IEE RADAR, SONAR, NAVIGATION AND AVIONICS SERIES 5

Series Editors: Professor E. D. R. Shearman
P. Bradsell

Strapdown inertial navigation technology

Strapdown inertial navigation technology

D. H. Titterton and J. L. Weston

Peter Peregrinus Ltd. on behalf of the Institution of Electrical Engineers

Published by: Peter Peregrinus Ltd., on behalf of
the Institution of Electrical Engineers, London,
United Kingdom

© 1997: Peter Peregrinus Ltd.

Peter Peregrinus Ltd.
The Institution of Electrical Engineers,
Michael Faraday House,
Six Hills Way, Stevenage,
Herts. SG1 2AY, United Kingdom

British Library Cataloguing in Publication Data

A CIP catalogue record for this book
is available from the British Library

ISBN 0 86341 260 2

Printed and bound by Antony Rowe Ltd, Eastbourne

Contents

Preface

Inertial navigation is widely used for the guidance of aircraft, missiles, ships and land vehicles as well as a number of more novel applications, such as surveying underground pipelines and bore holes during drilling operations. This book sets out to provide a clear and concise description of the physical principles of inertial navigation; there is also a more detailed treatise covering recent developments in inertial sensor technology and the techniques for implementing such systems.

It is intended that the book should provide an up-to-date guide to the techniques of inertial navigation, which will be of interest to both the practising engineer and the postgraduate student. The text, therefore, contains an introduction to the techniques involved and information on modern technological developments, as well as a more rigorous mathematical treatment for the reader wishing to explore the subject in greater depth.

In order to be of value to systems engineers and project managers, the text describes a range of technologies and evaluation techniques to enable informed judgements to be made about the suitability of competing technologies and sensors. Data are provided to give an indication of the range of performance that can be achieved from both component devices and systems. There is a detailed description of the techniques which may be used for the independent evaluation of such different technologies, and also included are illustrated examples to highlight the interaction between competing effects and the impact on performance.

At the end of the text there is a description of a system study undertaken to define an inertial navigation system using the data, equations and algorithms defined in the earlier chapters. This study should be of value to many different types of technologists, as it shows the methods that may be used to define a system that has to perform to a given specification; it also considers many of the interactions and compromises that have to be made during the design of a system.

A comprehensive glossary of terms has been included in order to give an accurate definition of the principal terms and expressions used in this subject.

The completion of a book of this form would not have been possible without the help of many people. The authors are very grateful to Mr S. G. Smith and Mr C. R. Milne, both formerly at RAE Farnborough, who read the manuscript and made many helpful suggestions. Grateful thanks are also expressed to the following people for their advice and assistance given during the preparation of the text: Mrs S. P. Potts of British Aerospace plc, formerly with ACO (ASWE), Slough; Mr G. Baker of Systron-Donner Inertial Division; Mr R. J. Chaplin and Dr A. R. Malvern of British Aerospace plc; Mr S. K. Davidson, Mr D. J. Miles and Mr R. Wright of the DRA, Farnborough, and Dr R. Coleman and Mr J. E. Krautmann of Beacon Consultants Limited.

A special thanks to our families for
their support and encouragement
during the writing of this book

Chapter 1

Introduction

1.1 Navigation

Navigation is a very ancient skill or art which has become a complex science. It is essentially about travel and finding the way from one place to another and there are a variety of means by which this may be achieved [1].

Perhaps one of the simplest forms of navigation is the following of directions, or instructions. For example, a person wishing to travel to a given destination from where they are at the moment may be instructed: turn right at the next junction, turn left at the Rose and Crown, keep to the right of a given landmark, . . . , it will be in front of you! Clearly, this method of navigation relies on the observation and recognition of known features or fixed objects in one's surroundings and moving between them. In technical narratives, the locations of these features are often referred to as way points.

An extension of this process is navigation by following a map. In this case, the navigator will determine his/her position by observation of geographical features such as roads, rivers, hills and valleys which are shown on the map. These features may be defined on the map with respect to a grid system or reference frame. For example, positions of terrain features are often defined with respect to the Earth's equator and the Greenwich meridian by their latitude and longitude. Hence the navigator is able to determine his/her position in that reference frame. As will become clear later in the text, the use of reference frames is fundamental to the process of navigation.

As an alternative method, the navigator may choose to observe other objects or naturally occurring phenomena to determine his/her position. An ancient and well established technique is to take sightings of certain of the fixed stars to which the navigator can relate his/her position. The fixed stars effectively define a reference frame which is fixed in space. Such a reference is commonly referred to as an inertial reference frame and star sightings enable an observer to determine his/her position with respect to that frame. Given knowledge of the motion of the Earth and the time of the observation, the navigator is able to use the celestial measurements to define his position on the surface of the Earth. Navigation systems of this type which rely on observation of the outside world are known as position fixing systems.

An alternative approach is to use the principle of dead reckoning by which one's present position may be calculated from knowledge of initial position and measurements of speed and direction. The process of dead reckoning is performed by taking last known position and the time at which it was obtained, and noting average speed and heading since that time and current time. The speed must be resolved though the heading angle to give velocity components north and east. Each is then multiplied by the time which has elapsed since the last position was obtained to give the change in position. Finally, the position changes are summed with initial position to obtain present position.

An equivalent process may be conducted using inertial sensors—gyroscopes and accelerometers—to sense rotational and translational motion with respect to an inertial reference frame. This is known as inertial navigation.

$$\int a\,dt = v$$
$$\int v\,dt = d$$

1.2 Inertial navigation

The operation of inertial navigation systems depends on the laws of classical mechanics as formulated by Newton. Newton's laws tell us that the motion of a body will continue uniformly in a straight line unless disturbed by an external force acting on the body. These laws also tell us that this force will produce a proportional acceleration of the body. Given the ability to measure that acceleration, it would be possible to calculate the change in velocity and position by performing successive mathematical integrations of the acceleration with respect to time. Acceleration can be determined using a device known as an accelerometer. An inertial navigation system usually contains three such devices, each of which is capable of detecting acceleration in a single direction. The accelerometers are commonly mounted with their sensitive axes perpendicular to one another.

In order to navigate with respect to our inertial reference frame, it is necessary to keep track of the direction in which the accelerometers are pointing. Rotational motion of the body with respect to the inertial reference frame may be sensed using gyroscopic sensors and used to determine the orientation of the accelerometers at all times. Given this information, it is possible to resolve the accelerations into the reference frame before the integration process takes place.

Hence, inertial navigation is the process whereby the measurements provided by gyroscopes and accelerometers are used to determine the position of the vehicle in which they are installed. By combining the two sets of measurements, it is possible to define the translational motion of the vehicle within the inertial reference frame and so to calculate its position within that frame.

Unlike many other types of navigation system, inertial systems are entirely self-contained within the vehicle, in the sense that they are not dependent on the transmission of signals from the vehicle or reception from an external source. However, inertial navigation systems do rely on the availability of accurate knowledge of vehicle position at the start of navigation. The inertial measurements are then used to obtain estimates of changes in position which take place thereafter.

1.3 Strapdown technology

Although the underlying principles of operation are common to all types of inertial navigation system, their implementation may take a variety of different forms. The original applications of inertial navigation technology used stable platform techniques. In such systems, the inertial sensors are mounted on a stable platform and are mechanically isolated from the rotational motion of the vehicle. Platform systems are still in common use, particularly for those applications requiring very accurate estimates of navigation data, such as ships and submarines.

Modern systems have removed most of the mechanical complexity of platform systems by having the sensors attached rigidly, or strapped down, to the body of the host vehicle. The potential benefits of this approach are lower cost, reduced size and greater reliability compared with equivalent platform systems. As a result, small, light weight and accurate inertial navigation systems may now be fitted to small guided missiles, for instance. The major penalties incurred are a substantial increase in computing complexity and the need to use sensors capable of measuring much higher rates of turn. However, recent advances in computer technology combined with the development of suitable sensors have allowed such designs to become a reality.

Inertial navigation systems of this type, usually referred to as strapdown inertial navigation systems, are the subject of this book. Although there are many books which described the older and well established platform technology, no similar book exists which deals explicitly with strapdown systems. It was this fact which provided the primary motivation for this publication.

This text describes the basic concepts of inertial navigation and the technological developments which have led to modern strapdown systems. It is intended to provide an introduction to the subject of strapdown inertial navigation which may be read at various levels by both suppliers of inertial sensors and systems and customers for such products and so encourage a more effective two-way dialogue.

By selective reading, the engineer new to the subject may obtain a background understanding of the subject. For those needing to become more closely involved in the various aspects of strapdown system technology, the text provides a more extensive description of system configurations, an appreciation of strapdown inertial sensors and computational requirements and an awareness of techniques which may be used to analyse and assess the performance of such systems. References are provided for those seeking more detailed information on different aspects of the subject.

Strapdown inertial navigation systems rely on complex technology and both technology specific terms and jargon are in common usage. Such terminology is defined in the glossary of terms.

Where appropriate, mathematical descriptions of the physical principles and processes involved are presented. The reader new to the subject, who perhaps wishes to gain an appreciation of physical principles without dwelling on the mathematical details of the processes involved, may merely wish to take note of the results of the more mathematical sections, or possibly to skip over these aspects altogether.

1.4 Layout of book

Chapter 2 introduces the underlying concepts of strapdown inertial navigation systems with the aid of simplified examples, and culminates in the definition of the basic functions which must be implemented within such a system. It is shown how the measurement of rotational and translation motion are fundamental to the operation of an inertial navigation system. There follows a brief review of the historical developments which have led to the current state of development of strapdown inertial navigation systems. This is accompanied by an outline discussion of system applications.

The way in which the measurements of rotational and translational motion are combined to form an inertial navigation system are addressed more fully in Chapter 3. This chapter deals at some length with attitude computation and the concept of the navigation equation, both of which are fundamental to the operation of strapdown inertial navigation systems. In addition, a number of possible system configurations are described.

Gyroscope and accelerometer technologies are discussed in some detail in Chapters 4 to 6. This part of the text provides descriptions of the various instrument types currently in use and some which are likely to become available within the foreseeable future. These include conventional angular momentum gyroscopes, optical rate sensors such as the ring laser gyroscope and the fibre optic gyroscope, pendulous force feedback accelerometers and solid state devices, as well as multi-sensors. The text covers mechanical and electronic aspects of the instruments, measurement accuracy, mathematical descriptions and applications. Chapter 5 is devoted to the description and performance of optical sensors. The testing and calibration of inertial sensors and systems is addressed in Chapter 7.

Chapter 8 describes the basic building blocks which combine to form a strapdown inertial system, drawing attention to alternative mechanisations.

A vital factor in the achievement of accurate navigation is the initialisation of the inertial navigation system before the commencement of navigation, prior to take-off in the case of an aircraft navigation system, for instance. This process involves the accurate determination of the position, velocity and attitude of the vehicle navigation system with respect to the chosen reference frame, and is usually referred to as inertial navigation system alignment. Such alignment may have to be undertaken in a vehicle which is moving, as in the case of the in-flight alignment of an airborne inertial navigation system. The difficulties of achieving an accurate alignment in various vehicle applications are highlighted, and techniques for alleviating such problems are described in Chapter 9.

The computer processing of the gyroscope and accelerometer measurements which must be carried out in order to complete the task of navigation is examined in Chapter 10. Computational algorithms are discussed in some detail during the course of this chapter.

Techniques for the analysis of inertial navigation system performance are presented in Chapter 11 to enable the designer to assess system performance. Attention is drawn to a number of errors which are of particular concern in strapdown systems and the use of simulation methods to assess system performance is highlighted.

It is common practice for many applications to combine the outputs of an inertial navigation system with some external measurement data to achieve an overall improvement in navigation accuracy. For example, independent position fixes may be used to aid an inertial navigation system, and so enhance navigation performance beyond that which may be obtained using either the position fixing system or the inertial navigator in isolation. Possible navigation aids are discussed and techniques are presented for mixing inertial and external measurement data to form a so-called integrated navigation system in Chapter 12.

The final chapter draws together much of the preceding text through the discussion of a design example. Because the background of the authors is predominantly in the field of guided missile systems, this part of the book is directed at such an application. The design example will be of particular interest to the engineer wishing to specify a system to meet a given requirement and to assess its potential performance.

The appendices provide descriptions of Kalman filtering techniques and inertial navigation error budgets. A glossary of principal terms used is given at the end of the book.

Reference

1. ANDERSON, E.W.: *'The principles of navigation'* (Hollis and Carter, 1966)

Chapter 2
Fundamental principles and historical development of inertial navigation

2.1 Basic concepts

The basic concepts of inertial navigation are outlined here with the aid of some simple examples.

A simple one-dimensional example of navigation involves the determination of the position of a train which is moving along a straight track between two locations on a perfectly flat plane. It is possible to determine the instantaneous speed of the train and the distance it has travelled from a known starting point by using measurements of its acceleration along the track. Sensors called accelerometers provide such information about their own movement. If an accelerometer is fixed in the train, it will provided information about the acceleration of the train. The time integral of the acceleration measurement provides a continuous estimate of the instantaneous speed of the train, provided its initial speed is known. A second integration yields the distance travelled with respect to a known starting point. The accelerometer together with a computer, or other suitable device capable of integration, therefore constitutes a simple one-dimensional navigation system.

In general, a navigation system is required to provide an indication of the position of a vehicle with respect to a known grid system or reference frame. For instance, it may be required to determine the location of a vehicle in terms of x and y co-ordinates in a Cartesian reference frame. Considering again the example of a train moving along a track, as depicted in Figure 2.1, it is now necessary to determine the position of the train with respect to the co-ordinate reference frame shown in the figure.

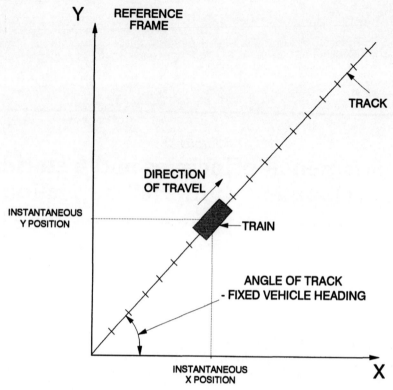

Figure 2.1   One-dimensional navigation

Given knowledge of the train's acceleration along the track, and the angle which the track makes with the reference frame, the x and y co-ordinate positions may be determined. This may be accomplished by resolving the measured acceleration in the reference frame to give x and y components, and by suitable integration of the resolved signals to yield the velocity and position of the train in reference axes. In this simple case, the angle of the track defines the heading of the train with respect to the reference frame.

With the more general situation illustrated in Figure 2.2, where the track curves, it is necessary to detect continuously both the translational motion of the train in two directions and changes in its direction of travel, i.e. to detect the rotations of the train about the perpendicular to the plane of motion as the train moves along the track.

Two accelerometers are now required to detect the translational motion in directions along and perpendicular to the track. One sensor suitable for the measurement of the rotational motion is a gyroscope.

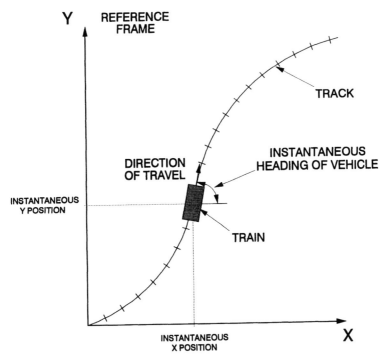

Figure 2.2   Two-dimensional navigation

Depending on the form of construction of this sensor, it may be used to provide either a direct measure of the train's heading with respect to the reference frame, or a measurement of the turn rate of the train. In the latter case, the angular orientation of the train may be calculated by the integration of this measurement, provided that the angle is known at the start of navigation. Given such information, it is possible to relate the measurements of acceleration, which are obtained in an axis set which is fixed in the train, to the reference frame. The instantaneous measurements of acceleration may therefore be resolved in the reference frame and integrated with respect to time to determine the instantaneous velocity and position of the vehicle with respect to that frame.

Clearly then, it is possible to construct a simple, two dimensional, navigation system using a gyroscope, two accelerometers and a computer. In practice, the inertial sensors may be mounted on a platform which is stabilised in space, and hence isolated from the rotation of the vehicle, or mounted directly on the vehicle to form a strapdown system. The measurements are processed in the computer to provide continuous estimates of the position, speed and the direction of travel or heading of the train. It must be stressed that inertial navigation is fundamentally dependent on an accurate knowledge of position, speed and heading being available prior to the start of navigation. This is because it uses dead reckoning which relies for its operation on the updating of the system's previous estimates of these navigational quantities, commencing with the initial values entered into the system at the start of navigation.

It will be apparent from the preceding discussion that successful navigation of a vehicle can be achieved by using the properties of suitable sensors mounted in the vehicle. In general, one needs to determine a vehicle's position with respect to a three-dimensional reference frame. Consequently, if single-axis sensors are used, three gyroscopes will be required to provide measurements of a vehicle's turn rate about three separate axes, whilst three accelerometers provide the components of acceleration which the vehicle experiences along these axes. For convenience and accuracy, the three axes are usually chosen to be mutually perpendicular.

In most applications the axis set defined by the sensitive axes of the inertial sensors is made coincident with the axes of the vehicle, or body, in which the sensors are mounted, usually referred to as the body axis set. The measurements provided by the gyroscopes are used to determine the attitude and heading of the body with respect to the reference frame within which it is required to navigate. The attitude and heading information is then used to resolve the accelerometer measurements into the reference frame. The resolved accelerations may then be integrated twice to obtain vehicle velocity and position in the reference frame.

Gyroscopes provide measurements of changes in a vehicle's attitude or its turn rate with respect to inertial space. Accelerometers, however, are unable to separate the total acceleration of the vehicle, the acceleration with respect to inertial space, from that caused by the presence of a gravitational field. These sensors do in fact provide measurements of the difference between the true acceleration in space and the acceleration due to gravity[1]. This quantity is the non-gravitational force per unit mass exerted on the instrument, referred to in this text for brevity as specific force [1].

Hence, the measurements provided by the accelerometers, especially when close to a large body such as the Earth, must be combined with knowledge of the gravitational field of that body in

---

[1]Algebraically, the sum of the acceleration with respect to inertial space and the acceleration due to gravitational attraction.

order to determine the acceleration of the vehicle with respect to inertial space. Using this information, vehicle acceleration relative to the body may be derived.

The navigational function is therefore fulfilled by combining the measurements of vehicle rotation and specific force with knowledge of the gravitational field to compute estimates of attitude, velocity and position with respect to a pre-defined reference frame. A schematic representation of such an inertial navigation system is shown in Figure 2.3.

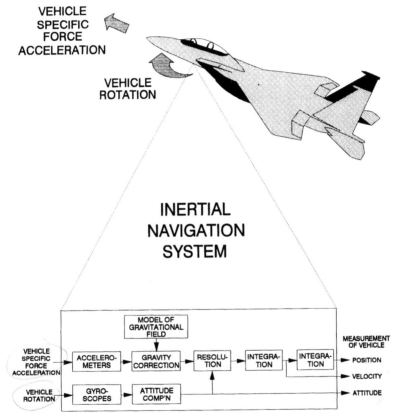

Figure 2.3   Functional components of an inertial navigation system

Summary

It follows from the introductory discussion that the essential functions which an inertial navigation system must perform may be defined as follows:

- use gyroscopic sensors to determine the angular motion of a vehicle from which its attitude relative to a reference frame may be derived;
- measure specific force using accelerometers;
- resolve the specific force measurements into the reference frame using the knowledge of attitude derived from the information provided by the gyroscopes;
- have access to a function representing the gravitational field—the gravitational attraction of the Earth in the case of systems operating in the vicinity of the Earth;
- integrate the resolved specific force measurements to obtain estimates of the velocity and position of the vehicle.

The later chapters describe the principles of inertial navigation in some depth and provide detailed information on system mechanisations, inertial sensor technology, computational aspects (including algorithms), design analysis and applications of such systems. However, prior to this, it is instructive to have a brief review of the historical developments which have led to the current state of development of present day inertial navigation systems and their technology.

2.2 Historical developments

From the earliest times, people have moved from one place to another by finding or knowing their way; this skill has required some form of navigation. There is an oblique reference to inertial navigation in the Bible [2]. Generally, as in the case of the biblical reference, the earliest applications were on land. Then as the desire developed to explore farther afield, instruments were developed for marine applications. More recently, there have been significant developments in inertial sensors and systems for inertial navigation on land, in the air, on or under the oceans as well as in space to the planets and beyond.

Our earliest ancestors travelled in search of food, usually on land. As they developed, they crossed rivers generally using landmarks, i.e. navigation by observation. Further development of position fixing techniques saw the Polynesians cross the Pacific Ocean about two millennia ago using their understanding of celestial bodies and landmarks. These techniques can only be used in clear weather conditions. During the thirteenth century the Chinese discovered the properties of lodestone and applied the principles of magnetism to fabricate a compass which they used to navigate successfully across the South China Sea. This device could be used irrespective of visibility but was difficult to use in rough weather. The other significant development to help the long distance traveller was the sextant, which enabled position fixes to be made accurately on land.

In the seventeenth century Sir Isaac Newton defined the laws of mechanics and gravitation, which are the fundamental principles on which inertial navigation is based. Despite this, it was to be about another two centuries before the inertial sensors were developed that would enable the demonstration of inertial navigation techniques. However, in the early eighteenth century, there were several significant developments; Serson demonstrating a stabilised sextant [3] and Harrison devising an accurate chronometer, the former development enabling sightings to be taken of celestial objects without reference to the horizon and the latter enabling an accurate determination of longitude. These instruments, when used with charts and reference tables of location of celestial bodies, enabled accurate navigation to be achieved, provided that the objects were visible.

Foucault is generally credited with the discovery of the gyroscopic effect in 1852. He was certainly the first to use the word. There were others, such as Bohneberger, Johnson and Lemarle, developing similar instruments. All of these people were investigating the rotational motion of the Earth and the demonstration of rotational dynamics. They were using the ability of the spin axis of a rotating disc to remain fixed in space. Later in the nineteenth century many fine gyroscopic instruments were made. In addition, there were various ingenious applications of the gyroscopic principle in heavy equipment such as the grinding mill.

A significant discovery was made in 1890 by Professor G.H. Bryan concerning the ringing of hollow cylinders, a phenomenon later applied to solid state gyroscopes.

The early years of the twentieth century saw the development of the gyrocompass for the provision of a directional reference. The basic principle of this instrument is the indication of true north by establishing the equilibrium between the effect of its pendulosity and the angular momentum of the rotating base carrying the compass. Initially, this instrument was sensitive to acceleration. Professor Max Schuler produced an instrument with a vertical erection system enabling an accurate vertical reference to be defined [4]. This instrument was tuned to the undamped natural period defined by

$2\pi\sqrt{(R/g)}$, approximately equal to 84 minutes, where R is the radius of the Earth and g is the acceleration caused by the Earth's gravitational field. Later, this technique became known as Schuler tuning [5], a phrase originated by Dr Walter Wrigley of Massachusetts Institute of Technology (MIT). This ingenious method produced a directional instrument insensitive to acceleration for use at sea. Elmer and Lawrence Sperry improved the design of the gyrocompass with further refinements by Brown and Perry. These instruments provided the first steps towards all weather, autonomous navigation. The Sperry brothers were also at the forefront of the application of the gyroscopic effect to control and guidance in the early twentieth century. They produced navigation and autopilot equipment for use in aircraft and gyroscopes for use in torpedoes.

Rate of turn indicators, artificial horizons and directional gyroscopes for aircraft were being produced in the 1920s. At a similar time, side slip sensors were being developed, early open loop accelerometers, and Schuler was demonstrating a north-seeking device for land use giving an accuracy of 22 seconds of arc! There was significant progress during the early part of the twentieth century with the development of stable platforms for fire control systems for guns on ships and the identification of the concept for an inertial navigation system. Boykow identified the use of accelerometers and gyroscopes to produce a full inertial navigation system. However, at this stage, the quality of the inertial sensors was not suitable for the production and demonstration of such a system.

World War II saw the demonstration of the principles of inertial guidance in the V2 rockets by German scientists, a prime step forward being the use of a system with feedback leading to accurate guidance. At this time there was much activity in various parts of the world devising new types of inertial sensors, improving their accuracy and, in 1949, the first publication suggesting the concept of the strapdown technique for navigation.

The pace of development and innovation quickened in the 1950s with many significant developments for seaborne and airborne applications. More accurate sensors were produced, with the accuracy of the gyroscope being increased substantially. The error in such sensors was reduced from about 15°/hour to about 0.01°/hour, Professor Charles Stark Draper and his co-workers at MIT being largely responsible for many technical advances with the demonstration of the floated rate integrating gyroscope [6]. It was also during the 1950s that the principle of force feedback was applied to the proof mass in an accelerometer to produce an accurate acceleration sensing instrument.

The early part of the 1950s saw the fabrication of a stabilised platform inertial navigation system followed by the first crossing of the United States of America by an aircraft using full inertial navigation. Inertial navigation systems became standard equipment in military aircraft, ships and submarines during the 1960s, all of these applications using the so-called stable platform technology. This era saw further significant developments with increases in the accuracy of sensors, the miniaturisation of these devices and the start of ring laser gyroscope development. Major projects of this period in which inertial system technology was applied were the ballistic missile programmes and the exploration of space.

Similar progress has taken place in the last two decades; one major advance being the application of the microcomputer and development of gyroscopes with large dynamic ranges enabling the strapdown principle to be realised. This has enabled the size and complexity of the inertial navigation system to be reduced significantly for very many applications. The use of novel methods has enabled small, reliable, rugged and accurate inertial sensors to be produced that are relatively inexpensive, thus enabling a very wide range of diverse applications as discussed below. This period has also seen significant advances in the development of solid state sensors such as optical fibre gyroscopes and silicon accelerometers.

The development of inertial navigation systems in recent years has been characterised by the gradual move from stable platform to strapdown technology as indicated in Figure 2.4. The figure gives an indication of the increasing application of strapdown systems which has resulted from advances in gyroscope technology. Milestones in this continuing development have occurred as a result of the development of the miniature rate integrating gyroscope, the dynamically tuned gyroscope and, more recently, ring laser and fibre optic rate sensors and vibratory gyroscopes, all of which are described in Chapters 4 and 5.

Strapdown systems are becoming widely used for aircraft and guided missile applications. In theory, there is no reason why such technology should not also be applied to ship and submarine applications as indicated in Figure 2.4. The diagram shows other applications, for strapdown technology, the accuracy range required from the gyroscopes being related to the position and size of the box in which it is mentioned.

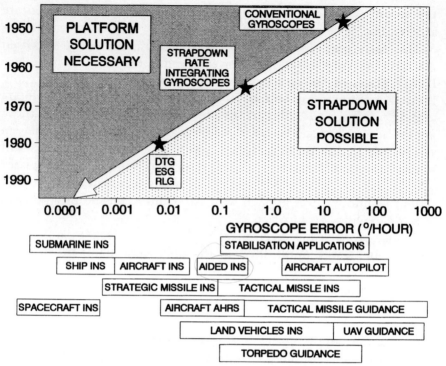

Figure 2.4   Strapdown sensor development and some applications

2.3 The modern-day inertial navigation system

From the preceding section it is clear that the range of applications in which inertial navigation systems can and are being used is very extensive, covering navigation of ships, aircraft, tactical and strategic missiles and spacecraft. In addition, there are some more novel applications in the fields of robotics, active suspension in racing or high performance motor cars and for surveying underground wells and pipelines.

Such diverse applications call for navigation systems having a very broad range of performance capabilities, as well as large differences in the periods of time over which they will be required to provide navigation data. For example, tactical missile applications may require inertial navigation and guidance to an accuracy of a few hundred metres for periods of minutes or even a few seconds, whilst other airborne systems are required to operate for several hours while maintaining knowledge of aircraft position to an accuracy of one or two nautical miles or better. In the cases of marine or space applications, such systems may be required to provide navigation data to similar accuracy over periods of weeks, months or even longer in the case of interplanetary exploration. One extreme example is the Voyager space craft which has been navigating through the solar system for more than ten years.

Although the basic principles of inertial navigation systems do not change from one application to another, it will come as no surprise to find that the accuracy of the inertial sensors and the precision to which the associated computation must be carried out varies dramatically over the broad range of applications indicated above. It follows therefore that the instrument technologies and the techniques used for the implementation of the navigation function in such diverse applications also vary greatly. Part of the function of this text is to provide some insight into the methods and technologies appropriate to some of the different types of inertial system application outlined above.

References

1 BRITTING, K.: '*Inertial Navigation System Analysis*', (Wiley Interscience, New York, 1971)
2 The Bible, Amos 7:7–9
3 SORG, H.W.: 'From Serson to Draper—two centuries of gyroscopic development', J. Inst. Navig., Winter 1976–77, **23** (4)
4 SCHULER, M.: 'Die Storung von Pendul-und Kreiselapparaten durch die Beschleunigung der Fahrzeuges', *Physikalische Zeitschrift*, 1923, B.24
5 WRIGLEY, W., HOLLISTER, W.M. and DENHARD, W.G.: '*Gyroscopic theory, design and instrumentation* (MIT Press, 1969)
6 DRAPER, C., WRIGLEY, W. and HOVORKA, J.: '*Inertial guidance*' (Pergamon Press, 1960)

Raytheon

EXPECT GREAT THINGS

Raytheon

EXPECT GREAT THINGS

Chapter 3
Basic principles of strapdown inertial navigation systems

3.1 Introduction

The previous chapter has provided some insight into the basic measurements that are necessary for inertial navigation. For the purposes of the ensuing discussion, it is assumed that measurements of specific force and angular rate are available along and about axes which are mutually perpendicular. Attention is focused on how these measurements are combined and processed to enable navigation to take place.

3.2 A simple 2-D strapdown navigation system

We begin this chapter by describing a simplified two-dimensional strapdown navigation system. Although functionally identical to the full three-dimensional system discussed later, the computational processes which must be implemented to perform the navigation task in two dimensions are much simplified compared with a full strapdown system. Therefore, through this introductory discussion, we hope to provide the reader with an appreciation of the basic processing tasks which must be implemented in a strapdown system without becoming too deeply involved in the intricacies and complexities of the full system computational tasks.

For the purposes of this discussion, it is assumed that a system is required to navigate a vehicle which is constrained to move in a single plane. A two-dimensional strapdown system capable of fulfilling this particular navigation task was introduced very briefly in Chapter 2 and is shown diagramatically in Figure 3.1.

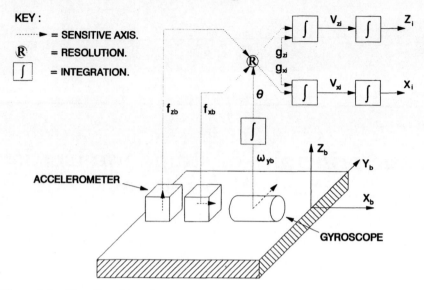

Figure 3.1   Two-dimensional strapdown navigation system

The system contains two accelerometers and a single rate gyroscope, all of which are attached rigidly to the body of the vehicle. The vehicle body is represented in the figure by the block on which the instruments shown are mounted. The sensitive axes of the accelerometers, indicated by the directions of the arrows in the diagram, are at right angles to one another and aligned with the body axes of the vehicle in the plane of motion; they are denoted as the x_b and z_b axes. The gyroscope is mounted with its sensitive axis orthogonal to both accelerometer axes allowing it to detect rotations about an axis perpendicular to the plane of motion; the y_b axis. It is assumed that navigation is required to take place with respect to a space fixed reference frame denoted by the axes x_i and z_i. The reference and body axis sets are shown in Figure 3.2, where θ represents the angular displacement between the body and reference frames.

Referring now to Figure 3.1, body attitude, θ, is computed by integrating the measured angular rate, ω_{yb}, with respect to time. This information is then used to resolve the measurements of specific force, f_{xb} and f_{zb}, into the reference frame. A gravity model, stored in the computer, is assumed to provide estimates of the gravity components in the reference frame, g_{xi} and g_{zi}. These quantities are combined with the resolved measurements of specific force, f_{xi} and f_{zi}, to determine true accelerations, denoted by $\dot{v}_{xi}$ and $\dot{v}_{zi}$. These derivatives are subsequently integrated twice to obtain estimates of vehicle velocity and position. The full set of equations which must be solved is given in Figure 3.3.

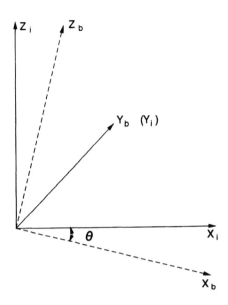

Figure 3.2   Reference frames for 2-D navigation

$$\dot{\theta} = \omega_{yb}$$
$$f_{xi} = f_{xb}\cos\theta + f_{zb}\sin\theta$$
$$f_{zi} = -f_{xb}\sin\theta + f_{zb}\cos\theta$$
$$\dot{v}_{xi} = f_{xi} + g_{xi}$$
$$\dot{v}_{zi} = f_{zi} + g_{zi}$$
$$\dot{x}_{i} = v_{xi}$$
$$\dot{z}_{i} = v_{zi}$$

Figure 3.3   2-D strapdown navigation system equations

Having defined the basic functions which must be implemented in a strapdown inertial navigation system, consideration is now given to the application of the two-dimensional system, described above, for navigation in a rotating reference frame. For instance, consider the situation where one needs to navigate a vehicle moving in a meridian plane around the Earth, as depicted in Figure 3.4. Hence, we are concerned here with a system which is operating in the vertical plane alone. Such a system would be required to provide estimates of velocity with respect to the Earth, position along the meridian and height above the Earth.

Whilst the system mechanisation as described could be used to determine such information, this would entail a further transformation of the velocity and position, derived in space fixed co-ordinates, to a geographic frame. An alternative and often used approach is to navigate directly in a local geographic reference frame, defined in this simplified case by the direction of the local vertical at the current location of the vehicle. In order to provide the required navigation information, it now becomes necessary to keep track of vehicle attitude with respect to the local geographic frame denoted by the axes x and z. This information can be extracted by differencing the successive gyroscopic measurements of body turn rate with respect to inertial space, and the current estimate of the turn rate of the reference frame with respect to inertial space. For a vehicle moving at a velocity, v_x, in a single plane around a perfectly spherical Earth of radius R_0, this rate is given by $v_x / (R_0 + z)$ where z is the height of the vehicle above the surface of the Earth. This is often referred to as the transport rate.

Figure 3.4 shows a modified two-dimensional strapdown system for navigation in the moving reference frame. As shown in the Figure, an

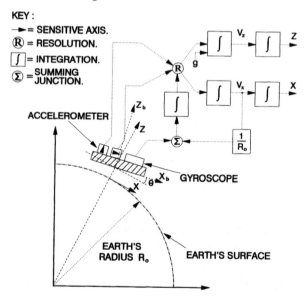

Figure 3.4   *Two-dimensional strapdown system for navigation in a rotating reference frame*

estimate of the turn rate of the reference frame is derived using the estimated component of horizontal velocity.

The equations which must be solved in this system are given in Figure 3.5.

$$\dot{\theta} = \omega_{yb} - v_x/(R_0 + z)$$
$$f_x = f_{xb}\cos\theta + f_{zb}\sin\theta$$
$$f_z = -f_{xb}\sin\theta + f_{zb}\cos\theta$$
$$\dot{v}_x = f_x + v_x v_z/(R_0 + z)$$
$$\dot{v}_z = f_z + g - v_x^2/(R_0 + z)$$
$$\dot{x} = v_x$$
$$\dot{z} = v_z$$

Figure 3.5   Simplified 2-D strapdown system equations1 for navigation in a rotating reference frame

Comparison with the equations given in Figure 3.3, relating to navigation with respect to a space fixed axis set, reveals the following differences. The attitude computation is modified to take account of the turn rate of the local vertical reference frame as described above. Consequently, the equation in θ is modified by the subtraction of the term $v_x/(R_0 + z)$ in Figure 3.4. The terms $v_x v_z/(R_0 + z)$ and $v_x^2/(R_0 + z)$ which appear in the velocity equations are included to take account of the additional forces acting as the system moves around the Earth (Coriolis forces, see Section 3.4). The gravity term (g) appears only in the v_z equation as it is assumed that the Earth's gravitational acceleration acts precisely in the direction of the local vertical.

This section has outlined the basic form of the computing tasks to be implemented in a strapdown navigation system using a much simplified two-dimensional representation. In the remainder of this chapter the extension of this simple strapdown system to three dimensions is described in some detail. It will be appreciated that this entails a substantial increase in the complexity of the computing tasks involved. In particular, attitude information in three dimensions can no longer be obtained by a simple integration of the measured turn rates.

3.3 Reference frames

Fundamental to the process of inertial navigation is the precise definition of a number of Cartesian co-ordinate reference frames. Each frame is an orthogonal, right handed, co-ordinate frame or axis set.

For navigation over the Earth, it is necessary to define axis sets which allow the inertial measurements to be related to the cardinal directions of the Earth, i.e. frames which have a physical significance when attempting to navigate in the vicinity of the Earth. Therefore, it is customary to consider an inertial reference frame which is stationary with respect to the fixed stars, the origin of which is located at the centre of the Earth. Such a reference frame is shown in Figure 3.6, together with an Earth fixed reference frame and a local geographic navigation frame defined for the purposes of terrestrial inertial navigation.

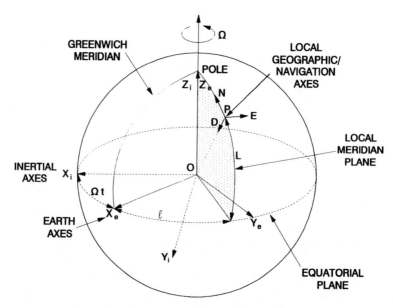

Figure 3.6   Frames of reference

The following co-ordinate frames are used in the text:

The inertial frame (*i*-frame) has its origin at the centre of the Earth and axes which are non-rotating with respect to the fixed stars, defined by the axes Ox_i, Oy_i, Oz_i, with Oz_i coincident with the Earth's polar axis (which is assumed to be invariant in direction).

The Earth frame (*e*-frame) has its origin at the centre of the Earth and axes which are fixed with respect to the Earth, defined by the axes Ox_e, Oy_e, Oz_e with Oz_e along the Earth's polar axis. The axis Ox_e lies along the intersection of the plane of the Greenwich meridian with the Earth's equatorial plane. The Earth frame rotates, with respect to the inertial frame, at a rate Ω about the axis Oz_i.

The navigation frame (*n*-frame) is a local geographic frame which has its origin at the location of the navigation system, point P, and axes aligned with the directions of north, east and the local vertical (down). The turn rate of the navigation frame, with respect to the Earth fixed frame, ω_{en}, is governed by the motion of the point P with respect to the Earth. This is often referred to as the transport rate.

The wander azimuth frame (*w*-frame) may be used to avoid the singularities in the computation which occur at the poles of the navigation frame. Like the navigation frame, it is locally level but is rotated through the wander angle about the local vertical. Its use is described in Section 3.5.

The body frame (*b*-frame), depicted in Figure 3.7, is an orthogonal axis set which is aligned with the roll, pitch and yaw axes of the vehicle in which the navigation system is installed.

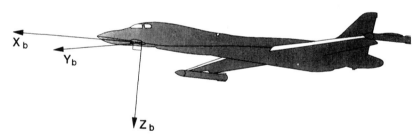

Figure 3.7   Illustration of a body reference frame

3.4 3-D strapdown navigation system—general analysis

3.4.1 Navigation with respect to a fixed frame

Consider the situation where one is required to navigate with respect to a fixed, or non-accelerating, and non-rotating set of axes. The measured components of specific force and estimates of the gravitational field are summed to determine components of acceleration with respect to a space fixed reference frame. These quantities can then be integrated twice, giving estimates of velocity and position in that frame.

This process may be expressed mathematically in the following manner[1]. Let **r** represent the position vector of the point P with respect to O, the origin of the reference frame shown in Figure 3.8.

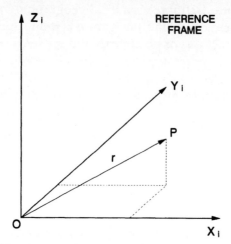

Figure 3.8　Position vector with respect to reference frame

The acceleration of P with respect to a space fixed axis set, termed the i-frame and denoted by the subscript $_i$, is defined by:

$$\mathbf{a}_i = \left.\frac{d^2}{dt^2}\mathbf{r}\right|_i \tag{3.1}$$

A triad of perfect accelerometers will provide a measure of the specific force acting at point P where:

$$\mathbf{f} = \left.\frac{d^2}{dt^2}\mathbf{r}\right|_i - \mathbf{g} \tag{3.2}$$

in which **g** is the mass attraction gravitation vector.

Rearranging equation (3.2) yields the following equation:

$$\left.\frac{d^2}{dt^2}\mathbf{r}\right|_i = \mathbf{f} + \mathbf{g} \tag{3.3}$$

---

[1]Vector and matrix notation is widely used throughout the text for the mathematical representation of strapdown inertial system processes. This notation is adopted both in the interests of brevity and to be consistent with other texts on the subject. Vector and matrix quantities are written in boldface type.

This is called the navigation equation since, with suitable integration, it yields the navigational quantities of velocity and position. The first integral gives the velocity of point P with respect to the i-frame:

$$\mathbf{v}_i = \frac{d}{dt}\mathbf{r}\bigg|_i \qquad\qquad (3.4)$$

whilst a second integration gives its position in that frame.

3.4.2 Navigation with respect to a rotating frame

In practice, one often needs to derive estimates of a vehicle's velocity and position with respect to a rotating reference frame, as when navigating in the vicinity of the Earth. In this situation, additional apparent forces will be acting which are functions of reference frame motion. This results in a revised form of the navigation equation which may be integrated to determine the ground speed of the vehicle, $\mathbf{v}_e$, directly. Alternatively, $\mathbf{v}_e$ may be computed from the inertial velocity, $\mathbf{v}_i$, using the theorem of Coriolis, as follows,

$$\mathbf{v}_e = \frac{d}{dt}\mathbf{r}\bigg|_e = \mathbf{v}_i - \boldsymbol{\omega}_{ie} \times \mathbf{r} \qquad\qquad (3.5)$$

where $\boldsymbol{\omega}_{ie} = [\,0\ \ 0\ \ \Omega\,]^T$ is the turn rate of the Earth frame with respect to the i-frame and $\times$ denotes a vector cross product.

Revised forms of the navigation equation suitable for navigation with respect to the Earth are the subject of Section 3.5.

3.4.3 The choice of reference frame

The navigation equation, eqn. 3.3, may be solved in any one of a number of reference frames. If the Earth frame is chosen, for example, then the solution of the navigation equation will provide estimates of velocity with respect to either the inertial frame or the Earth frame, expressed in Earth co-ordinates, denoted $\mathbf{v}_i^e$ and $\mathbf{v}_e^e$ respectively[1].

In Section 3.5 a number of different strapdown system mechanisations for navigating with respect to the Earth are described. In each case it will be shown that the navigation equation is expressed in a different manner depending on the choice of reference frame.

---

[1]Superscripts attached to vector quantities denote the axis set in which the vector quantity co-ordinates are expressed.

3.4.4 Resolution of accelerometer measurements

The accelerometers usually provide a measurement of specific force in a body fixed axis set, denoted f^b. In order to navigate, it is necessary to resolve the components of the specific force in the chosen reference frame. In the event that the inertial frame is selected, this may be achieved by pre-multiplying the vector quantity f^b by the direction cosine matrix, C_b^i, using:

$$f^i = C_b^i f^b \tag{3.6}$$

where C_b^i is a 3×3 matrix which defines the attitude of the body frame with respect to the i-frame. The direction cosine matrix C_b^i may be calculated from the angular rate measurements provided by the gyroscopes using the following equation:

$$\dot{C}_b^i = C_b^i \Omega_{ib}^b \tag{3.7}$$

where Ω_{ib}^b is the skew symmetric matrix:

$$\Omega_{ib}^b = \begin{pmatrix} 0 & -r & q \\ r & 0 & -p \\ -q & p & 0 \end{pmatrix} \tag{3.8}$$

This matrix is formed from the elements of the vector $\omega_{ib}^b = [p, q, r]^T$, which represents the turn rate of the body with respect to the i-frame as measured by the gyroscopes. Eqn. 3.7 is derived in Section 3.6.

The attitude of the body with respect to the chosen reference frame, which is required to resolve the specific force measurements into the reference frame, may be defined in a number of different ways. For the purposes of the discussion of navigation system mechanisations in this and the following section, the direction cosine method will be adopted. Direction cosines and some alternative attitude representations are described in some detail in Section 3.6.

3.4.5 System example

Consider the situation in which one needs to navigate with respect to inertial space and the solution of the navigation equation takes place in the i-frame. Eqn. 3.3 may be expressed in i-frame co-ordinates as follows:

$$\frac{d^2}{dt^2} r \Big|_i = f^i + g^i$$

$$= C_b^i f^b + g^i \tag{3.9}$$

It is clear from the preceding discussion that the integration of the navigation equation involves the use of information from both the gyroscopes and the accelerometers contained within the inertial navigation system. A block diagram representation of the resulting navigation system is given in Figure 3.9.

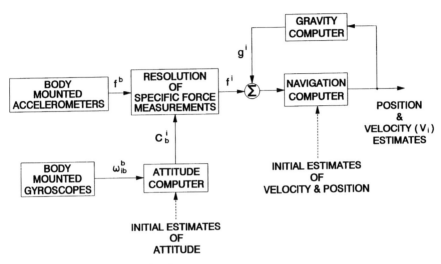

Figure 3.9   Strapdown navigation system

The diagram displays the main functions to be implemented within a strapdown navigation system: the processing of the rate measurements to generate body attitude, the resolution of the specific force measurements into the inertial reference frame, gravity compensation and the integration of the resulting acceleration estimates to determine velocity and position.

3.5 Strapdown system mechanisations

Attention is focused here on inertial systems which may be used to navigate in the vicinity of the Earth. It has been shown in Section 3.4 how estimates of position and velocity are derived by integrating a navigation equation of the form given in eqn. 3.3. In systems of the type described below, in which one needs to derive estimates of vehicle velocity and position with respect to an Earth fixed frame, additional apparent forces will be acting which are functions of the reference frame motion. In this section, further forms of the navigation equation are derived, corresponding to different choices of reference frame [1].

The resulting system mechanisations are described, together with their applications. As will become apparent, the variations in the mechanisations described here are in the strapdown computational algorithms and not in the arrangement of the sensors or the mechanical layout of the system.

3.5.1 Inertial frame mechanisation

In this system, one needs to calculate a vehicle's speed with respect to the Earth, the ground speed, in inertial axes, denoted by the symbol v_e^i. This may be accomplished by expressing the navigation eqn. 3.3 in inertial axes and deriving an expression for $\left.\dfrac{d^2}{dt^2}r\right|_e$ in terms of ground speed and its time derivatives with respect to the inertial frame.

Inertial velocity may be expressed in terms of ground speed using the Coriolis equation:

$$\left.\frac{d}{dt}\mathbf{r}\right|_i = \left.\frac{d}{dt}\mathbf{r}\right|_e + \boldsymbol{\omega}_{ie} \times \mathbf{r} \tag{3.10}$$

Differentiating this expression and writing $\left.\dfrac{d}{dt}\mathbf{r}\right|_e = \mathbf{v}_e$, we have:

$$\left.\frac{d^2}{dt^2}\mathbf{r}\right|_i = \left.\frac{d}{dt}\mathbf{v}_e\right|_i + \left.\frac{d}{dt}\left[\boldsymbol{\omega}_{ie} \times \mathbf{r}\right]\right|_i \tag{3.11}$$

Applying the Coriolis equation in the form of eqn. 3.10 to the second term in eqn. 3.11 gives:

$$\left.\frac{d^2}{dt^2}\mathbf{r}\right|_i = \left.\frac{d}{dt}\mathbf{v}_e\right|_i + \boldsymbol{\omega}_{ie} \times \mathbf{v}_e + \boldsymbol{\omega}_{ie} \times \left[\boldsymbol{\omega}_{ie} \times \mathbf{r}\right] \tag{3.12}$$

In generating the above equation, it is assumed that the turn rate of the Earth is constant, hence $(d/dt)\boldsymbol{\omega}_{ie} = 0$.

Combining eqns. 3.3 and 3.12 and re-arranging yields:

$$\left.\frac{d}{dt}\mathbf{v}_e\right|_i = \mathbf{f} - \boldsymbol{\omega}_{ie} \times \mathbf{v}_e - \boldsymbol{\omega}_{ie} \times \left[\boldsymbol{\omega}_{ie} \times \mathbf{r}\right] + \mathbf{g} \tag{3.13}$$

In this equation, $\mathbf{f}$ represents the specific force acceleration to which the navigation system is subjected, whilst $\boldsymbol{\omega}_{ie} \times \mathbf{v}_e$ is the acceleration caused by its velocity over the surface of a rotating Earth, usually referred to as the Coriolis acceleration. The term $\boldsymbol{\omega}_{ie} \times [\boldsymbol{\omega}_{ie} \times \mathbf{r}]$, in eqn. 3.13, defines the centripetal acceleration experienced by the system owing to the rotation of the Earth, and is not separately distinguishable from the gravitational acceleration which arises through mass attraction, **g**. The sum of the accelerations caused by the mass attraction force and the centripetal force constitutes what is known as the local gravity vector, the vector to which a plumb bob would align itself when held above the Earth. This is denoted here by the symbol $\mathbf{g}_l$, i.e.:

$$\mathbf{g}_l + \mathbf{g} - \boldsymbol{\omega}_{ie} \times [\boldsymbol{\omega}_{ie} \times \mathbf{r}] \tag{3.14}$$

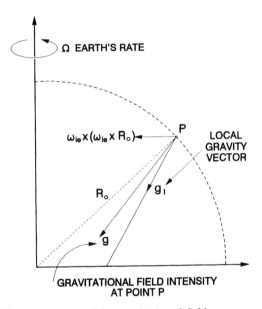

Figure 3.10 The components of the gravitational field

Combining eqns. 3.13 and 3.14 gives the following form of the navigation equation:

$$\frac{d}{dt}\mathbf{v}_e\Big|_i = \mathbf{f} - \boldsymbol{\omega}_{ie} \times \mathbf{v}_e + \mathbf{g}_l \tag{3.15}$$

This equation may be expressed in inertial axes, as follows, using the superscript notation mentioned earlier.

$$\dot{\mathbf{v}}_e^i = \mathbf{f}^i - \boldsymbol{\omega}_{ie}^i \times \mathbf{v}_e^i + \mathbf{g}_l^i \tag{3.16}$$

The measurements of specific force provided by the accelerometers are in body axes, as denoted by the vector quantity $\mathbf{f}^b$. In order to set up the navigation eqn. 3.16, the accelerometer outputs must be resolved into inertial axes to give $\mathbf{f}^i$. This may be achieved by pre-multiplying the measurement vector $\mathbf{f}^b$ by the direction cosine matrix $\mathbf{C}^i_b$, as described in Section 3.4.4 (eqn. 3.6). Given knowledge of the attitude of the body at the start of navigation, the matrix $\mathbf{C}^i_b$ is updated using eqns. 3.7 and 3.8 based on measurements of the body rates with respect to the i-frame which may be expressed as follows:

$$\boldsymbol{\omega}^b_{ib} = \begin{bmatrix} p & q & r \end{bmatrix}^T \tag{3.17}$$

Substituting for $\mathbf{f}^i$, from eqn. 3.6 in eqn. 3.16 gives the following form of the navigation equation:

$$\dot{\mathbf{v}}^i_e = \mathbf{C}^i_b \mathbf{f}^b - \boldsymbol{\omega}^i_{ie} \times \mathbf{v}^i_e + \mathbf{g}^i_l \tag{3.18}$$

The final term in this equation represents the local gravity vector expressed in the inertial frame.

A block diagram representation of the resulting inertial frame mechanisation is shown in Figure 3.11.

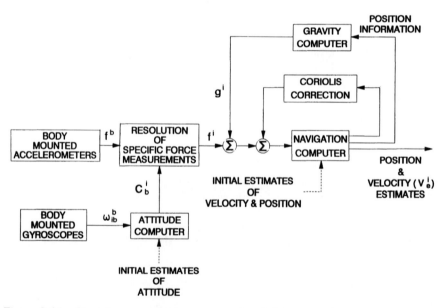

Figure 3.11   Strapdown navigation system—inertial frame mechanisation

3.5.2 Earth frame mechanisation

In this system, ground speed is expressed in an Earth fixed co-ordinate frame to give $\mathbf{v}^e_e$. It follows from the Coriolis equation, that the rate of change of $\mathbf{v}_e$, with respect to Earth axes, may be expressed in terms of its rate of change in inertial axes using:

$$\frac{d}{dt}\mathbf{v}_e\bigg|_e = \frac{d}{dt}\mathbf{v}_e\bigg|_i - \boldsymbol{\omega}_{ie} \times \mathbf{v}_e \tag{3.19}$$

Substituting for $\dfrac{d}{dt}\mathbf{v}_e\bigg|_i$ from eqn. 3.15 we have:

$$\frac{d}{dt}\mathbf{v}_e\bigg|_e = \mathbf{f} - 2\boldsymbol{\omega}_{ie} \times \mathbf{v}_e + \mathbf{g}_l \tag{3.20}$$

This may be expressed in Earth axes as follows:

$$\dot{\mathbf{v}}_e^e = \mathbf{C}_b^e \mathbf{f}^b - 2\boldsymbol{\omega}_{ie}^e \times \mathbf{v}_e^e + \mathbf{g}_l^e \tag{3.21}$$

where $\mathbf{C}_b^e$ is the direction cosine matrix used to transform the measured specific force vector into Earth axes. This matrix propagates in accordance with the following equation:

$$\dot{\mathbf{C}}_b^e = \mathbf{C}_b^e \boldsymbol{\Omega}_{eb}^b \tag{3.22}$$

where $\boldsymbol{\Omega}_{eb}^b$ is the skew symmetric form of $\boldsymbol{\omega}_{eb}^b$, the body rate with respect to the Earth fixed frame. This is derived by differencing the measured body rates, $\boldsymbol{\omega}_{ib}^b$, and estimates of the components of Earth's rate, $\boldsymbol{\omega}_{ie}^e$, expressed in body axes as follows:

$$\boldsymbol{\omega}_{eb}^b = \boldsymbol{\omega}_{ib}^b - \mathbf{C}_e^b \boldsymbol{\omega}_{ie}^e \tag{3.23}$$

in which $\mathbf{C}_e^b = \mathbf{C}_b^{eT}$, the transpose of the matrix $\mathbf{C}_b^e$.

A block diagram representation of the Earth frame mechanisation is shown in Figure 3.12.

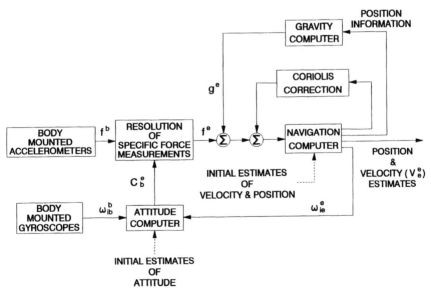

Figure 3.12   Strapdown navigation system—Earth frame mechanisation

A variation on this system may be used when one needs to navigate over relatively short distances, with respect to a fixed point on the Earth. A mechanisation of this type may be used for a tactical missile application in which navigation is required with respect to a ground based tracking station. In such a system, target tracking information provided by the ground station may need to be combined with the outputs of an onboard inertial navigation system to provide missile midcourse guidance commands. In order that the missile may operate in harmony with the ground systems, all information must be provided in a common frame of reference.

In this situation, an Earth fixed reference frame may be defined, the origin of which is located at the tracking station, its axes aligned with the local vertical and a plane which is tangential to the Earth's surface, as illustrated in Figure 3.13.

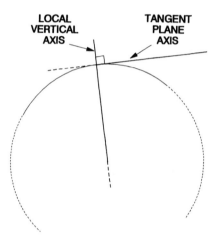

Figure 3.13   Tangent plane axis set

For very short term navigation, as required for some tactical missile applications, further simplifications to this system mechanisation may be permitted. For instance, where the navigation period is short, typically ten minutes or less, the effects of the rotation of the Earth on the attitude computation process can sometimes be ignored, and Coriolis corrections are no longer essential in the velocity equation to give sufficiently accurate navigation. In this situation, attitude is computed solely as a function of the turn rates measured by the gyroscopes, and eqn. 3.21 reduces to the following:

$$\dot{\mathbf{v}}_e^e = \mathbf{C}_b^e \mathbf{f}^b + \mathbf{g}_l^e \tag{3.24}$$

It is stressed that such simplifications can only be allowed in cases where the navigation errors, induced by the omission of Earth rate and Coriolis terms, lie within the error bounds in which the navigation system is required to operate. This situation arises when the permitted gyroscopic errors are in excess of the rotation rate of the Earth, and allowable accelerometer biases are in excess of the acceleration errors introduced by ignoring the Coriolis forces.

3.5.3 Navigation frame mechanisation

In order to navigate over large distances around the Earth, navigation information is most commonly required in the local geographic or navigation axis set described earlier, i.e. in terms of north and east velocity components, latitude, longitude and height above the Earth. Although such information can be computed using the position estimates provided by the inertial or Earth frame mechanisations described above, this involves a further transformation of the vector quantities $\mathbf{v}_e^i$ or $\mathbf{v}_e^e$. Further, difficulties arise in representing the Earth's gravitational field precisely in a computer. For these reasons, the navigation frame mechanisation, described here, is often used when navigating around the Earth.

In this mechanisation, ground speed is expressed in navigation coordinates to give $\mathbf{v}_e^n$. The rate of change of $\mathbf{v}_e^n$ with respect to navigation axes may be expressed in terms of its rate of change in inertial axes as follows:

$$\frac{d}{dt}\mathbf{v}_e\Big|_n = \frac{d}{dt}\mathbf{v}_e\Big|_i - (\boldsymbol{\omega}_{ie} + \boldsymbol{\omega}_{en}) \times \mathbf{v}_e \tag{3.25}$$

Substituting for $\dfrac{d}{dt}\mathbf{v}_e\Big|_i$ from eqn. 3.15 we have :

$$\frac{d}{dt}\mathbf{v}_e\Big|_n = \mathbf{f} - [2\boldsymbol{\omega}_{ie} + \boldsymbol{\omega}_{en}] \times \mathbf{v}_e + \mathbf{g}_l \tag{3.26}$$

This may be expressed in navigation axes as follows:

$$\dot{\mathbf{v}}_e^n = \mathbf{C}_b^n \mathbf{f}^b - [2\boldsymbol{\omega}_{ie}^n + \boldsymbol{\omega}_{en}^n] \times \mathbf{v}_e^n + \mathbf{g}_l^n \tag{3.27}$$

where $\mathbf{C}_b^n$ is a direction cosine matrix used to transform the measured specific force vector into navigation axes. This matrix propagates in accordance with the following equation:

$$\dot{\mathbf{C}}_b^n = \mathbf{C}_b^n \Omega_{nb}^b \tag{3.28}$$

where Ω_{nb}^b is the skew symmetric form of $\boldsymbol{\omega}_{nb}^b$, the body rate with respect to the navigation frame. This is derived by differencing the measured body rates, ω_{ib}^b, and estimates of the components of navigation frame rate, ω_{in}. The latter term is obtained by summing the Earth's rate with respect to the inertial frame and the turn rate of the navigation frame with respect to the Earth, i.e. $\omega_{in} = \omega_{ie} + \omega_{en}$. Therefore :

$$\omega_{nb}^b = \omega_{ib}^b - C_n^b\left[\omega_{ie}^n + \omega_{en}^n\right] \tag{3.29}$$

A block diagram representation of the navigation frame mechanisation is shown in Figure 3.14.

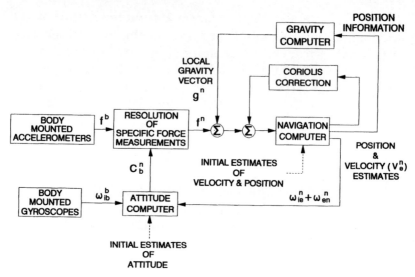

Figure 3.14   Strapdown navigation system—navigation frame mechanisation

It is instructive to consider the physical significance of the various terms in the navigation eqn. 3.27. From this equation, it can be seen that the rate of change of the velocity, with respect to the surface of the Earth, is made up of the following terms:

(i)   The specific force acting on the vehicle, as measured by a triad of accelerometers mounted within it;

(ii) A correction for the acceleration caused by the vehicle's velocity over the surface of a rotating Earth, usually referred to as the Coriolis acceleration. The effect in two dimensions is illustrated in Figure 3.15. As the point P moves away from the axis of rotation, it traces out a curve in space as a result of the Earth's rotation;

(iii) A correction for the centripetal acceleration of the vehicle, resulting from its motion over the Earth's surface. For instance, a vehicle moving due east over the surface of the Earth will trace out a circular path with respect to inertial axes. To follow this path, the vehicle is subject to a force acting towards the centre of the Earth of magnitude equal to the product of its mass, its linear velocity and its turn rate with respect to the Earth;

(iv) Compensation for the apparent gravitational force acting on the vehicle. This includes the gravitational force caused by the mass attraction of the Earth, and the centripetal acceleration of the vehicle resulting from the rotation of the Earth. The latter term arises even if the vehicle is stationary with respect to the Earth, since the path which its follows in space is circular.

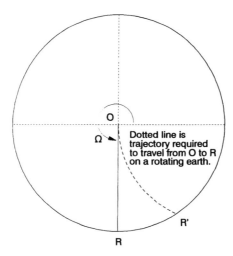

Figure 3.15   Illustration of the effect of Coriolis acceleration

A simple example serves to illustrate the importance of the Coriolis effect. Consider a vehicle launched from the North Pole with the intention of flying to New York. The vehicle is assumed to travel at an average speed of 3600 miles/hour. During the flight, of approximately one hour, the Earth will have rotated by about 15 degrees, a distance of approximately 900 miles at the latitude of New York. Consequently, if no Coriolis correction is made to the onboard inertial guidance system during the course of the flight, the vehicle will arrive in the Chicago area rather than New York as originally intended.

3.5.4 Wander azimuth mechanisation

In the navigation frame mechanisation described in the previous section, singularities arise in the equations as a result of vector quantities being expressed in navigation axes. For example, the turn rate of the navigation frame, the transport rate, may be expressed in component form as:

$$\boldsymbol{\omega}_{en}^{n} = \left[\frac{v_E}{\left(R_0 + h \right)} \frac{-v_N}{\left(R_0 + h \right)} \frac{-v_N \tan L}{\left(R_0 + h \right)} \right]^{T}$$

(3.30)

where v_N = north velocity
 v_E = east velocity
 R_0 = radius of the Earth
 L = latitude
 h = height above ground

It will be seen that the third component of the transport rate becomes indeterminate at the geographic poles. One way of avoiding such singularities, and so providing a navigation system with worldwide capability, is to adopt a wander azimuth mechanisation.

A wander axis system is a locally level frame which moves over the Earth's surface with the vehicle. However, as the name implies, the azimuth angle between true north and the x axis of the wander axis frame varies with vehicle position on the Earth. This variation is chosen in order to avoid discontinuities in the orientation of the wander frame with respect to the Earth as the vehicle passes over either the North or South Pole.

A navigation equation for a wander azimuth system, which is similar in form to eqn. 3.27, may be constructed as follows:

$$\dot{\mathbf{v}}_e^w = \mathbf{C}_b^w \mathbf{f}^b - \left[2\mathbf{C}_e^w \boldsymbol{\omega}_{ie}^e + \boldsymbol{\omega}_{ew}^w \right] \times \mathbf{v}_e^w + \mathbf{g}_l^w$$

(3.31)

This equation is integrated to generate estimates of a vehicle's ground speed in the wander azimuth frame, $\mathbf{v}_e^w$. This is then used to generate the turn rate of the wander frame with respect to the Earth, $\boldsymbol{\omega}_{ew}^w$. The direction cosine matrix which relates the wander frame to the Earth frame, $\mathbf{C}_e^w$, may be updated using the equation:

$$\dot{\mathbf{C}}_e^w = \mathbf{C}_e^w \Omega_{ew}^e$$

(3.32)

where Ω_{ew}^w is a skew symmetric matrix formed from the elements of the angular rate vector $\boldsymbol{\omega}_{ew}^w$. This process is implemented iteratively and enables any singularities to be avoided. Further details concerning wander azimuth systems and the mechanisations described earlier appear in Reference 1.

3.5.5 Summary of strapdown system mechanisations

This section has provided outline descriptions of a number of possible strapdown inertial navigation system mechanisations. Further details are given in Reference 1. The choice of mechanisation is dependent on the application. Although any of the schemes described may be used for navigation close to the Earth, the navigation frame mechanisation is commonly employed for navigation over large distances. The wander azimuth system provides a worldwide navigation capability. These mechanisations provide navigation data in terms of north and east velocity, latitude and longitude and allow a relatively simple gravity model to be used. For navigation over shorter distances, an Earth fixed reference system may be applicable.

3.6 Strapdown attitude representations

3.6.1 Introductory remarks

Consider now ways in which a set of strapdown gyroscopic sensors may be used to instrument a reference co-ordinate frame within a vehicle, which is free to rotate about any direction. The attitude of the vehicle with respect to the designated reference frame may be stored as a set of numbers in a computer within the vehicle. The stored attitude is updated as the vehicle rotates using the measurements of turn rate provided by the gyroscopes.

The co-ordinate frames referred to during the course of the discussion which follows are orthogonal, right handed axis sets in which positive rotations about each axis are taken to be in a clockwise direction looking along the axis from the origin, as indicated in the Figure 3.16. A negative rotation acts in an opposite sense, i.e. in an anticlockwise direction. This convention is used throughout this book.

It is important to remember that the change in attitude of a body, which is subjected to a series of rotations about different axes, is not only a function of the angles through which it rotates about each of those axes, but the order in which the rotations occur. The illustration given in Figure 3.17, although somewhat extreme, shows quite clearly that the order in which a sequence of rotations occurs is most important.

Rotations are defined here with respect to the orthogonal right handed axis set, *Oxyz*, indicated in the Figure. The sequence of

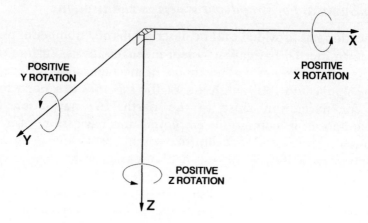

Figure 3.16   Definition of axis rotations

rotations shown in the left half of the figure is made up of a 90° pitch, or y axis rotation, followed by a 90° yaw, or z axis rotation, and a further pitch rotation of −90°. On completion of this sequence of turns it can be seen that a net rotation of 90° about the roll (x) axis has taken place. In the right hand half of the figure, the order of the rotations has been reversed. Although the body still ends up with its roll axis aligned in the original direction, it is seen that a net roll rotation of −90° has now taken place. Hence, individual axis rotations are said to be non-commutative. It is clear that failure to take account of the order in which rotations arise can lead to a substantial error in the computed attitude.

Various mathematical representations can be used to define the attitude of a body with respect to a co-ordinate reference frame. The parameters associated with each method may be stored within a computer and updated as the vehicle rotates using the measurements of turn rate provided by the strapdown gyroscopes. Three attitude representations are described here, namely:

(i)   **Direction cosines.** The direction cosine matrix, introduced in Section 3.5, is a 3×3 matrix, the columns of which represent unit vectors in body axes projected along the reference axes.

(ii)   **Euler angles**. A transformation from one co-ordinate frame to another is defined by three successive rotations about different axes taken in turn. The Euler angle representation is perhaps one of the simplest techniques in terms of physical appreciation. The three angles correspond to the angles which would be

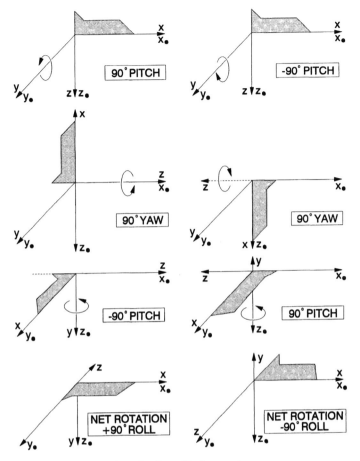

Figure 3.17 Illustration of effect of order of body rotations

measured between a set of mechanical gimbals[1], which is supporting a stable element, where the axes of the stable element represent the reference frame, and with the body being attached via a bearing to the outer gimbal.

(iii) **Quaternions.** The quaternion attitude representation allows a transformation from one co-ordinate frame to another to be effected by a single rotation about a vector defined in the reference frame. The quaternion is a four element vector representation, the elements of which are functions of the orientation of this vector and the magnitude of the rotation.

---

[1] A gimbal is a rigid mechanical frame which is free to rotate about a single axis to isolate it from angular motion in that direction. A stable platform can be isolated from body motion if supported by three such frames with their axes of rotation nominally orthogonal to each other.

In the following sections, each of these attitude representations is described in some detail.

3.6.2 Direction cosine matrix

3.6.2.1 Introduction

The direction cosine matrix, denoted here by the symbol $\mathbf{C}_b^n$, is a 3×3 matrix, the columns of which represent unit vectors in body axes projected along the reference axes. $\mathbf{C}_b^n$ is written here in component form as follows:

$$\mathbf{C}_b^n = \begin{pmatrix} c_{11} & c_{12} & c_{13} \\ c_{21} & c_{22} & c_{23} \\ c_{31} & c_{32} & c_{33} \end{pmatrix} \tag{3.33}$$

The element in the ith row and the jth column represents the cosine of the angle between the i axis of the reference frame and the j axis of the body frame.

3.6.2.2 Use of direction cosine matrix for vector transformation

A vector quantity defined in body axes, $\mathbf{r}^b$, may be expressed in reference axes by pre-multiplying the vector by the direction cosine matrix as follows:

$$\mathbf{r}^n = \mathbf{C}_b^n \mathbf{r}^b \tag{3.34}$$

3.6.2.3 Propagation of direction cosine matrix with time

The rate of change of $\mathbf{C}_b^n$ with time is given by:

$$\begin{aligned} \dot{\mathbf{C}}_b^n &= \lim_{\delta t \to 0} \frac{\delta \mathbf{C}_b^n}{\delta t} \\ &= \lim_{\delta t \to 0} \frac{\mathbf{C}_b^n(t + \delta t) - \mathbf{C}_b^n(t)}{\delta t} \end{aligned} \tag{3.35}$$

where $\mathbf{C}_b^n(t)$ and $\mathbf{C}_b^n(t + \delta t)$ represent the direction cosine matrix at times t and $t + \delta t$ respectively. $\mathbf{C}_b^n(t + \delta t)$ can be written as the product of two matrices as follows:

$$\mathbf{C}_b^n(t + \delta t) = \mathbf{C}_b^n(t)\,\mathbf{A}(t) \tag{3.36}$$

where $\mathbf{A}(t)$ is a direction cosine matrix which relates the b-frame at time t to the b-frame at time $t + \delta t$. For small angle rotations, $\mathbf{A}(t)$ may be written as follows:

$$\mathbf{A}(t) = [\mathbf{I} + \delta\Psi] \qquad (3.37)$$

where $\mathbf{I}$ is a 3×3 identity matrix and

$$\delta\Psi = \begin{pmatrix} 0 & -\delta\psi & \delta\theta \\ \delta\psi & 0 & -\delta\phi \\ -\delta\theta & \delta\phi & 0 \end{pmatrix} \qquad (3.38)$$

in which $\delta\psi$, $\delta\theta$ and $\delta\phi$ are the small rotation angles through which the b-frame has rotated over the time interval δt about its yaw, pitch and roll axes respectively. In the limit as δt approaches zero, small angle approximations are valid and the order of the rotations becomes unimportant.

Substituting for $\mathbf{C}_b^n(t + \delta t)$ in eqn. 3.35 we obtain:

$$\dot{\mathbf{C}}_b^n = \mathbf{C}_o^n \lim_{\delta t \to 0} \frac{\delta\Psi}{\delta t} \qquad (3.39)$$

In the limit as $\delta t \to 0$, $\delta\Psi / \delta t$ is the skew symmetric form of the angular rate vector $\boldsymbol{\omega}_{nb}^b = [\omega_x \; \omega_y \; \omega_z]^T$, which represents the turn rate of the b-frame with respect to the n-frame expressed in body axes, i.e.

$$\lim_{\delta t \to 0} \frac{\delta\Psi}{\delta t} = \Omega_{nb}^b \qquad (3.40)$$

Substituting in eqn. 3.39 gives:

$$\dot{\mathbf{C}}_b^n = \mathbf{C}_b^n \Omega_{nb}^b \qquad (3.41)$$

where

$$\Omega_{nb}^b = \begin{pmatrix} 0 & -\omega_z & \omega_y \\ \omega_z & 0 & -\omega_x \\ -\omega_y & \omega_x & 0 \end{pmatrix} \qquad (3.42)$$

An equation of the form of eqn. 3.41 may be solved within a computer in a strapdown inertial navigation system to keep track of body attitude with respect to the chosen reference frame. It may be expressed in component form as follows:

$$\dot{c}_{11} = c_{12}\omega_z - c_{13}\omega_y \quad \dot{c}_{12} = c_{13}\omega_x - c_{11}\omega_z \quad \dot{c}_{13} = c_{11}\omega_y - c_{12}\omega_x$$

$$\dot{c}_{21} = c_{22}\omega_z - c_{23}\omega_y \quad \dot{c}_{22} = c_{23}\omega_x - c_{21}\omega_z \quad \dot{c}_{23} = c_{21}\omega_y - c_{22}\omega_x$$

$$\dot{c}_{31} = c_{32}\omega_z - c_{33}\omega_y \quad \dot{c}_{32} = c_{33}\omega_x - c_{31}\omega_z \quad \dot{c}_{33} = c_{31}\omega_y - c_{32}\omega_x \qquad (3.43)$$

3.6.3 Euler angles

3.6.3.1 Introduction

A transformation from one co-ordinate frame to another can be carried out as three successive rotations about different axes. For example, a transformation from reference axes to a new co-ordinate frame may be expressed as follows:

> rotate through angle ψ about reference z axis
>
> rotate through angle θ about new y axis
>
> rotate through angle ϕ about new x axis

where ψ, θ and ϕ are referred to as the Euler rotation angles. This type of representation is popular because of the   physical significance of the Euler angles which correspond to the angles which would be measured by angular pick-offs between a set of three gimbals in a stable platform inertial navigation system.

3.6.3.2 Use of Euler angles for vector transformation

The three rotations may be expressed mathematically as three separate direction cosine matrices as defined below:

$$\text{rotation } \psi \text{ about } z \text{ axis, } \mathbf{C}_1 = \begin{pmatrix} \cos\psi & \sin\psi & 0 \\ -\sin\psi & \cos\psi & 0 \\ 0 & 0 & 1 \end{pmatrix} \tag{3.44}$$

$$\text{rotation } \theta \text{ about } y \text{ axis, } \mathbf{C}_2 = \begin{pmatrix} \cos\theta & 0 & -\sin\theta \\ 0 & 1 & 0 \\ \sin\theta & 0 & \cos\theta \end{pmatrix} \tag{3.45}$$

$$\text{rotation } \phi \text{ about } x \text{ axis, } \mathbf{C}_3 = \begin{pmatrix} 1 & 0 & 0 \\ 0 & \cos\phi & \sin\phi \\ 0 & -\sin\phi & \cos\phi \end{pmatrix} \tag{3.46}$$

Thus a transformation from reference to body axes may be expressed as the product of these three separate transformations as follows:

$$\mathbf{C}_n^b = \mathbf{C}_3\mathbf{C}_2\mathbf{C}_1 \tag{3.47}$$

Similarly, the inverse transformation from body to reference axes is given by:

$$\mathbf{C}_b^n = \mathbf{C}_n^{bT} = \mathbf{C}_1^T \mathbf{C}_2^T \mathbf{C}_3^T \tag{3.48}$$

$$\mathbf{C}_b^n = \begin{pmatrix} \cos\psi & -\sin\psi & 0 \\ \sin\psi & \cos\psi & 0 \\ 0 & 0 & 1 \end{pmatrix} \begin{pmatrix} \cos\theta & 0 & \sin\theta \\ 0 & 1 & 0 \\ -\sin\theta & 0 & \cos\theta \end{pmatrix} \begin{pmatrix} 1 & 0 & 0 \\ 0 & \cos\phi & -\sin\phi \\ 0 & \sin\phi & \cos\phi \end{pmatrix}$$

$$= \begin{pmatrix} \cos\theta\cos\psi & -\cos\phi\sin\psi + \sin\phi\sin\theta\cos\psi & \sin\phi\sin\psi + \cos\phi\sin\theta\cos\psi \\ \cos\theta\sin\psi & \cos\phi\cos\psi + \sin\phi\sin\theta\sin\psi & -\sin\phi\cos\psi + \cos\phi\sin\theta\sin\psi \\ -\sin\theta & \sin\phi\cos\theta & \cos\phi\cos\theta \end{pmatrix} \tag{3.49}$$

This is the direction cosine matrix given by eqn. 3.33 expressed in terms of Euler angles.

For small angle rotations, $\sin\phi \to \phi$, $\sin\theta \to \theta$, $\sin\psi \to \psi$ and the cosines of these angles approach unity. Making these substitutions in eqn. 3.49 and ignoring products of angles which also become small, the direction cosine matrix expressed in terms of the Euler rotations reduces approximately to the skew symmetric form:

$$\mathbf{C}_b^n \approx \begin{pmatrix} 1 & -\psi & \theta \\ \psi & 1 & -\phi \\ -\theta & \phi & 1 \end{pmatrix} \tag{3.50}$$

This form of matrix is used in Chapter 10 to represent the small change in attitude which occurs between successive updates in the real time computation of body attitude, and in Chapters 9 and 11 to represent the error in the estimated direction cosine matrix.

3.6.3.3 Propagation of Euler angles with time

Following the gimbal analogy mentioned above, ϕ, θ and ψ are the gimbal angles and $\dot{\phi}$, $\dot{\theta}$ and $\dot{\psi}$ are the gimbal rates. The gimbal rates are related to the body rates, ω_x, ω_y and ω_z as follows:

$$\begin{pmatrix} \omega_x \\ \omega_y \\ \omega_z \end{pmatrix} = \begin{pmatrix} \dot{\phi} \\ 0 \\ 0 \end{pmatrix} + \mathbf{C}_3 \begin{pmatrix} 0 \\ \dot{\theta} \\ 0 \end{pmatrix} + \mathbf{C}_3\mathbf{C}_2 \begin{pmatrix} 0 \\ 0 \\ \dot{\psi} \end{pmatrix} \tag{3.51}$$

This equation can be rearranged and expressed in component form as follows:

$$\dot{\phi} = \left(\omega_y \sin\phi + \omega_z \cos\phi\right)\tan\theta + \omega_x$$

$$\dot{\theta} = \omega_y \cos\phi - \omega_z \sin\phi \tag{3.52}$$

$$\dot{\psi} = \left(\omega_y \sin\phi + \omega_z \cos\phi\right)\sec\theta$$

Equations of this form may be solved in a strapdown system to update the Euler rotations of the body with respect to the chosen reference frame. However, their use is limited since the solution of the $\dot{\phi}$ and $\dot{\psi}$ equations become indeterminate when $\theta = \pm 90°$.

3.6.4 Quaternions

3.6.4.1 Introduction

The quaternion attitude representation is a four parameter representation based on the idea that a transformation from one co-ordinate frame to another may be effected by a single rotation about a vector μ defined with respect to the reference frame. The quaternion, denoted here by the symbol q, is a four element vector, the elements of which are functions of this vector and the magnitude of the rotation:

$$q = \begin{pmatrix} a \\ b \\ c \\ d \end{pmatrix} = \begin{pmatrix} \cos(\mu/2) \\ (\mu_x/\mu)\sin(\mu/2) \\ (\mu_y/\mu)\sin(\mu/2) \\ (\mu_z/\mu)\sin(\mu/2) \end{pmatrix} \tag{3.53}$$

where μ_x, μ_y, μ_z = components of the angle vector μ

and μ = magnitude of μ

The magnitude and direction of μ are defined in order that the reference frame may be rotated into coincidence with the body frame by rotating about μ through an angle μ.

A quaternion with components a, b, c and d may also be expressed as a four parameter complex number with a real component a, and three imaginary components, b, c and d, as follows:

$$q = a + ib + jc + kd \tag{3.54}$$

This is an extension of the more usual two parameter complex number form with one real component and one imaginary component, $x = a + ib$, with which the reader is more likely to be familiar.

The product of two quaternions,

$$q = a + ib + jc + kd \text{ and } p = e + if + jg + kh$$

may then be derived as shown below applying the usual rules for products of complex numbers:

$$\mathbf{i} \cdot \mathbf{i} = -1 \qquad \mathbf{i} \cdot \mathbf{j} = \mathbf{k} \qquad \mathbf{j} \cdot \mathbf{i} = -\mathbf{k} \dots \text{etc.}$$

Hence $\mathbf{q} \cdot \mathbf{p} = (a + ib + jc + kd)(e + if + jg + kh)$

$$= ea - bf - cg - dh$$
$$+ (af + be + ch - dg)\mathbf{i}$$
$$+ (ag + ce - bh + df)\mathbf{j}$$
$$+ (ah + de + bg - cf)\mathbf{k} \qquad (3.55)$$

Alternatively, the quaternion product may be expressed in matrix form as:

$$\mathbf{q} \cdot \mathbf{p} = \begin{pmatrix} a & -b & -c & -d \\ b & a & -d & c \\ c & d & a & -b \\ d & -c & b & a \end{pmatrix} \begin{pmatrix} e \\ f \\ g \\ h \end{pmatrix} \qquad (3.56)$$

3.6.4.2 Use of quaternions for vector transformation

A vector quantity defined in body axes, $\mathbf{r}^b$, may be expressed in reference axes as $\mathbf{r}^n$ using the quaternion directly. First define a quaternion, $\mathbf{r}^{b'}$ in which the complex components are set equal to the components of $\mathbf{r}^b$, and with a zero scalar component, i.e. if:

$$\mathbf{r}^b = \mathbf{i}x + \mathbf{j}y + \mathbf{k}z$$
$$\mathbf{r}^{b'} = 0 + \mathbf{i}x + \mathbf{j}y + \mathbf{k}z$$

This is expressed in reference axes as $\mathbf{r}^{n'}$ using:

$$\mathbf{r}^{n'} = \mathbf{q}\,\mathbf{r}^{b'}\,\mathbf{q}^* \qquad (3.57)$$

where $\mathbf{q}^* = (a - ib - jc - kd)$, the complex conjugate of $\mathbf{q}$.

Hence

$$\mathbf{r}^{n'} = (a + ib + jc + kd)(0 + ix + jy + kz)(a - ib - jc - kd)$$

$$= 0$$
$$+\{(a^2 + b^2 - c^2 - d^2)x + 2(bc - ad)y + 2(bd + ac)z\}\mathbf{i}$$
$$+\{2(bc + ad)x + (a^2 - b^2 + c^2 - d^2)y + 2(cd - ab)z\}\mathbf{j}$$
$$+\{2(bd - ac)x + 2(cd + ab)y + (a^2 - b^2 - c^2 + d^2)z\}\mathbf{k} \qquad (3.58)$$

Alternatively, $\mathbf{r}^{n'}$ may be expressed in matrix form as follows:

$$\mathbf{r}^{n'} = \mathbf{C}'\mathbf{r}^{b'}$$

where $\quad \mathbf{C}' = \begin{pmatrix} 0 & 0 \\ 0 & \mathbf{C} \end{pmatrix} \quad \mathbf{r}^{b'} = \begin{pmatrix} 0 \\ \mathbf{r}^b \end{pmatrix}$

and $\mathbf{C} = \begin{pmatrix} \left(a^2 + b^2 - c^2 - d^2\right) & 2(bc - ad) & 2(bd + ac) \\ 2(bc + ad) & \left(a^2 - b^2 + c^2 - d^2\right) & 2(cd - ab) \\ 2(bd - ac) & 2(cd + ab) & \left(a^2 - b^2 - c^2 + d^2\right) \end{pmatrix}$

$$(3.59)$$

which is equivalent to writing:

$$\mathbf{r}^n = \mathbf{C}\,\mathbf{r}^b$$

Comparison with eqn. 3.34 reveals that $\mathbf{C}$ is equivalent to the direction cosine matrix $\mathbf{C}_b^n$.

3.6.4.3 Propagation of a quaternion with time

The quaternion, $\mathbf{q}$, propagates in accordance with the following equation:

$$\dot{\mathbf{q}} = 0.5\mathbf{q} \cdot \mathbf{p}_{nb}^b \qquad (3.60)$$

This equation may be expressed in matrix form as a function of the components of $\mathbf{q}$ and $\mathbf{p}_{nb}^b = [0, \boldsymbol{\omega}_{nb}^{bT}]^T$ as follows:

$$\dot{\mathbf{q}} = \begin{pmatrix} \dot{a} \\ \dot{b} \\ \dot{c} \\ \dot{d} \end{pmatrix} = 0.5 \begin{pmatrix} a & -b & -c & -d \\ b & a & -d & c \\ c & d & a & -b \\ d & -c & b & a \end{pmatrix} \begin{pmatrix} 0 \\ \omega_x \\ \omega_y \\ \omega_z \end{pmatrix} \qquad (3.61)$$

i.e. $\quad \dot{a} = -0.5\left(b\omega_x + c\omega_y + d\omega_z\right)$

$\qquad \dot{b} = 0.5\left(a\omega_x - d\omega_y + c\omega_z\right)$

$\qquad \dot{c} = 0.5\left(d\omega_x + a\omega_y - b\omega_z\right) \qquad (3.62)$

$\qquad \dot{d} = -0.5\left(c\omega_x - b\omega_y - a\omega_z\right)$

Equations of this form may be solved in a strapdown navigation system to keep track of the quaternion parameters which define body orientation. The quaternion parameters may then be used to compute an equivalent direction cosine matrix, or used directly to transform the measured specific force vector into the chosen reference frame (see eqn. 3.57).

3.6.5 Relationships between direction cosines, Euler angles and quaternions

As shown in the preceding sections, the direction cosines may be expressed in terms of Euler angles or quaternions:

$$
\mathbf{C}_b^n = \begin{pmatrix} c_{11} & c_{12} & c_{13} \\ c_{21} & c_{22} & c_{23} \\ c_{31} & c_{32} & c_{33} \end{pmatrix}
$$

$$
= \begin{pmatrix} \cos\theta\cos\psi & -\cos\phi\sin\psi + \sin\phi\sin\theta\cos\psi & \sin\phi\sin\psi + \cos\phi\sin\theta\cos\psi \\ \cos\theta\sin\psi & \cos\phi\cos\psi + \sin\phi\sin\theta\sin\psi & -\sin\phi\cos\psi + \cos\phi\sin\theta\sin\psi \\ -\sin\theta & \sin\phi\cos\theta & \cos\phi\cos\theta \end{pmatrix}
$$

$$
= \begin{pmatrix} \left(a^2 + b^2 - c^2 - d^2\right) & 2(bc - ad) & 2(bd + ac) \\ 2(bc + ad) & \left(a^2 - b^2 + c^2 - d^2\right) & 2(cd - ab) \\ 2(bd - ac) & 2(cd + ab) & \left(a^2 - b^2 - c^2 + d^2\right) \end{pmatrix} \tag{3.63}
$$

By comparing the elements of the above equations, the quaternion elements may be expressed directly in terms of Euler angles or direction cosines. Similarly, the Euler angles may be written in terms of direction cosines or quaternions. Some of these relationships are summarised below.

3.6.5.1 Quaternions expressed in terms of direction cosines

For small angular displacements, the quaternion parameters may be derived using the following relationships:

$$
a = \frac{1}{2}\left(1 + c_{11} + c_{22} + c_{33}\right)^{1/2}
$$

$$
b = \frac{1}{4a}\left(c_{32} - c_{23}\right)
$$

$$
c = \frac{1}{4a}\left(c_{13} - c_{31}\right)
$$

$$
d = \frac{1}{4a}\left(c_{21} - c_{12}\right) \tag{3.64}
$$

A more comprehensive algorithm for the extraction of quaternion parameters from the direction cosines, which takes account of the relative magnitudes of the direction cosine elements, is described by Shepperd [2]

3.6.5.2 Quaternions expressed in terms of Euler angles

$$a = \cos\frac{\phi}{2}\cos\frac{\theta}{2}\cos\frac{\psi}{2} + \sin\frac{\phi}{2}\sin\frac{\theta}{2}\sin\frac{\psi}{2}$$

$$b = \sin\frac{\phi}{2}\cos\frac{\theta}{2}\cos\frac{\psi}{2} - \cos\frac{\phi}{2}\sin\frac{\theta}{2}\sin\frac{\psi}{2}$$

$$c = \cos\frac{\phi}{2}\sin\frac{\theta}{2}\cos\frac{\psi}{2} + \sin\frac{\phi}{2}\cos\frac{\theta}{2}\sin\frac{\psi}{2}$$

$$d = \cos\frac{\phi}{2}\cos\frac{\theta}{2}\sin\frac{\psi}{2} + \sin\frac{\phi}{2}\sin\frac{\theta}{2}\cos\frac{\psi}{2} \qquad (3.65)$$

3.6.5.3 Euler angles expressed in terms of direction cosines

The Euler angles may be derived directly from the direction cosines as described below. For conditions where θ is not equal to 90° the Euler angles can be determined using:

$$\phi = \arctan\left\{\frac{c_{32}}{c_{33}}\right\}$$

$$\theta = \arcsin\left\{-c_{31}\right\}$$

$$\psi = \arctan\left\{\frac{c_{21}}{c_{11}}\right\} \qquad (3.66)$$

For situations in which θ approaches $\pi/2$ radians, the equations in ϕ and ψ become indeterminate because the numerator and the denominator approach zero simultaneously. Under such conditions, alternative solutions for ϕ and ψ are sought based upon other elements of the direction cosine matrix. This difficulty may be overcome by using the direction cosine elements c_{12}, c_{13}, c_{22} and c_{23}, which do not appear in eqn. 3.66, to derive the following relationships:

$$c_{23} + c_{12} = (\sin\theta - 1)\sin(\psi + \phi)$$

$$c_{13} - c_{22} = (\sin\theta - 1)\cos(\psi + \phi)$$

$$c_{23} - c_{12} = (\sin\theta + 1)\sin(\psi - \phi)$$

$$c_{13} + c_{22} = (\sin\theta + 1)\cos(\psi - \phi) \qquad (3.67)$$

For θ near $+\pi/2$:

$$\psi - \phi = \arctan\left\{\frac{c_{23} - c_{12}}{c_{13} + c_{22}}\right\}$$

For θ near $-\pi/2$:

$$\psi + \phi = \arctan\left\{\frac{c_{23} + c_{12}}{c_{13} - c_{22}}\right\} \qquad (3.68)$$

Eqns. 3.67 and 3.68 provide values for the sum and difference of ϕ and ψ under conditions where θ approaches $\pi/2$. Separate solutions for ϕ and ψ cannot be obtained when $\theta = \pm\pi/2$ because both become measures of angle about parallel axes (about the vertical), i.e. a degree of rotational freedom is lost. This is equivalent to the gimbal lock condition which arises in a set of mechanical gimbals when the pitch, or inner, gimbal is rotated through 90°.

When θ approaches $\pm\pi/2$, either ϕ or ψ may be selected arbitrarily to satisfy some other condition whilst the unspecified angle is chosen to satisfy eqn. 3.68. To avoid jumps in the values of ϕ or ψ between successive calculations when θ is in the region of $\pm\pi/2$, one approach would be to freeze one angle, ϕ for instance, at its current value and to calculate ψ in accordance with eqn. 3.68. At the next iteration, ψ would be frozen and ϕ determined using eqn. 3.68. This process of updating ϕ or ψ alone at successive iterations would continue until θ is no longer in the region of $\pm\pi/2$.

3.7 Detailed navigation equations

3.7.1 Navigation equations expressed in component form

For a terrestrial navigation system, operating in the local geographic reference frame, it has been shown (Section 3.4) that the navigation equation may be expressed as follows:

$$\dot{\mathbf{v}}_e^n = \mathbf{f}^n - \left(2\boldsymbol{\omega}_{ie}^n + \boldsymbol{\omega}_{en}^n\right) \times \mathbf{v}_e^n + \mathbf{g}_l^n \qquad (3.69)$$

where, $\mathbf{v}_e^n$ represents velocity with respect to the Earth expressed in the local geographic frame defined by the directions of true north, east and the local vertical. In component form:

$$\mathbf{v}_e^n = \begin{bmatrix} v_N & v_E & v_D \end{bmatrix}^T \qquad (3.70)$$

$\mathbf{f}^n$ is the specific force vector as measured by a triad of accelerometers and resolved into the local geographic reference frame:

$$\mathbf{f}^n = \begin{bmatrix} v_N & v_E & v_D \end{bmatrix}^T \qquad (3.71)$$

$\boldsymbol{\omega}_{ie}^n$ is the turn rate of the Earth expressed in the local geographic frame:

$$\boldsymbol{\omega}_{ie}^n = \begin{bmatrix} \Omega \cos L & 0 & -\Omega \sin L \end{bmatrix}^T \qquad (3.72)$$

$\boldsymbol{\omega}_{en}^n$ represents the turn rate of the local geographic frame with respect to the Earth fixed frame; the transport rate. This quantity may be expressed in terms of the rate of change of latitude and longitude as follows:

$$\boldsymbol{\omega}_{en}^{n} = \left[\dot{\lambda}\cos L \quad -\dot{L} \quad -\dot{\lambda}\sin L\right]^{T} \tag{3.73}$$

Writing $\dot{\lambda} = v_E / (R_0 + h)\cos L$ and $\dot{L} = v_N / (R_0 + h)$ yields :

$$\boldsymbol{\omega}_{en}^{n} = \left[v_E / (R_0 + h) \quad -v_N / (R_0 + h) \quad -v_E \tan L / (R_0 + h)\right]^{T} \tag{3.74}$$

where R_0 is the radius of the Earth and h is the height above the surface of the Earth.

$\mathbf{g}_l^n$ is the local gravity vector which includes the combined effects of the mass attraction of the Earth ($\mathbf{g}$) and the centripetal acceleration caused by the Earth's rotation ($\boldsymbol{\omega}_{ie} \times \boldsymbol{\omega}_{ie} \times \mathbf{R}$). Hence, we may write:

$$\mathbf{g}_l^n = \mathbf{g} - \boldsymbol{\omega}_{ie} \times \left[\boldsymbol{\omega}_{ie} \times \mathbf{R}\right] \tag{3.75}$$

$$= \mathbf{g} - \frac{\Omega^2(R_0 + h)}{2}\begin{pmatrix} \sin 2L \\ 0 \\ (1 + \cos 2L) \end{pmatrix}$$

The navigation equation may be expressed in component form as follows:

$$\dot{v}_N = f_N - v_E\left(2\Omega + \dot{\lambda}\right)\sin L + v_D\dot{L} + \xi g$$

$$= f_N - 2\Omega v_E \sin L + \frac{\left(v_N v_D - v_E^2 \tan L\right)}{(R_0 + h)} + \xi g \tag{3.76}$$

$$\dot{v}_E = f_E + v_N\left(2\Omega + \dot{\lambda}\right)\sin L + v_D\left(2\Omega + \dot{\lambda}\right)\cos L - \eta g$$

$$= f_E - 2\Omega(v_N \sin L + v_D \cos L) + \frac{v_E}{(R_0 + h)}(v_D + v_N \tan L) - \eta g \tag{3.77}$$

$$\dot{v}_D = f_D - v_E\left(2\Omega + \dot{\lambda}\right)\cos L - v_N\dot{L} + g$$

$$= f_D - 2\Omega v_E \cos L - \frac{\left(v_E^2 + v_N^2\right)}{(R_0 + h)} + g \tag{3.78}$$

where ξ and η represent angular deflections in the direction of the local gravity vector with respect to the local vertical due to gravity anomalies, as discussed in Section 3.7.3.

Latitude, longitude and height above the surface of the Earth are given by:

$$\dot{L} = \frac{v_N}{(R_0 + h)} \tag{3.79}$$

$$\dot{\lambda} = \frac{v_E \sec L}{(R_0 + h)} \tag{3.80}$$

$$\dot{h} = -v_D \tag{3.81}$$

It is assumed, in the equations given above, that the Earth is perfectly spherical in shape. Additionally, it is assumed that there is no variation in the Earth's gravitational field with changes in the position of the navigation system on the Earth or its height above the surface of the Earth.

The modifications which must be applied to the navigation equations in order to take account of the errors introduced by these assumptions and so permit accurate navigation over the surface of the Earth are summarised briefly in the following sections. The reader requiring a more detailed analysis of these effects is referred to the texts by Britting [3] and Steiler and Winter [4] in which such aspects are discussed in detail.

3.7.2 The shape of the Earth

In order to determine position on the Earth using inertial measurements, it is necessary to make some assumptions regarding the shape of the Earth. The spherical model assumed so far is not sufficiently representative for very accurate navigation. Owing to the slight flattening of the Earth at the poles, it is customary to model the Earth as a reference ellipsoid, which approximates more closely to the true geometry. Terrestrial navigation involves the determination of velocity and position relative to a navigational grid which is based on the reference ellipsoid.

In accordance with this model, illustrated in section in Figure 3.18, the following parameters are defined:

length of the semi-major axis,	R	$= 6378137.0$ m
length of the semi-minor axis,	$r = R(1 - f)$	$= 6356752.3142$ m
flattening of the ellipsoid,	$f = (R - r)/R$	$= 1/298.257223563$
major eccentricity of the ellipsoid,	$e = [f(2 - f)]^{1/2}$	$= 0.0818191908426$
Earth's rate,	Ω	$= 7.292115 \times 10^{-5}$ rad/s $(15.041067°/\text{hour})$

The figures given above are as defined by the World Geodetic System Committee in 1984, the WGS-84 system,[5].

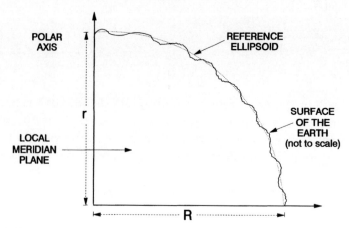

Figure 3.18   Reference ellipsoid

By modelling the Earth in accordance with the reference ellipsoid defined above, the rates of change of latitude and longitude may be expressed in terms of a meridian radius of curvature (R_N) and a transverse radius of curvature (R_E) as follows:

$$\dot{L} = \frac{v_N}{\left(R_N + h\right)} \qquad \dot{\lambda} = \frac{v_E \sec L}{\left(R_E + h\right)} \tag{3.82}$$

where $R_N = R(1 - e^2)/(1 - e^2 \sin^2 L)^{3/2}$

$R_E = R/(1 - e^2 \sin^2 L)^{1/2}$

The mean radius of curvature used in the earlier equations, $R_0 = (R_E R_N)^{1/2}$. Similarly, the transport rate now takes the following form:

$$\omega_{en}^n = \left(\frac{v_E}{\left(R_E + h\right)} \frac{-v_N}{\left(R_N + h\right)} \frac{-v_E \tan L}{\left(R_N + h\right)}\right)^T \tag{3.83}$$

3.7.3 Variation of gravitational attraction over the Earth

As described earlier, accelerometers provide measurements of the difference between the acceleration with respect to inertial space and the gravitational attraction acting at the location of the navigation system. In order to extract the precise estimates of true acceleration needed for very accurate navigation in the vicinity of the Earth, it is necessary to model accurately the Earth's gravitational field. This, of course, is also true for navigation close to any other body with a gravitational field.

It is assumed in the earlier derivation of the navigation equation that the gravity vector acts vertically downwards, i.e. normal to the reference ellipsoid. In practice, both the magnitude and the direction of the gravity vector vary with position on the Earth's surface and height above it. Variations occur because of the variation between the mass attraction of the Earth and gravity vector; the centrifugal acceleration being a function of latitude. In addition, gravity varies with position on the Earth because of the inhomogenous mass distribution of the Earth. Such deviations in the magnitude and direction of the gravity vector from the calculated values are known as gravity anomalies.

Mathematical representations of the Earth's gravitational field are discussed in some depth by Britting [3]. The deflection of the local gravity vector from the vertical may be expressed as angular deviations about the north and east axes of the local geographic frame as follows:

$$\mathbf{g}_l = [\xi g, -\eta g, g]^T \tag{3.84}$$

where ξ is the meridian deflection and η is the deflection perpendicular to the meridian. The deflection in the meridian plane is illustrated in Figure 3.19.

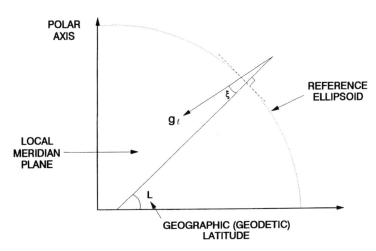

Figure 3.19 Deflections of local vertical owing to gravity anomalies

The resulting deviation of the vertical over the surface of the Earth varies by up to 30 arc seconds.

Precise knowledge of the gravity vector becomes important for certain high accuracy applications, such as for marine navigation where the deflection of the vertical becomes an important factor. Exact knowledge of the magnitude of gravity is also vital for the testing of very precise accelerometers, i.e. sensors having a measurement bias of less than 10^{-5} g. Similarly, it is important for surveying and gravity gradiometry, where attempts are made to measure the gravity vector very accurately.

Various international models for the variation of gravity with latitude are given in the literature. Steiler and Winter [4] give the following expressions for the variation of the magnitude of the gravity vector with latitude at sea level ($h = 0$) and its rate of change with height above ground:

$$g(0) = 9.780318\left(1 + 5.3024.10^{-3}\sin^2 L - 5.9.10^{-6}\sin^2 2L\right)\mathrm{m/s}^2 \quad (3.85)$$

$$\frac{d}{dh}g(0) = -0.0000030877\left(1 - 1.39.10^{-3}\sin^2 L\right)\mathrm{m/s}^2/\mathrm{m} \quad (3.86)$$

For many applications, precise knowledge of gravity is not required and it is sufficient to assume that the variation of gravity with altitude is as follows:

$$g(h) = g(0)/(1 + h/R_0)^2 \quad (3.87)$$

where $g(0)$ is derived from eqn. 3.85.

References

1 WRAY, G.L. and FLYNN, D.J.: 'An assessment of various solutions of the navigation equation for a strapdown inertial system'. Royal Aircraft Establishment, technical report 79017, January 1979
2 SHEPPERD, S.W.: 'Quaternion from rotation matrix', *AIAA Journal of Guidance and Control*, 1978, 1 (3)
3 BRITTING, K.: *'Inertial Navigation System Analysis'* (Wiley Interscience, New York, 1971)
4 STEILER, B. and WINTER, H.: 'AGARD flight test instrumentation volume 15 on gyroscopic instruments and their application to flight testing'. AGARD-AG-160-VOL.15, September 1982
5 'Department of Defense World Geodetic System 1984: its definition and relationship with local geodetic systems'. DMA TR 8350.2, September 1987

Chapter 4
Gyroscope technology 1

4.1 Introduction

Gyroscopes are used in various applications to sense either the angle turned through by a vehicle or structure (displacement gyroscopes) or, more commonly, its angular rate of turn about some defined axis (rate gyroscopes). The sensors are used in a variety of roles such as:

- flight path stabilisation;
- autopilot feedback;
- sensor or platform stabilisation;
- navigation.

It is possible with modern gyroscopes for a single sensor to fulfil each of the above tasks, but often two or more separate clusters of sensors are used.

The most basic and the original form of gyroscope makes use of the inertial properties of a wheel, or rotor, spinning at high speed. Many people are familiar with the child's toy which has a heavy metal rotor supported by a pair of gimbals [1]. When the rotor is spun at high speed, the rotor axis continues to point in the same direction despite the gimbals being rotated. This is a crude example of a mechanical, or conventional, displacement gyroscope.

Examples of mechanical spinning wheel gyroscopes used in strapdown applications are the single-axis rate integrating gyroscope and twin-axis tuned or flex gyroscopes. An alternative class designation for gyroscopes that cannot be categorised in this way is, not surprisingly, called unconventional sensors, some of which are solid state devices. The very broad and expanding class of unconventional sensors includes devices such as:

rate transducers which include mercury sphere and magneto-hydrodynamic sensors;

vibratory gyroscopes;

nuclear magnetic resonance (NMR) gyroscopes;

electrostatic gyroscopes (ESGs);

optical rate sensors which include ring laser gyroscopes (RLGs) and fibre optic gyroscopes (FOGs).

Although many of the sensors in this class are strictly angular rate sensors and not gyroscopes in the sense that they do not rely on the dynamical properties of rotating bodies, it has become accepted that all such devices be referred to as gyroscopes since they all provide measurements of body rotation.

In this chapter, some conventional sensors are described followed by sections which outline the principles of operation and performance of some of the other gyroscope technologies noted above. Finally, a brief mention is made of other forms of instrument or novel techniques that may be used to sense rotational motion. Optical gyroscope technology is discussed separately in Chapter 5.

Throughout this and the following chapter emphasis is placed on those sensors which are used, or have the potential to be used, in strapdown inertial systems. It is for this reason that optical sensors are described in some detail, such sensors being seen as the technology of the future with wide application in strapdown systems. Details of fabrication of various types of gyroscopes can be found in References 2 and 3.

4.2 Conventional sensors

4.2.1 Introduction

Conventional gyroscopes make use of the inertial properties of a wheel, or rotor, spinning at high speed [2, 3]. A spinning wheel tends to maintain the direction of its spin axis in space by virtue of its angular momentum vector, the product of its inertia and spin speed, and so defines a reference direction. The development of the mechanical gyroscope owes much to the excellent work of Professor C.S. Draper and his co-workers, at the Massachusetts Institute of Technology. The performance which may be achieved using gyroscopes of this type varies from the precision devices with error rates of less than 0.001°/hour, to less accurate sensors with error rates of tens of degrees per hour. Many devices of this type have been developed for strapdown applications, being able to measure angular rates up to about 500°/second. Some designs are very rugged, having characteristics which allow them to operate in harsh environments such as guided weapons.

4.2.2 Fundamental principles

There are several phenomena on which the operation of the conventional spinning mass gyroscope depends, namely gyroscopic inertia, angular momentum and precession. In the case of two-degree of freedom gyroscopes, there are also the phenomena of nutation, gimbal lock and tumbling. These are considered in turn.

4.2.2.1 Gyroscopic inertia

Gyroscopic inertia is fundamental to the operation of all spinning mass gyroscopes, as it defines a direction in space which remains fixed in the inertial reference frame, i.e. fixed in relation to a system of co-ordinates which do not accelerate with respect to the fixed stars. The establishment of a fixed direction enables rotation to be detected, by making reference to this fixed direction. The rotation of an inertial element generates an angular momentum vector which is coincident with the axis of spin of the rotor, or wheel. It is the direction of this vector which remains fixed in space, given perfection in the construction of the gyroscope.

A practical reference instrument may be designed by having the rotor supported in a set of frames, or gimbals, which are free to rotate with respect to one another as shown in Figure 4.1. This is an external gimbal type gyroscope. The orientation of the case of the instrument with respect to the direction of the spin axis may be measured with angle pick-off devices mounted on the gimbals.

4.2.2.2 Angular momentum

The angular momentum (H) of a rotating body is the product of its moment of inertia (I) and its angular velocity (ω_s) referred to the same axis of rotational motion, i.e.:

$$H = I\omega_s \tag{4.1}$$

where I is the sum of the products of the mass elements that make up the rotor and the square of their distances from the given axis.

Angular momentum is defined by the distribution of mass on a rotor as well as by its angular velocity. For many applications, the

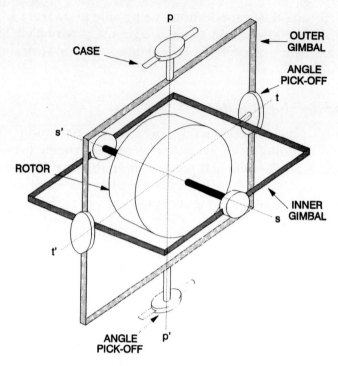

Figure 4.1   Schematic diagram of a two-axis gyroscope

angular momentum is chosen to be very high, so that the undesired torques that can act on a rotor and cause errors are virtually insignificant. This, of course, given good design and fabrication techniques, results in a gyroscope with little movement of the direction of the spin axis. Any undesired movement of the direction of the spin axis is usually referred to as drift. Clearly, one technique for producing a very high angular momentum is to have the majority of the mass of the rotor at its edge owing to the dependence of the moment of inertia on the square of the distance of its mass element from the centre of rotation.

Careful consideration must be given to the value of the angular momentum selected for a gyroscope to be used in a given application. The choice of a very high angular momentum should result in negligible drift, but there could be some considerable penalties. The gyroscope would almost certainly be relatively large and heavy, and it may take many seconds, if not minutes, for the rotor to reach its operating speed. Further, when used in a strapdown mode, the associated control system may not be capable of recording, or capturing, angular rates beyond a few tens of degrees per second.

Hence, many compromises have to be made when selecting a gyroscope for a given application.

4.2.2.3 Precession

Because the motion of a spinning mass occurs in a way that does not coincide with common sense expectations, it has acquired a confusing aura of mystery. Some simple explanation may help.

First, think of the gyroscope rotor mounted in bearings in a gimbal, as shown in Figure 4.2. The gimbal axis system has one axis through the bearing axes, *ss'*, and two mutually orthogonal axes through the centre of mass of the rotor, *tt'* and *pp'*.

spin is the rotation of the gyroscope rotor relative to the gimbal;

precession is rotation of the gimbal, relative to inertial space. In the case of a freely spinning body, such as the Earth (or the rotor of an electrostatic gyroscope, see Section 4.6), there is not a material frame with spin bearings. In this case, the precession must be considered to be that of the axis system which an imaginary gimbal would have—one axis through the North and South poles, and two mutually orthogonal in the plane of the Equator.

Consider now the disc shown in Figure 4.2 spinning about the axis *ss'*. If the disc is acted upon by a couple, i.e. a torque, the torque being about the axis *tt'*, the spin axis of the disc will be forced to turn about the axis *pp'*. This turning is the precession. Note that the precession axis, *pp'*, is orthogonal to the torque axis, *tt'*. It is the unexpectedness of this result which causes confusion. However, if Newton's laws are applied carefully, the result can be explained, both qualitatively and quantitatively.

The disc is spinning about the axis *ss'* in an anti-clockwise direction looking from *s* to *s'*. Suppose that the disc is rigid with all the mass in the rim, and that the rim has a peripheral speed, *u*. Consider an element of mass at the highest point, P_1. Apply an impulsive couple *FF'*, as shown in the Figure in an anti-clockwise direction looking from *t* to *t'*.

The instantaneous velocity of the mass is changed by adding the velocity *w* in the same sense as the couple *FF'*. The resultant velocity, *v*, is now in a different direction. It is noted that the other elements of the rim change their velocities in proportion to their distance from the axis *tt'*. After the disc has spun through 90 degrees, the element of mass arrives at the point P_2, which is not in the expected line *tt'*, but in a plane which has precessed about the axis *pp'*.

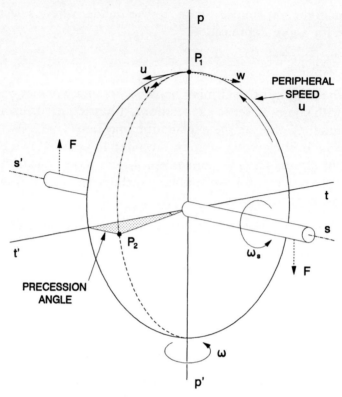

Figure 4.2   Simple explanation of precession

This simple picture indicates how the spinning disc reacts to the impulsive couple, and shows the axis and sense of the precession. Of course, the process is normally continuous, not impulsive. The dynamics can be analysed using co-ordinate geometry and applying Newton's Laws. The result agrees with eqn. 4.5, below which is arrived at using vectors.

The particles making up a spinning body undergo:

1 accelerations caused by accelerations of the centre of mass of the body;
2 centripetal accelerations caused by the spinning of the body;
3 Coriolis accelerations as a result of the precession of the body.

The Coriolis accelerations are simply the additional accelerations experienced by a mass moving relative to an axis system when that axis system is itself rotating in inertial space. The precession torque is simply the torque necessary to produce the sum of the particle masses multiplied by their Coriolis accelerations.

Mathematical description of precession

Consider a heavy spinning disc, as shown in cross section in Figure 4.3, with angular momentum H defined by the vector **OA**, i.e. H**a**, where **a** is a unit vector.

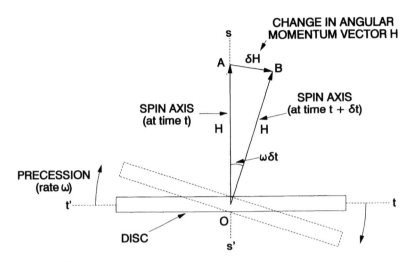

Figure 4.3 Illustration of precession

From Newton's First Law, applied to angular motion, the angular momentum vector **H** remains constant unless the disc is acted on by a torque. Let us suppose that a torque T is applied to the disc which causes it to precess at a rate ω ($= \omega$**c**, where **c** is also a unit vector) about an axis which will lie in the plane of the disc and may be taken to be normal to the plane of the paper. Over a short period of time δt the disc will have precessed through an angle $\omega \delta t$ about **c**, and the angular momentum vector will have changed to **OB**, i.e. to $(H + \delta H)$**b**, where **b** = **a** + $\omega \delta t$ (**c** $\times$ **a**).

The change in angular momentum over this time is represented by the vector **AB** and may be expressed as:

$$\delta\mathbf{H} = (H + \delta H)\mathbf{b} - H\mathbf{a}$$

$$= H(\mathbf{b} - \mathbf{a}) + \delta H \mathbf{b}$$

i.e. $\delta\mathbf{H} = H\omega \delta t\ (\mathbf{c} \times \mathbf{a}) + \delta H \mathbf{b}$ (4.2)

Thus in the limit, as $\delta t \to 0$, the rate of change of angular momentum is given by:

$$\frac{d}{dt}\mathbf{H} = H\omega(\mathbf{c} \times \mathbf{a}) + \frac{d}{dt}H\mathbf{b}$$

i.e. $\frac{d}{dt}\mathbf{H} = \omega \times \mathbf{H} + \frac{d}{dt}H\mathbf{b}$ (4.3)

From Newton's Second Law, the rate of change of angular momentum is equal to the torque **T** applied to the body, hence:

$$\mathbf{T} = \omega \times \mathbf{H} + \frac{d}{dt}H\mathbf{b}$$ (4.4)

Thus the component of the torque which is along the spin axis **b** gives rise to an acceleration in the spin rate. In a practical gyroscope, this is normally negligible and countered by the effect of the spin motor. The component normal to the spin axis gives rise to a precession ω about an axis which is normal to both the torque and the spin axes, and from inspection of the figure, the direction of the precession is such as to try to align the spin axis with the torque axis.

Neglecting the component along the spin axis, in vector terms we may write:

$$\mathbf{T} = \omega \times \mathbf{H}$$ (4.5)

and in magnitude terms:

$$T = \omega H$$ (4.6)

This is sometimes known as the Law of Gyroscopics.

The application of the precession principle

The principle of precession can be exploited to provide a very accurate measure of angular rotation or rotation rate. Since a spinning wheel, or rotor, will precess if a torque is applied to it, a rotor suspended in an instrument case by gimbals will maintain its spin axis in a constant direction in space. Changes in the angles of the gimbals will then reflect any changes in orientation of the case with reference to the spin axis direction.

Alternatively, if controlled torques are applied to the rotor to keep its spin axis aligned with a direction defined by the case of the instrument, then the measurement of these torques will provide measurements of the angular velocity of the instrument, and hence of the angular velocity of any body to which the instrument is rigidly attached.

Note that when a torque is applied to the rotor, which responds by precessing, then there is an equal but opposite reaction torque from the rotor to the application mechanism. However, if precession is prevented, as happens when the supporting gimbal hits a stop, then the reaction torque disappears and the rotor and gimbal act as a non-gyroscopic body about this axis.

In the single axis rate gyroscope, shown schematically in Figure 4.4, the gyroscope's rotor is supported by a single gimbal whose axis is normal to the spin axis. The gimbal is restrained about its axis by a spring attached to the case, and there is an angular pick-off which measures the displacement of the gimbal about its axis from a null, or zero, position.

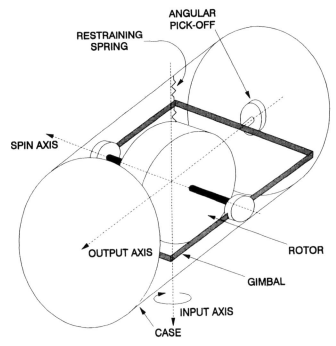

Figure 4.4   A single axis gyroscope

In the case where the instrument is rotating about the input axis and the rotor is not precessing at the same rate, the difference in rates will result in elastic compression of the gimbal pivots. This gives a torque on the gimbal (and a reaction torque on the case) which is applied to the rotor about the input axis. This torque causes the rotor to precess initially about the output axis, about which it is free to turn. The resulting displacement about the output axis causes a torque to be produced about this axis by the restraining spring.

The spring torque on both the gimbal and the rotor about the output axis results in the rotor precessing about the input axis until, in the steady state, it is precessing at the same rate as the case is turning, with the deflection of the restraining spring providing just that amount of torque needed to keep the case and rotor in alignment.

Assuming that the restraining spring is linear, the deflection of the gimbal is proportional to the torque required to keep the rotor precessing with the case, and so to the turn rate of the case.

In practice, it is quite difficult to measure angular displacements accurately without resorting to sophisticated and, consequently, expensive equipment. However, it is quite easy to measure accurately a fixed or defined position, particularly a zero deflection or null position. Hence, if the spin axis of a rotor is made to precess back to the null position by the application of a suitable torque, there is potential for very accurate angular measurement, provided that the torque required to null the deflection can be generated and measured.

This is achieved in practice by replacing the restraining spring with an electromagnetic torque generator, which produces a torque to cause precession of the rotor in a direction opposite to that caused by rotation of the case about its input axis. The current required can be measured very accurately by simple techniques, and when the system is balanced this current is directly proportional to the applied angular rate. This technique is commonly called nulling, and is fundamental to the use of strapdown techniques. Hence, application of the reverse of the precession principle enables very accurate measurements to be made of the angular displacement or rate of turn of the case of the rate gyroscope.

A conventional single-axis gyroscope can be considered to have three orthogonal axes: its rotor or spin axis, an input precession axis and an output or torque axis. i.e. torque is always applied about the output axis to cause precession about the input axis to keep it in alignment with the case. In the case of the so-called two or dual-axis gyroscopes, these sensors have a spin axis with two orthogonal input axes. In this case, angular motion of the case of the sensor with respect to the rotor is sensed by pick-off angle sensors on the gimbals, as indicated in Figure 4.1, and each gimbal is provided with a torque mechanism.

Even the most accurate of gyroscopes will appear to drift, or have their spin axis precess. This is because the angular momentum vector is fixed with respect to space axes, not the co-ordinate system defined by the Earth. Hence, for some orientations on Earth, it is necessary to apply corrective torques to precess the gyroscope if it is to be used as an Earth reference.

4.2.2.4 Nutation

This is a natural phenomenon that occurs with so-called two-degree of freedom gyroscopes, such as those in which the rotor is supported by a gimbal structure. Nutation is simply a wobbling of the spin axis of the rotor. It is a self sustaining oscillation which physically represents a continuous transfer of energy from one degree of freedom to the other and back again. In contrast to precession, this motion does not need any external torques to sustain it. This motion has a natural frequency ω_R, commonly known as the nutation frequency, given by:

$$\omega_R = \frac{H}{\sqrt{\left(I_{ig} I_{og}\right)}} \tag{4.7}$$

where H is the angular momentum of the rotor, I_{ig} is the moment of inertia of the rotor and inner gimbal about the inner axis and I_{og} is the moment of inertia of the rotor, inner gimbal and outer gimbal about the outer axis.

In a frictionless system, nutation would be self perpetuating. However, friction in the gimbal bearings or deliberately applied viscous drag damps out this undesirable motion. Energy dissipation varies in proportion to the nutation frequency. Therefore, in order to minimise the occurrence of nutation, it is necessary to increase ω_R. It is usual for the rotor to have as large an angular momentum as possible, combined with gimbals having low moments of inertia. This is achieved through the use of light but stiff materials such as beryllium in the construction of gimbals.

4.2.2.5 Gimbal lock

Gimbal lock is an effect which prevents a two degree-of-freedom gyroscope having 360° of freedom about both its inner and outer gimbal axes. Gimbal lock occurs when the spin axis of the rotor coincides with the outer gimbal axis owing to a 90° rotation about the inner axis. At this point, the gyroscope loses one degree-of-freedom. Application of motion about an axis perpendicular to the plane containing the outer gimbal causes the outer gimbal to spin. Once this spinning motion has begun, the spin axis of the rotor and the axis of the outer gimbal remain permanently coaxial. The only method of separating them is to stop the rotation of the inertial element to allow the two axes to be reset. This undesirable effect is prevented by using mechanical stops to limit the motion of the inner gimbal. These stops usually permit up to ±85° of motion by the inner gimbal. Gimbal lock can also occur in stable platforms with three gimbals.

4.2.2.6 Tumbling

Tumbling is a consequence of using mechanical stops to prevent gimbal lock. This phenomenon occurs when the inner gimbal hits one of the mechanical stops, this causing the outer gimbal to turn through 180° about its own axis. This motion of the outer gimbal is known as tumbling. Once tumbling occurs, the reference is lost.

4.2.3 Components of a mechanical gyroscope

The basic components of mechanical gyroscopes are as follows:

1 **The instrument case**: the case in which the other elements are housed and which provides the structure by which the instrument is mounted in a vehicle.
2 **The rotor (or inertial) element**: essentially a flywheel rotated at high angular velocity. The rotor usually has the majority of its mass at the outer edge, as the moment of inertia is the sum of the products of the individual masses (m_i) and the square of their distance (r_i) from the axis of rotation, $\Sigma m_i r_i^2$. This enables a high angular momentum (H) to be achieved for a given angular velocity, since H is the product of moment of inertia (I) and angular velocity (ω_s) as described in the previous section. This approach also allows a high angular momentum to be achieved with the lowest overall mass. A low rotor mass is desirable to minimise vibration and shock effects.
3 **Gimbals**: support frames on which the rotor or another gimbal is mounted to isolate the rotor from rotational motion, by allowing freedom of angular movement of these frames about the rotor. In the case of a two-gimbal sensor, the axes of rotation of the two gimbals and the rotor are arranged to be mutually orthogonal as shown in Figure 4.1.
4 **Pick-off**: a device used to detect relative motion between the rotor and the gimbals or, in some cases, between the rotor and the instrument case. The pick-off produces an electrical signal, indicating the direction and amplitude of the motion from a reference position. There are three basic forms of pick-off technology commonly used with mechanical gyroscopes operating in torque rebalance mode:

   (i)   moving coil—using a small receiver coil and an a.c. excitation coil, so that any relative movement between the two modifies the flux sensed by the receiver coil;
   (ii) variable reluctance—the excitation and receiver coils are

fixed to the case of the gyroscope, with a soft iron assembly attached to the moving component so that it is in the flux return path between the excitation and receiver coils. Motion of the soft iron assembly causes a change in its orientation in the excitation field, thereby modifying the return flux to the receiver coil;

(iii) capacitive—there is a stationary plate close to the rotor, or moving component, whilst the rotor acts as the other plate of the capacitor. Movement of the rotor about its input axis, or axes for a two axis sensor, causes a change in separation between the two plates of the capacitor and hence there is a change in capacitance.

Open loop gyroscopes, such as simple rate sensors, often use a potentiometer to sense angular displacement of a gimbal. Generally, this form of sensor is not used for navigation purposes.

5 **Torque motor or electromagnetic torquer**: when a gyroscope is used in a closed loop or torque rebalance mode, it is necessary to generate a torque on the rotor in order to return the rotor to the null, or zero, position. This is achieved using a torque generator, which usually takes one of two common forms:

(i) permanent magnet—this type relies on the interaction between the field generated by a permanent magnet and that of an electromagnetic coil. Particularly with single axis sensors, a coil cup is fixed to the moving element and the permanent magnet attached to the case. This has several advantages such as reducing the sensitivity to external magnetic fields and allowing the magnet to be outside of the flotation fluid. However, it does require a pair of flexible leads to the coil which can generate error torques. In general, as a result of other constraints, dynamically tuned gyroscopes have the opposite configuration with the permanent magnet fixed to the moving element.

(ii) electromagnet—a soft iron component is attached to the sensing element and a coil is fixed to the case. When a current is applied to the coil, a magnetic field is produced that interacts with the soft iron producing a torque on the sensing element.

6 **Rebalance loop**: this is the term given to the electronic circuitry that receives and uses the signals from the pick-off assembly. It interprets these signals in terms of the current required in the

torquer coils to return the inertial element to its null position. The rebalance loop electronics can either be analogue or digital. In the case of the analogue rebalance loop, a continuously variable current is passed through the coil to return the inertial element to its null position. When there is no displacement, then there is no current flow. A digital rebalance system generates precision current pulses of particular duration to force the inertial element back to its null position. With some implementations, pulses of equal amplitude but opposite sign are passed into the torque generator even when there is no displacement of the rotor. Imbalance in the number of pulses applied in each direction gives rise to a net torque.

7 **Spin motor**: the motor used to rotate the inertial element and give it the angular momentum vital for the operation of the mechanical gyroscope. Usually the spin motor is either a hysteresis motor or an inductive device. Some gyroscopes that have a short run time use a blast of air or a small explosive charge to spin the inertial element, and for cheap and crude applications, a d.c. electric motor may be used.

8 **Float**: rate integrating gyroscopes, as discussed in Section 4.2.5, have their rotor and spin motor sealed in a can which is immersed in a fluid to reduce the load on the gimbal bearings. This can, with its encapsulated components, is known as the float. Careful choice of the flotation fluid can reduce the load of the rotor assembly on the gimbal to zero. In such a design, bellows are used to compensate for changes in the volume of the fluid when the temperature inside the case of the gyroscope changes. The centre of buoyancy is arranged to be close to the centre of gravity of the float and along the output axis in order to minimise acceleration sensitive errors. There is further consideration of the effect of this on sensor performance in Section 4.2.4 .

9 **Flotation fluid**: the fluid in the gyroscope that gives buoyancy to the float in a floated rate integrating gyroscope. It also provides damping of the motion of the float which gives rise to the integration function for the single-axis rate integrating gyroscope.

10 **Bearings**: for the spin axis of the rotor, most gyroscopes use ball bearings in a race with a retainer, which are chosen to have low noise characteristics. This form of bearing needs a suitable lubricant with the following characteristic:

- it should not separate into solid and liquid components;
- it should have small or negligible change in viscosity over the temperature range of the sensor;

- it should not leak out from the bearing;
- it should retain its physical and chemical properties for the required shelf life of the gyroscope.

Lubricants can severely limit the environmental performance and shelf life of a sensor. An alternative form of bearing that overcomes the well known problems of ball bearings is the gas bearing. This form of bearing can be either self acting or externally pumped. In the former case, the bearing draws a gas, usually air, into a series of grooves which, owing to the viscosity of the fluid, supports the structure in the other part of the bearing. In the latter case, the fluid is pumped into the grooves to support the structure. The drawbacks with gas bearings are the need for very tight tolerances in manufacture, of the order of a micrometre or better, and the use of very hard materials such as boron carbide as the two bearing surfaces touch and rub during both starting and stopping of the rotor. However, they are very low noise bearings and can last a very long time, particularly if the bearing runs continuously. A schematic diagram of a self pumping gas bearing is shown in Figure 4.5.

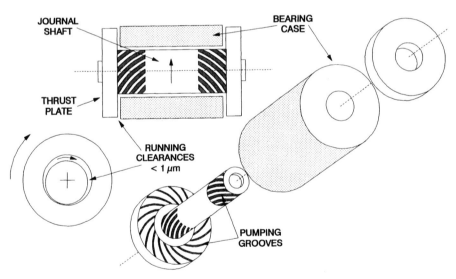

Figure 4.5   Self pumping gas bearing

4.2.4 Sensor errors

All gyroscopic sensors are subject to errors which limit the accuracy to which the angle of rotation or applied turn rate can be measured. Spurious and undesired torques, caused by design limitations and constructional deficiencies, act on the rotors of all mechanical gyroscopes. These imperfections give rise to precession of the rotor, which manifests itself as a drift in the reference direction defined by the spin axis of the rotor. In a free gyroscope, i.e. one which measures angular displacements from a given direction, it is customary to describe the performance in terms of an angular drift rate. For a restrained gyroscope, i.e. one operating in a nulling or rebalance loop mode to provide a measure of angular rate, any unwanted torques act to produce a bias on the measurement of angular rate.

The terms drift and bias are commonly used interchangeably. In this book, we reserve the term drift for the motion of the spin axis in a free gyroscope, whereas bias is used with nulled sensors. In practice, the way in which the errors are quoted often depends on the accuracy band of the sensor rather than whether the gyroscope is used with its spin axis fixed in space or restrained in some way.

The major sources of error which arise in mechanical gyroscopes are itemised below. Further details relating to specific types of gyroscope will be given later in the Chapter where the physical effects which give rise to each type of error are discussed more fully.

Fixed bias (g-independent): the sensor output which is present even in the absence of an applied input rotation. It may be a consequence of a variety of effects, including residual torques from flexible leads within the sensor, spurious magnetic fields and temperature gradients. The size of the bias is independent of any motion to which the gyroscope may be subjected and is sometimes referred to as the acceleration (or g) independent bias. It is usually expressed in units of °/hour, or for the less accurate sensors in °/second.

Acceleration dependent bias (g-dependent): biases which are proportional to the magnitude of the applied acceleration. Such errors arise in spinning mass gyroscopes as a result of mass unbalance in the rotor suspension, i.e. non-coincidence of the rotor centre of gravity and the centre of the suspension mechanism. The relationship between these components of bias and the applied acceleration can be expressed by means of coefficients having units of °/hour/g. In general, such terms relate accelerations in each of the principal axes of the gyroscope, i.e. accelerations which act both along and orthogonal to the sensitive axis of the sensor, to errors in the measurement of turn rate. In the presence of a steady acceleration, a fixed bias in the measured rate occurs.

Anisoelastic bias (g²-dependent): biases which are proportional to the product of acceleration along orthogonal pairs of axes. Such biases arise in spinning mass gyroscopes because the gyroscope rotor suspension structure, particularly the bearings, has finite compliances which are unequal in different directions. The anisoelastic coefficients have units of $°/\text{hour}/\text{g}^2$.

Anisoinertia errors: such errors arise in spinning mass gyroscopes and introduce biases owing to inequalities in a gyroscope's moments of inertia about different axes. Anisoinertia is frequency sensitive if the rotor is driven by a hysterisis motor. This is a consequence of the elastic coupling between the magnetic ring on the rotor and the rotating magnetic field. The resulting biases are proportional to the product of angular rates applied about pairs of orthogonal axes. The anisoinertia coefficients may be expressed in units of $°/\text{hour}/(\text{radians}/\text{second})^2$.

Scale factor errors: errors in the ratio relating the change in the output signal to a change in the input rate which is to be measured. Scale factor error is commonly expressed as a ratio of output error to input rate, in parts per million (ppm), or as a percentage figure for the lower performance class of sensor. Additional errors arise as a result of scale factor non-linearity and scale factor asymmetry. Scale factor non-linearity refers to the systematic deviations from the least squares straight line or non-linear function fitted to the measurements, which relates the output signal to the applied angular rate. The latter term includes differences in the magnitude of the output signal for equal rotations of the sensor in opposite directions. In spinning mass gyroscopes, scale factor non-linearity relates to thermal changes that result in changes of the magnetic flux.

Cross-coupling errors: erroneous gyroscope outputs resulting from gyroscope sensitivity to turn rates about axes normal to the input axis. Such errors arise through non-orthogonality of the sensor axes and may also be expressed as parts per million or a percentage of the applied angular rate.

Angular acceleration sensitivity: this error is also known as the gyroscopic inertial error. All mechanical gyroscopes are sensitive to angular acceleration owing to the inertia of the rotor. Such errors become important in wide bandwidth applications. This error increases with increasing frequency of input motion. Hence, it is necessary to compensate for this error if accuracy is to be preserved. A detailed analysis is given by Edwards in Reference 4 for both the rate integrating gyroscope and the dynamically tuned gyroscope, which are described in Sections 4.2.5 and 4.2.6.

It is important to realise that each of the errors described will, in general, include some or all of the following components:

- fixed or repeatable terms;
- temperature induced variations;
- switch-on to switch-on variations;
- in-run variations.

For instance, the measurement of angular rate provided by a gyroscope will include:

(i)   a bias component which is predictable and is present each time the sensor is switched on and can therefore be corrected;
(ii) a temperature dependent bias component which can be corrected with suitable calibration;
(iii) a random bias which varies from gyroscope switch-on to switch-on but is constant for any one run;
(iv) an in-run random bias which varies throughout a run; the precise form of this error varies from one type of sensor to another.

The fixed components of error, and to a large extent the temperature induced variations, can be corrected to leave residual errors attributable to switch-on to switch-on variation and in-run effects, i.e. the random effects caused by instabilities within the gyroscope. Assuming that the systematic errors are compensated, it is mainly the switch-on to switch-on and in-run variations which influence the performance of the inertial system in which the sensors are installed. Compensation techniques are discussed further in Chapter 7.

A number of different types of mechanical gyroscope of interest in strapdown applications are now described.

4.2.5 Rate integrating gyroscope

4.2.5.1 Introductory remarks

The design of this type of mechanical gyroscope was conceived in the late 1950s for use on stabilised platforms, early examples appearing at the start of the 1960s. This basic concept is capable of achieving a wide spectrum of performance from a very small gyroscope that fits into a cylinder of diameter 25 mm (1 inch) and length 50 mm (2 inches). Typically, the drift performance of the miniature versions of this type of sensor is in the 1°/hour to 10°/hour class, although substantially better than 0.01°/hour can be achieved with the larger top of the range sensors. The smaller sensors are able to measure turn rates typically of the order of 400°/second or better. This type of sensor has found many different applications as a result of this wide spectrum of performance, including navigation systems in aircraft, ships and guided weapons.

4.2.5.2 Detailed description of sensor

A rate integrating gyroscope has one input axis and so is known as a single-axis gyroscope. Besides the case, it has three main component parts, as illustrated in Figure 4.6:

(i)   the float, which contains the rotor and its motor. It is supported in precision bearings to allow rotation about an axis perpendicular to the spin axis of the rotor;
(ii) the angle pick-off which senses rotation of the float assembly;
(iii) the torque motor, which is used to apply precise torques to return
(iv) the float to its null position.

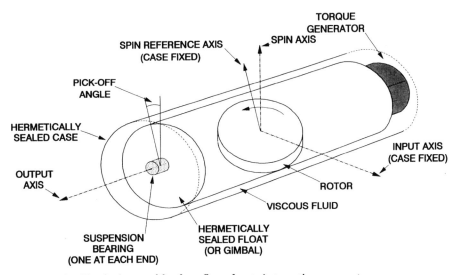

Figure 4.6   Single degree-of-freedom floated rate integrating gyroscope

These components are sealed into a case and the small gap between the float and the case is filled with a highly viscous liquid. This liquid provides some support for the float in its bearings, thus reducing undesired torques and, in some very particular instances, it provides total buoyancy. The flotation fluid also provides viscous damping between the float and the case. Electrical signals and power are transmitted between the case and the float via delicate flexible (flex) leads.

When an angular rate is applied about the input axis, the float develops a precessional rate about the output axis shown in Figure 4.6. As a result of the damping fluid which supports the float, the output axis rate gives rise to a viscous torque about the output axis. This torque causes the float to precess about the input axis at the input rate and so follow the case rotation. The output axis rate therefore becomes proportional to the input rate. The gyroscope operates in this manner, as a precision rate integrating gyroscope, i.e. the output which is sensed by the pick-off is proportional to the integral of the input axis rate, i.e. to the change in input angle.

If an additional torque is applied electrically via the torque motor, the pick-off angle rate becomes proportional to the difference between the input rate and the precessional rate induced by the torque motor. Hence the pick-off angle becomes proportional to the integral of the difference between the input and torque motor rates. For strapdown operation, the pick-off angle is nulled by feeding back the pick-off output to the torque motor. In this situation, the time integral of the difference between the input and torque motor rates becomes zero. It follows that the current applied to the torque motor to maintain the null position is proportional to the applied input rate. This gyroscope is used as a closed loop sensor as this leads to a better definition of the input axis and more accurate measurement of rotation.

Figure 4.7 shows the components of a rate integrating gyroscope in more detail.

4.2.5.3 Sources of error

The major error processes that influence the performance of this type of gyroscope are shown outlined below:

g-independent bias, resulting from a variety of causes which includes residual flex lead torques, thermal gradients across the sensor which result in fluid flow around the float assembly and pivot stiction.

g-dependent bias caused by:

(a) float mass unbalance relative to the pivots of the gimbal along the spin motor axis—principally the result of rotor movement along the spin axis caused by spin motor bearing compliance.

(b) float mass unbalance along the input axis.

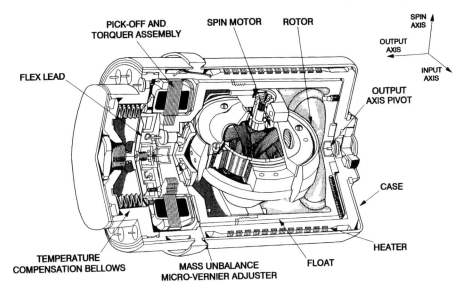

Figure 4.7 Rate integrating gyroscope

anisoelastic bias (g^2-dependent), which results from unequal compliance of the gyroscope's float assembly along the input and spin axes.

scale factor error, caused by imperfections and temperature fluctuations in the pick-off and nulling components, which may be expressed as the sum of a fixed error and a set of non-linear components.

cross-coupling which arises through imperfections in the construction of the sensor.

zero-mean random bias caused by instabilities in the gyroscope which have short correlation times, variations in pivot friction and random movements of the rotor along the spin axis, for example.

This sensor is intended to measure angular rates, but unfortunately it is also sensitive to linear and angular accelerations and vibrations and these can give rise to errors in measurements. Careful shielding is required to eliminate errors resulting from stray magnetic fields interacting with the torque generator. Changes in temperature alter the characteristics of the magnetic materials within the sensor. Without at least approximate compensation, these changes in temperature give rise to scale factor errors. Generally, heating effects in conjunction with magnetic imperfections give rise to first, second and third order scale factor errors. The significant error sources are usually systematic and can be readily corrected.

The angular rate measurement ($\tilde{\omega}_x$) provided by a rate integrating gyroscope may be expressed in terms of the true input rate and the error terms as follows:

$$\tilde{\omega}_x = (1 + S_x)\omega_x + M_y\omega_y + M_z\omega_z + B_{fx} + B_{gx}a_x + B_{gz}a_z + B_{axz}a_xa_z + n_x \quad (4.8)$$

where ω_x is the turn rate of the gyroscope about its input axis, ω_y and ω_z are the turn rates of the gyroscope about its output and spin axes respectively and a_x and a_z are the accelerations of the gyroscope along its input and spin axes respectively.

B_{fx} = g-insensitive bias
B_{gx}, B_{gz} = g-sensitive bias coefficients
B_{axz} = anisoelastic bias coefficient
n_x = zero-mean random bias
M_y, M_z = cross-coupling coefficients
S_x = scale factor error which may be expressed as a polynomial in ω_x to represent scale factor non-linearities.

4.2.5.4 Typical performance characteristics

Typical 1σ values for the major error sources are:

g-independent bias	0.05–10°/hour
g-dependent/mass unbalance bias	1–10°/hour/g
anisoelastic bias	1–2°/hour/g²
scale factor error (uncompensated temperature effects)	up to 400 ppm/°C
scale factor non-linearities (at high rotation rates)	0.01–0.1%
bandwidth	up to 60 Hz
maximum input rate	up to 400°/second

In certain applications, other systematic error effects may become important, but generally, those given above are dominant.

4.2.6 Dynamically tuned gyroscope

4.2.6.1 Introductory remarks

This sensor is sometimes also called the tuned rotor gyroscope, or dry tuned gyroscope. It has two input axes which are mutually orthogonal and which lie in a plane which is perpendicular to the spin axis of the gyroscope. Work to demonstrate this form of technology was underway at the Royal Aircraft Establishment, Farnborough (now part of the Defence Research Agency) by Philpot and Mitchell [5] during the late 1940s. Although demonstration of the tuning phenomenon took place in the early 1950s, it is only since the 1970s that this type of gyroscope has been fully developed. The original concept was developed for stabilised platform applications, but has been applied to strapdown systems since the mid to late 1970s in many types of vehicle.

Generally, the performance of these gyroscopes is very similar to that achieved by the rate integrating gyroscope. Miniature instruments of this type developed for strapdown applications are typically about 30 mm in diameter and 50 mm long. Sub-miniature devices have also been produced, with some slight degradation in performance, which are about 15 mm by 35 mm. These gyroscopes have found many applications similar to the floated rate integrating gyroscope.

4.2.6.2 Detailed description of sensor

The sensor consists of three major sub-assemblies as indicated in Figure 4.8:

(i)   the body block, which consists of the spin motor and angle pick-off arrangement;
(ii) the rotor assembly, which also includes the torque generator magnets and the Hooke's joint suspension;
(iii) the case and torque generator coil assembly.

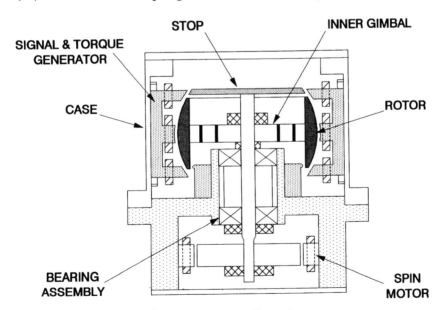

Figure 4.8   Typical tuned rotor gyroscope configuration

The rotor is connected to the drive shaft by a pairs of flexure hinges to an inner gimbal ring. This inner gimbal is also connected to the drive shaft by a pair of flexure hinges, the two axes of freedom being orthogonal as illustrated schematically in Figure 4.9. This is often called a Hooke's joint or a Cardan joint and allows torsional flexibility. This is an internal type of gimbal and is far more compact than the external gimbal shown in Figure 4.1. At the other end of the drive shaft is a synchronous motor.

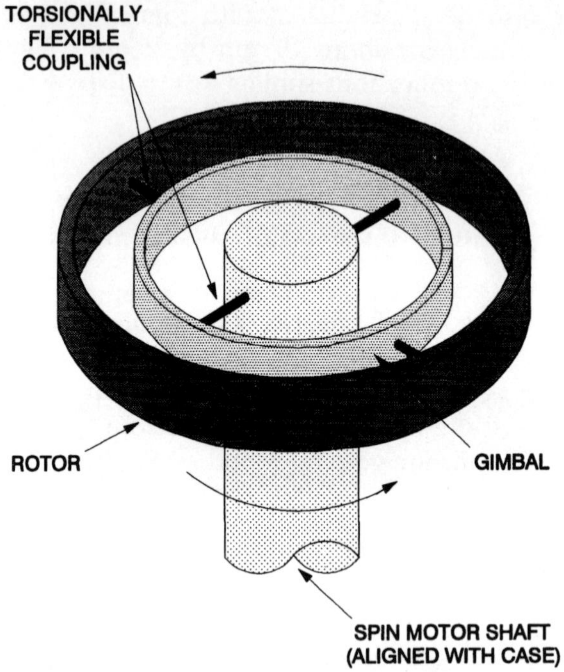

Figure 4.9   Dynamically tuned gyroscope rotor and drive shaft assembly

Rotation of the gimbal causes a reaction at the rotor that is equivalent to a negative torsional spring stiffness. This effect occurs when the angular momentum of the shaft does not coincide with that of the rotor, the angular momentum of the gimbal jumping between that of the shaft and the rotor at twice the speed of the rotor. Thus careful selection of the torsional stiffness of the gimbal components and the rotational speed of the rotor allows the rotor suspension to have a net zero spring stiffness at a particular rotor speed, known as the tuned speed. Under these conditions, the rotor is decoupled from the motion of the rest of the sensor and hence is free. In practice, this condition is usually adjusted or trimmed by the use of screws set into the inner gimbal ring which allow minor changes in the mass properties of the gimbal.

Normally, the decoupling of the rotor is not complete or perfect and residual elastic restraints restrict the useful angular range of movement of the rotor. Therefore, the sensor is usually used in a torque rebalance mode allowing only very small deflections of the rotor. Deflections of the rotor are sensed about two orthogonal axes, and are directly proportional to the motion of the gyroscope case about the respective axes in inertial space.

A figure of merit [6] is sometimes used for describing the quality of a dynamically tuned gyroscope. The figure of merit relates the inertias of the rotor to the inertias of the gimbal.

$$\text{figure of merit} = \frac{C}{I_g + J_g - K_g}$$

where $\quad C \quad$ = spin inertia of the rotor
$\quad I_g, J_g$= gimbal transverse inertias
$\quad K_g \quad$ = gimbal polar inertia

Typical values for figure of merit for a moderate performance instrument are in the region of 50.

4.2.6.3 Sources of error

It will be noticed that the errors are of similar form to those given for the rate integrating gyroscope errors.

g-independent bias, principally the result of stray internally generated magnetic fields which interact with the torque motor magnet mounted on the rotor plus rebalance loop biases. The effects of tuning errors and gimbal damping are often included in this error.
g-dependent bias, caused by mass unbalance of the rotor assembly and geometrical imperfections in the torsional elements. The flexures can also generate a torque when loaded axially, leading to an acceleration sensitive bias about the axis opposite to that axis along which the acceleration is acting. This is known as quadrature mass unbalance.
anisoelastic (g^2-dependent) bias which results from unequal compliance of the rotor assembly in the *x, y* and *z* directions.
anisoinertia bias results from differences in rotor inertias in the *x, y* and *z* directions, and is frequency sensitive.
scale factor errors mainly caused by thermally induced changes in magnets and coils used in the rebalance system.

zero-mean random bias caused, for example, by error torques resulting from changes in spin motor shaft orientation owing to variations in the bearing preload.

As in the case of the rate integrating gyroscope, this sensor is sensitive to linear and angular accelerations, vibratory motion, stray magnetic fields and temperature changes, all of which give rise to errors in measurements. This type of sensor is sensitive to vibrations at integer multiples of the spin speed, not only at the spin frequency, as in the single degree-of-freedom gyroscope, but also vibrations at twice this frequency. Vibration about the input axis interacts with the gimbal angular momentum, and is rectified to give a fixed bias.

The angular rate measurements provided by the sensor ($\tilde{\omega}_x$ and $\tilde{\omega}_y$) may be expressed mathematically as follows:

$$\tilde{\omega}_x = \left(1 + S_x\right)\omega_x + M_y\omega_y + M_z\omega_z + B_{fx} + B_{gx}a_x + B_{gy}a_y + B_{axz}a_xa_z + n_x$$

$$\tilde{\omega}_y = \left(1 + S_y\right)\omega_y + M_x\omega_x + M_z\omega_z + B_{fy} + B_{gy}a_y - B_{gx}a_x + B_{ayz}a_ya_z + n_y$$

$$(4.9)$$

where ω_x and ω_y are the turn rates of the gyroscope about its input axes, a_x and a_y are the accelerations along its input axes and a_z is the acceleration along its spin axis.

$$
\begin{aligned}
B_{fx}, B_{fy} &= \text{g-independent bias coefficients} \\
B_{gx}, B_{gy} &= \text{g-dependent bias coefficients} \\
B_{axz}, B_{ayz} &= \text{anisoelastic bias coefficients} \\
n_x, n_y &= \text{zero-mean random bias} \\
S_x, S_y &= \text{scale factor errors} \\
M_x, M_y, M_z &= \text{cross-coupling coefficients}
\end{aligned}
$$

4.2.6.4 Typical performance characteristics

Typical values for the significant error sources and performance parameters are given below:

g-independent bias	0.05–10°/hour
g-dependent/mass unbalance bias	1–10°/hour/g
anisoelastic bias	0.1–0.5°/hour/g²
scale factor error	up to 400 ppm/°C
(uncompensated temperature effects)	
scale factor non-linearities	0.01–0.1%
(at high rotation rates)	
bandwidth	up to 100 Hz
maximum input rate	up to 1000°/second

The dynamically tuned gyroscope offers a number of significant advantages for many applications, when compared with the rate integrating gyroscope. These are usually quoted as fewer parts, a fluid free suspension, no flex-lead torques, simplified spin motor bearing design and a fast warm-up characteristic. Of course, it offers the ability to measure angular motion about two axes, and additionally, its construction allows the sensor either to be reworked more easily, or to have its performance optimised before final sealing of the case. One potential drawback is its susceptibility to disturbances and oscillations at the tuned frequency and harmonics of this frequency. Its suspension is analogous to a mass on a spring. For this reason, careful design is required to ensure that mechanical resonances do not interact with the suspension and destroy it. For reliable performance in a harsh environment, careful design of the suspension and mounting is crucial. The rate integrating gyroscope is generally more resilient in this type of environment owing to its inherently rugged design. Figure 4.10 shows a typical arrangement of the various components of a dynamically tuned gyroscope.

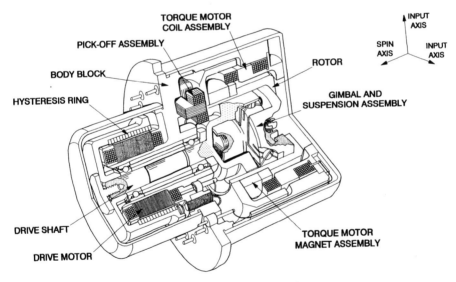

Figure 4.10   Sectional diagram of a dynamically tuned gyroscope

4.2.7 Flex gyroscope

4.2.7.1 Introductory remarks

This sensor bears a close resemblance to the dynamically tuned gyroscope and operates in a similar manner, as the rotor acts as a free inertial element. It also has two sensitive input axes. Development of this inertial instrument has progressed dramatically since the mid 1970s. The form of construction allows a very small instrument to be made, typically about 20 mm in diameter and 30 mm long. These sensors have found many applications in aerospace and industrial applications.

4.2.7.2 Detailed description of sensor

The major difference in construction between the flex gyroscope and the dynamically tuned gyroscope is that the flex device does not have a Hooke's joint type of flexure pivot arrangement but has a flexible pivot where the drive shaft is reduced in diameter, as shown in Figure 4.11.

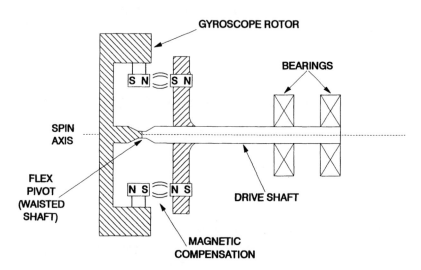

Figure 4.11   Shaft assembly of flex gyroscope

The rotor is attached to the main shaft usually using a spider and strut arrangement. Flexible joint torques arising from this form of suspension are compensated by small permanent magnets attached to the rim of the rotor which attract a set of high permeability screws mounted on a plate attached to the shaft. This use of magnetic forces to balance the flex pivot torques has the effect of decoupling the rotor from the drive shaft, as depicted in Figure 4.11 and is sometimes known as magnetic tuning. Generally, magnetic shielding is crucial with this sensor to ensure effective decoupling of the rotor. A schematic diagram of such a sensor is given in Figure 4.12.

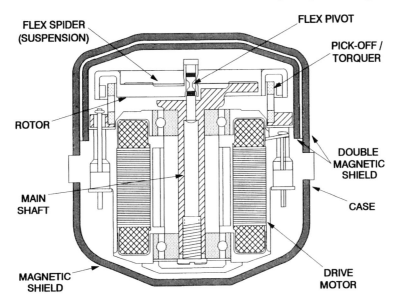

Figure 4.12   Flex gyroscope

4.2.7.3 Sources of error

The error mechanisms associated with this sensor are very similar to the dynamically tuned gyroscope already described and will not be repeated here. The outputs may be expressed mathematically in the same form as those for the dynamically tuned gyroscope in eqn. 4.9. Use of magnetic tuning allows the option of running the rotor at different speeds to fulfil different needs or applications. Additionally, this type of suspension gives very good resilience to vibratory inputs.

4.2.7.4 Typical performance characteristics

Typical values for the significant error sources and performance parameters are given below:

g-independent bias	1–5°/hour
g-dependent/mass unbalance bias	1–10°/hour/g
anisoelastic bias	0.5–0.25°/hour/g²
scale factor error (uncompensated temperature effects)	up to 400 ppm/°C
scale factor non-linearities (at high rotation rates)	0.01–0.1%
bandwidth	up to 100 Hz
maximum input rate	>500°/second

It can be seen that the error parameters are very similar to those quoted for the dynamically tuned gyroscope in Section 4.2.6.4. Typically, the drift performance of such a device is in the range 1 to 50°/hour with the capability to capture rotation rates up to at least 500°/second. Additionally, the anisoelasticity is often slightly smaller, typically by a factor of two to five.

4.3 Rate sensors

There is a class of mechanical sensors designed to sense angular rate using various physical phenomena which are suitable for use in some strapdown applications. Such devices resemble conventional gyroscopes in that they make use of the principles of gyroscopic inertia and precession described in Section 4.2.2. They are suitable for some lower accuracy strapdown applications, particularly those that do not require navigational data, but stabilisation. These devices tend to be rugged and to be capable of measuring rotation rates up to about 500°/second with typical drift accuracies of a few hundred degrees per hour. A number of devices of this type are discussed below.

4.3.1 Dual-axis rate transducer (DART)

4.3.1.1 Introductory remarks

Development of this type of gyroscope started in the United States during the 1960s. It has, as its name implies, the ability to sense angular rate about two orthogonal axes. Its basic performance is certainly sub-inertial, typically having a drift in the region of 0.5°/second or less. Its size is somewhat smaller than the rate integrating gyroscope being about 18 mm in diameter and 40 mm long.

4.3.1.2 Detailed description of sensor

The inertial element in this form of gyroscope is a sphere of heavy liquid, such as mercury, contained in a spherical cavity. This cavity is rotated at high speed about an axis along the case in order to give high angular momentum to the fluid sphere. There is an assembly of paddles, rigidly mounted to the inside of this spherical cavity. These paddles have piezoelectric crystals attached to them as shown in Figure 4.13. The instrument is sensitive to angular rates of the case about two orthogonal axes normal to the spin axis.

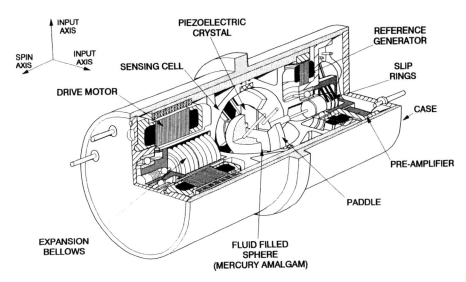

Figure 4.13 Dual-axis rate transducer

A simplified explanation of the operation of the complex dynamical interaction of this sensor is as follows. As the case of the sensor is rotated about either of its two sensitive axes, the spin axis of the mercury tends to lag behind that of the spherical cavity which moves with the rotation of the case. As a result of viscous coupling, a torque is applied to the rotating sphere of fluid in such a way as to make it precess at the input rate. This motion of the fluid causes a deflection of the paddles within the spherical cavity, bending the piezoelectric crystals and generating an a.c. electric signal which is proportional to the applied angular rate. The phase of this signal relative to the reference generator on the rotor shaft gives the axis of the applied rate.

4.3.1.3 Typical performance characteristics

Typical values for the significant error sources and performance parameters are as follows:

g-independent bias (including temperature effects)	0.1–0.4°/second
g-dependent bias	0.03–0.05°/second/g
g^2-dependent bias	~0.005°/second/g^2
scale factor error (sensitivity over operating temperature)	~5%
scale factor non-linearity (at high rotation rates)	~0.5% of maximum rate
bandwidth	>80 Hz
maximum input rate	up to 800°/second

This form of sensor is very rugged owing to the form of its fabrication. Its error processes tend to be similar to those of the dynamically tuned gyroscope. The temperature sensitivity is quite a complex function and can be difficult to correct exactly. In general, accuracy is usually somewhat less than that of the rate integrating and dynamically tuned gyroscopes, and so it is not usually used for inertial navigation applications. However, it does have many applications such as seeker stabilisation and provision of signals for autopilot feedback.

Derivatives of this sensor have also been produced which do not use any liquid within the sphere. The accuracy of such devices is somewhat less than the mercury filled device.

4.3.2 Magneto-hydrodynamic sensor

4.3.2.1 Introductory remarks

The development of this dual-axis rate sensor also has its origins in the United States and has taken place in parallel with the development of the dual-axis rate transducer described in Section 4.3.1. It is of similar size to the dual-axis rate transducer and has comparable performance capability, the g-independent bias being in the region of 0.05°/second to 0.5°/second.

4.3.2.2 Detailed description of sensor

This device does not rely upon the angular momentum of a spinning mass, but uses a rotating angular accelerometer to sense angular rates about two mutually perpendicular axes of the sensor. The rotating angular accelerometer acts as an integrator and provides an electrical signal directly proportional to the applied angular rate.

The sensor consists of the angular accelerometer and a synchronous motor, as illustrated in Figure 4.14. A slip ring assembly is required to access the electrical signals produced by the rotation of the angular accelerometer. The case is usually a high permeability alloy that provides the necessary magnetic shielding.

The principle by which the sensor operates is as follows. When an angular accelerometer is rotated at a constant rate (ω_a) about an axis perpendicular to its sensitive axis, and a steady rotation rate (ω_i) is applied about an axis perpendicular to this axis of rotation, then the instantaneous angular rate (ω_o) about the input axis of the angular accelerometer given by:

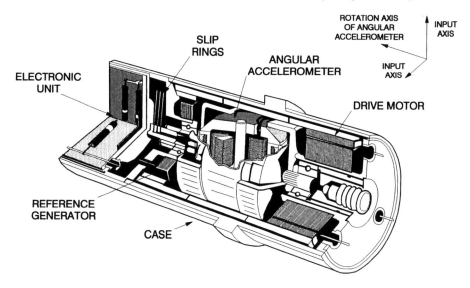

Figure 4.14 Magneto-hydrodynamic sensor

$$\omega_o = \omega_i \sin \omega_a t$$

Hence the angular acceleration is :

$$\dot{\omega}_o = \omega_a \omega_i \cos \omega_a t$$

Consequently, the input rate is changed to a time varying angular acceleration. The rotating angular accelerometer produces an alternating signal, the amplitude of which is directly proportional to the applied angular rate whilst the frequency is equal to the rate of turn of the angular accelerometer. This output signal can be resolved to give the applied angular rate about two orthogonal axes, both of which are mutually perpendicular to the axis of rotation of the angular accelerometer.

A diagrammatic representation of the angular accelerometer arrangement within the sensor is shown in Figure 4.15.

The angular accelerometer usually has an annular ring of mercury between the radially oriented permanent magnet and the magnetic case, which provides a path for the magnetic field. The presence of an input rate results in relative motion of the magnetic field with respect to the torus of mercury. As a result of the magneto-dynamic effect, the motion of the magnetic field produces a voltage gradient across the mercury, mutually at right angles to the motion and the magnetic field. The presence of the transformer windings, as shown in Figure 4.15, results in a voltage appearing in the secondary winding.

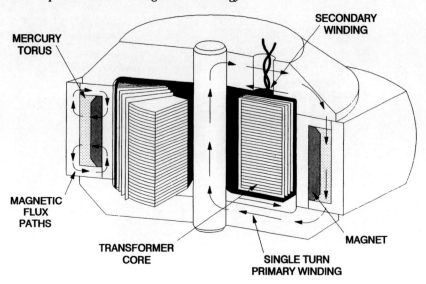

Figure 4.15   Magneto-hydrodynamic active element

4.3.2.3 Typical performance characteristics

The performance figures for the magneto-hydrodynamic (MHD) sensor are typically as shown below:

g-independent bias (including temperature effects)	0.05–0.5°/second
g-dependent bias	0.05°/second/g
g^2-dependent bias	~0.001°/second/g^2
scale factor error (sensitivity over operating temperature)	~4%
scale factor non-linearity (at maximum rotation rates)	~0.1% of maximum rate
bandwidth	100 Hz
maximum input rate	up to 400°/second

This sensor is very rugged and capable of surviving in very harsh environments. The performance capability appears to be that of a good rate sensor particularly suited to stabilisation applications. The error equation used to define performance may be expressed in a form similar to that used for the conventional mechanical gyroscopes as discussed in Section 4.2.

4.4 Vibratory gyroscopes

4.4.1 Introduction

The origins of this type of gyroscope may possibly be considered to be in the middle of the nineteenth century. Foucault demonstrated that a vibrating rod would maintain its plane of vibration whilst it was being rotated in a lathe. Later that century, Bryan [7] demonstrated that angular rate sensing, as well as linear acceleration sensing, could be achieved using this principle.

It was during the 1950s that work started to develop the principle of a vibrating element for use in sensing angular rate, the majority of the effort being in the United States. The vibrating element has taken various forms such as a string, a hollow cylinder, a rod, a tuning fork, a beam and a hemispherical dome. This form of gyroscope occurs in nature as laterae in flying insects. One of the earliest forms of gyroscope using a vibrating element was produced by the Sperry Gyroscope Company. It was based on the tuning fork principle and was known as the gyrotron.

The basic principle of operation of such sensors is that the vibratory motion of part of the instrument creates an oscillatory linear velocity. If the sensor is rotated about an axis orthogonal to this velocity, a Coriolis acceleration is induced. This acceleration modifies the motion of the vibrating element and provided that this can be detected, it will indicate the magnitude of the applied rotation.

The most common design technology for these sensors has generally used a stable quartz resonator with piezoelectric driver circuits. Some designs have produced sensors with small biases, in the region of 0.1°/hour. However, the smaller sensors have tended to produce biases in the region of 0.1 to 1°/second. Typical limitations for this type of technology, for use in inertial navigation systems, have been high drift rates, resonator time constants and sensitivity to environmental effects, particularly temperature changes and vibratory motion. However, these sensors can be made to be extremely rugged, including the capability of withstanding applied accelerations of many tens of thousands of g.

These sensors are usually quite small, usually with a diameter of something less than 15 mm and a length of about 25 mm. Others are significantly smaller than this and are packaged in rectangular cases. These sensors have been used in many applications, particularly to provide feedback for stabilisation or angular position measurement tasks.

As there are many similarities in the performance characteristics of vibratory gyroscopes, such aspects are covered in a single section following general descriptions of the operating principles for different types of design.

4.4.2 Vibrating wine glass sensor

This sensor is synonymous with the vibrating cylinder and vibrating dome gyroscopes. These devices usually have three basic components inside a sealed case:

1 A resonant body in the shape of a hemisphere or a cylinder with a high Q factor to maintain a stable resonance. It is often made from ceramic, quartz or steel.
2 A forcing or driving mechanism, commonly made from a piezo-electric material.
3 A pick-off device to sense the modified motion, also usually a piezoelectric device.

These components are shown schematically in Figure 4.16(a) together with the resonant vibration patterns in the static and rotating cases in Figure 4.16(b).

The resonant body, usually a hemisphere or cylinder, is forced to vibrate at its resonant frequency by four equally spaced piezoelectric 'driving crystals' firmly attached to its circumference. One opposite pair of crystals is driven with an oscillatory signal to distort the resonant body so that modes appear in the distortion pattern on its circumference. The other pair of crystals is used as feedback sensors to control the nodes in the induced motion. When the cylinder is stationary, the nodes in the vibratory motion are positioned exactly between the driving crystals, the anti-node axes *A* and *B* being shown in the Figure. If the resonant body is rotated at an angular rate about an axis orthogonal to the plane containing the vibratory motion of this body, the pattern of vibration is modified by the Coriolis acceleration. The effect is to add a tangential force to the vibratory force along the diameter of the resonant body. Consequently, there is a change in the motion at the points mid way between the driving crystals as the vibration pattern has moved through an angle relative to the case. Hence, the pick-off transducer crystals now sense movement of the resonant body, the amplitude of displacement being directly proportional to the applied rotation rate.

By demodulating the signal from the pick-off transducers, with respect to the waveform used to power the driving crystals that vibrate

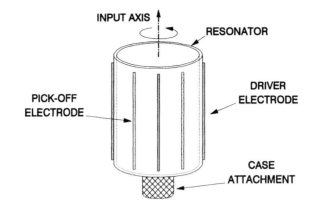

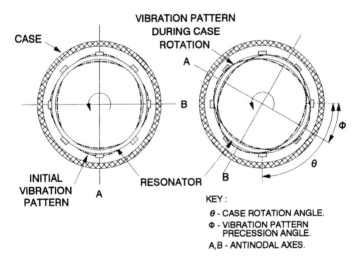

Figure 4.16 (a) *Vibratory gyroscope schematic diagram*
(b) *Resonator vibration patterns*

the cylinder, a d.c. signal is produced. Its magnitude is proportional to the applied rotation rate and its sign indicates the sense of rotation. The second pair of piezoelectric crystals, that is nominally at the nodal position, can be used to modify the vibration characteristics of the cylinder in order to enhance the bandwidth of this sensor. These crystals are driven by a feedback signal derived from the signal produced by the pick-off transducers.

An alternative configuration of this form of sensor is to fabricate the resonant body from a ceramic material and then deposit metal electrodes on to the ceramic. This design has some advantages in terms of reliability as the wires can be attached at points of zero movement. An alternative method of making the cylinder vibrate is to use a magnetically driven ferromagnetic cylinder. Capacitive pick-offs can be used thus reducing the damping of the resonance produced by attaching leads to the resonator.

It is very important that the vibrating shape, such as the shell of the hemispherical resonator gyroscope, is machined to have a wall thickness which is as uniform as possible and that it is then dynamically balanced to compensate for material inhomogenities and machining errors. A non-uniform shell is not sensitive to small rotations as the nodes do not move when the case is rotated at low rates about the input axis.

Sensors of this type can be very rugged and have been demonstrated to withstand accelerations or shocks well in excess of 20 000 g. Additionally, these devices can be activated very rapidly, but great care is required in the choice of resonant material to achieve a form of temperature sensitivity which does not mask its rate sensitivity. This form of device generally does not show any significant acceleration sensitivity as such a response only results from deformation of the resonant body.

Vibrating wine glass sensors can be operated as either open loop or closed loop devices. In the open loop configuration, the electric signal merely increases as the angular rate increases. When used in a closed loop configuration, the second set of crystals is used to null any displacement sensed by the pick-off crystals, this secondary drive signal being proportional to the detected rate. This latter technique leads to a far more linear relationship between the output signal and the input stimulus.

4.4.3 Vibrating disc sensor

An alternative configuration has been developed by British Aerospace based on a planar metal disc [8]. The resonator is formed from a metal alloy disc, which is machined to form a ring that is supported by rigid spokes, as shown in Figure 4.17. This ring is forced into resonant sinusoidal oscillation in the plane of the ring, using an alternating magnetic field, creating distortions in the shape of the ring. This motion of the ring is detected using capacitive techniques to measure the distance between a fixed plate and the edge of the ring.

The operation of this type of gyroscope is identical to that of the sensor described in the previous section. The vibration pattern remains fixed with respect to the ring whilst the sensor is stationary.

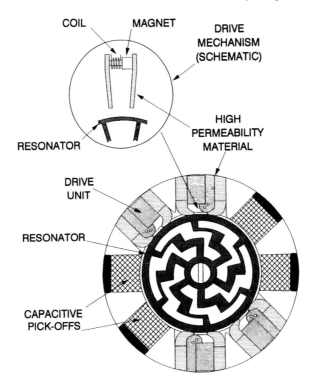

Figure 4.17   A section through a vibrating disc gyroscope

However, the positions of the nodes and anti-nodes of this vibratory motion are displaced through an angle when the sensor is rotated about an axis perpendicular to the plane containing the resonator. The magnitude of the angular displacement of the vibration pattern is proportional to the applied angular rate, and is measured using the capacitive pick-offs arranged around the edge of the resonator.

It has been suggested that the performance of this configuration is superior to the performance produced by the resonant cylinder sensor owing to the improved stability properties of the metal alloy used.

4.4.4 Tuning fork sensor

This form of device is very similar to the wine glass sensor described above. The sensing element is two vibrational structures mounted in parallel on a single base, each structure having a mass positioned at the end of a flexible beam. When the two structures are excited to vibrate in opposition, the effect is analogous to the motion of the tines of a tuning fork. When rotated about an axis parallel to the length of the beams, the effect of the Coriolis acceleration is to produce a torque couple about this input axis. The torque is oscillatory and is in phase with the tine mass velocity. The amplitude of the oscillation is proportional to the applied rate.

A schematic diagram showing the principle of operation the tuning fork sensor is given in Figure 4.18.

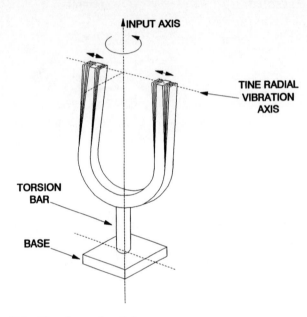

Figure 4.18   Principle of a tuning fork sensor

Two specific problems delayed the development of this type of sensor:

1 variation in the bending and torsional elastic modulii of materials with temperature;
2 bias instabilities caused by the lateral displacement of the tine mass centres.

Use of crystalline quartz tine forks has alleviated many of these problems.

A typical implementation of this form of technology is to use a pair of piezoelectric vibrating beams, each pair consisting of two piezoelectric bender elements mounted end to end. The element that is firmly attached to the base is driven resonantly so that the second element swings but does not bend. This element senses the angular motion. When there is angular motion about the sensitive axis of this tuning fork, there is a momentum transfer to the perpendicular plane as a result of the Coriolis acceleration. This sensing element now bends as a consequence of this momentum transfer and an electrical signal is produced that is proportional to the applied angular rate.

4.4.5 Quartz rate sensor

The quartz rate sensor is a direct application of the tuning fork principle. It is a single degree-of-freedom, open loop, solid-state sensor. In this device, quartz is formed into an H fork configuration, where one pair of tines has an array of electrodes. These tines are driven at their resonant frequency of about 10 kHz.

When the sensor is rotated at a given rate about the input axis, a Coriolis force is produced, which oscillates in phase with the tine mass velocity. This torque produces a walking motion of the pick-off tines, perpendicular to the vibrating plane of the driven tines. The time varying displacement of the tines, which is proportional to the applied rate, is detected with a capacitive sensor. It is vital that the mount is strong so that it supports the quartz element, but sufficiently isolated in order to maximise the Coriolis coupling torque into the pick-off tines. Drive and pick-off signals are routed through the mount. A general arrangement of this sensor is shown in Figure 4.19.

This sensor, like other solid-state devices, can have various rate sensitivities and the full scale output can be modified by changing the electronic gain control. These parameters are functions of the signal processor which controls the input range and the signal bandwidth. Additionally, the design of the vibrating tuning fork can have an almost infinite combination of size, thickness and electrode pattern, enabling flexibility of performance to be achieved.

High performance is possible from this configuration, giving a g-dependent bias of about 25°/hour/g.

4.4.6 Silicon sensor

The material silicon has many properties that make it suitable for the fabrication of very small components and intricate monolithic devices. It is inexpensive, very elastic, non-magnetic, it has a high strength to weight ratio and possesses excellent electrical properties allowing component formation from diffusion or surface deposition. Additionally, it can be electrically or chemically etched to very precise tolerances, of the order of micrometres.

A team at the Charles Stark Draper Laboratories [9] has used chemical etching techniques to make a very small gyroscope from a

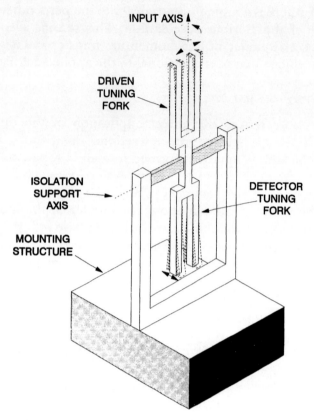

INPUT AXIS

DRIVEN
TUNING
FORK

ISOLATION
SUPPORT
AXIS

DETECTOR
TUNING
FORK

MOUNTING
STRUCTURE

Figure 4.19   Principle of operation of a quartz rate sensor

wafer of single crystal silicon. The sensor does not have any continuously rotating parts, but part of its structure is vibrated at very high frequency. A schematic representation of this sensor is shown in Figure 4.20.

The sensor comprises a double gimbal structure with a vertical member electro-plated with gold mounted on the inner gimbal. The gimbals are each supported by a set of orthogonal flexure pivots as indicated in the Figure. These pivots allow each gimbal a small amount of torsional freedom about their respective support axes whilst remaining rigid in other directions. The outer gimbal is forced to oscillate through a small angle by applying an oscillatory electrostatic torque using pairs of electrodes mounted above and below the outer gimbal structure. When the structure is rotated about an axis perpendicular to the plane of the sensor, as indicated in the Figure, the inner gimbal also starts to oscillate. The inner gimbal vibrates at the same frequency as the outer gimbal, but with an amplitude

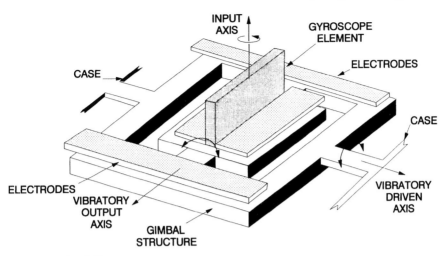

INPUT AXIS

GYROSCOPE ELEMENT

ELECTRODES

CASE

CASE

ELECTRODES

VIBRATORY OUTPUT AXIS

VIBRATORY DRIVEN AXIS

GIMBAL STRUCTURE

Figure 4.20   Silicon gyroscope

proportional to the applied angular rate. This motion is sensed electrostatically by a pair of bridging electrodes. The sensitivity of the device is determined largely by the geometrical arrangement of the structure.

In order to achieve high sensitivity and accuracy, the gyroscope is operated in a closed loop rebalance mode. The inner gimbal is torqued electrostatically in order to maintain it at a null position, the torquer drive signal being proportional to the angular displacement sensed by the electrostatic pick-off, and hence to the applied angular rate. The pick-off and rebalance signals pass through the same electrodes, but use different frequencies. This method of operation allows the gyroscope to tolerate variations in the frequency of the vibratory motion. It also allows the amplitude of the vibratory motion to be increased, without cross-coupling interactions becoming unacceptably large, thus enabling an increase in the signal to noise ratio of the output signal. The electronic loops which control the operation of the gyroscope also allow compensation for imperfections in the device fabrication and changes in temperature to be applied.

The gyroscope is packaged in a sealed case to maintain a vacuum. This enables a high Q (quality factor) to be achieved in the resonant structure, enhancing further the sensitivity of the device. It is anticipated that this type of device should be capable of achieving an in-run drift performance of better than 100°/hour, and be capable of measuring very high rotation rates. The device can be substantially less than 1 mm long and can be used with silicon accelerometers (discussed in Section 6.4.3) to make a very small inertial measurement unit.

Silicon vibrating disc sensor

It is possible that the resonator for the vibrating disc gyroscope outlined in Section 4.4.3 could also be manufactured from silicon. This material possesses many characteristics which would be ideal for this type of component: low cost, elasticity and high strength to weight ratio, for example. However, if silicon were to be used, it would be necessary to use an alternative technique to force the structure to vibrate, such as the use of piezoelectric devices. One clear advantage that silicon could offer would be a size reduction, probably with reduced cost of manufacture.

4.4.7 Vibrating wire rate sensor

This device, also known as the vibrating string gyroscope, has three fundamental components within its case, as shown in Figure 4.21:

1 the vibrating element in the form of a taut conductor;
2 the drive magnet;
3 the signal or pick-off magnet.

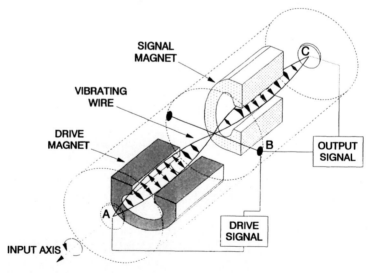

Figure 4.21   Illustration of a vibrating wire rate sensor

The principle of operation is similar to that of Foucault's pendulum. If a wire or string is oscillating in a plane and the supports of the vibrating element turn through an angle, then the plane of vibration remains fixed in space despite the fact that the string rotates with its supports.

An alternating current at a selected drive frequency is applied to the wire between the points *A* and *B*, as indicted in Figure 4.21. The interaction between the magnetic field around the wire and that of the drive magnet sets up a standing wave vibration of the wire. A second magnet, the signal magnet, is arranged with its magnetic field at right angles to the drive magnet as indicated in the Figure.

Consider now the effects of rotations about an axis that passes between the points *A* and *C*. When the gyroscope is not being rotated about this axis, the vibration of the wire between the poles of the signal magnet will not induce any change in the current in the wire. Thus, when the signal emerging from the point *C* is compared with the applied signal in a suitable demodulator, there will be no resultant signal. In the situation where the device is rotated about the axis *AC*, a rotation of the signal magnet with respect to the plane of vibration will arise. This causes the signal magnet to modify the current flowing in the wire and thus to modulate the carrier. Comparison of the drive and output signals now yields a resultant signal which is a measure of the applied angular rate about the axis *AC*.

It is usual to choose the natural frequency of both the wire and the drive frequency to be in the region of 20 kHz or more, so that the vibrations are well above those likely to be produced by environmental vibrations. This prevents synchronous vibration of the sensor's case along the pick off axis being interpreted as an input rotation.

4.4.8 General characteristics of vibratory sensors

All vibrating sensors tend to have a very short reaction time, i.e. rapid start-up capability, and some designs are very rugged. Significant sources of error with these devices are their sensitivity to changes in ambient temperature and the potential for cross talk between similar sensors mounted on the same structure. Careful design can minimise these effects and the errors they introduce into the output signal. These devices are usually termed solid-state sensors and offer good shelf life and good dormancy characteristics as they do not have bearings, lubricants or any other fluid within their case. Good reliability is possible because of the need for only one bonded joint and the power leads can be connected, with suitable design, at a point which does not move. The form of design also ensures low power consumption.

Vibratory sensors are subject to biases and scale factor errors equivalent to those which arise in conventional gyroscopes. Typical performance of the miniature vibratory sensors is not compatible with the requirements of inertial navigation systems, but have much to offer for control and stabilisation processes.

The performance range of such miniature devices is as follows:

g-independent bias[1] (including temperature effects)	0.1–1°/second
g-dependent bias	0.01–0.05°/second/g
scale factor temperature dependence (sensitivity over operating temperature)	0.01–0.05%/°C
scale factor non-linearity (at maximum rotation rates)	0.03–0.3%
bandwidth	60–500 Hz
shock resistance	>20 000 g

In addition, such sensors can be sensitive to vibration although, with careful design, such effects can be minimised. The error equation used to define performance may be expressed in a form similar to that used for the conventional mechanical gyroscopes, as discussed in Section 4.2.

4.5 Cryogenic devices

4.5.1 Nuclear magnetic resonance gyroscope

Investigation started in the 1960s into the application of the phenomenon of nuclear magnetic resonance to the sensing of angular rotation. The nuclear magnetic resonance gyroscope has many attractions, particularly as it will not have any moving parts. Its performance will be governed by the characteristics of the atomic material and will not demand the ultimate in accuracy from precision engineering techniques. Hence, in theory, it offers the prospect of a gyroscope with no limit to either its dynamic range or linearity and therefore, potentially, is an ideal sensor for use in strapdown inertial navigation.

Nuclear magnetic resonance (NMR) [10] is a physical effect arising from the interaction between the nuclei of certain elements and an external magnetic field. Generally, nuclei possess spin angular momentum and, associated with it, a magnetic dipole moment. In the

---

[1]The hemispherical resonator gyroscope (HRG) has a performance in the 0.01°/hour class.

presence of a magnetic field, H, the spinning nuclei are subjected to a torque which results in a precession of the nuclear spin axis about the direction of the magnetic field. This is known as the Larmor precession and has a characteristic angular frequency, ω_L, given by the relation:

$$\omega_L = \gamma H \tag{4.10}$$

where γ is the ratio of the magnetic dipole moment to the angular momentum, known as the gyromagnetic ratio peculiar to any species of nuclei.

When an angular rate, Ω, is applied to a cell containing the precessing nuclei, then the read-out mechanism is in a rotating axis frame, resulting in an apparent change in the precessing frequency of the atoms. This rotation of the cell is equivalent to applying a torque to an inertial element in a conventional gyroscope as the precessing nuclei act as an inertial element. Thus, the observed precessional frequency, ω_{obs}, becomes:

$$\omega_{obs} = \omega_L + \Omega \tag{4.11}$$

or

$$\omega_{obs} = \gamma H + \Omega$$

Therefore, the determination of the applied rate Ω is dependent on establishing a constant magnetic field at the sensing element and the measurement of a nuclear precessional frequency.

Several techniques [11, 12] employing optical pumping and optical read-out systems have been investigated to allow the small frequency shift induced by the rotation of the sensor to be detected. Optical pumping techniques will transfer an assembly of spins from an equilibrium to a non-equilibrium state. Light of the right frequency directed along the direction of the magnetic field will excite magnetic substances so that the chances of observing the transition of spins from one state to another is enhanced. Transitions are brought about by applying a weak oscillating field at 90° to the static field, H. When the frequency of the oscillating field is close to the Larmor frequency of the excited spins, the orientation of the nuclear spins is reversed. This effect can be detected by circularly polarised light directed at 90° to the direction of the static field. The orientation of the magnetic moments affects the plane of polarisation of this light and hence, after passing through an analyser, the resulting intensity modulation may be picked up by a photodetector.

Techniques for compensating for instability of the external field H involve the measurement of the Larmor frequencies of two magnetic substances contained within one sample cell.

The American company Litton produced a design for an experimental NMR gyroscope as shown in Figure 4.22. The NMR cell contains rubidium vapour, krypton and xenon. These gaseous materials are used as they have suitably long relaxation times enabling good sensitivity to be achieved from the very accurate determination of the observed precession frequencies. This design employs one light beam for both pumping and detection.

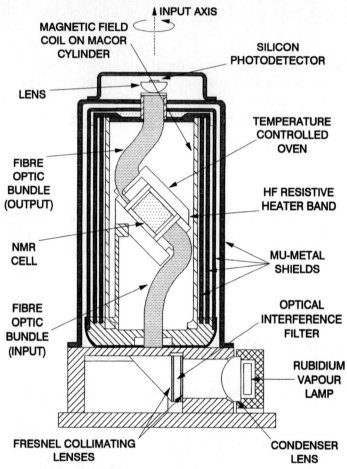

Figure 4.22   Experimental NMR gyroscope configuration

Clearly, a practical device requires a very uniform and constant magnetic field over the active sample of material. One way, and possibly the only practical technique, is to apply the unique properties of superconductors, such as the cryogenic Meissner effect [13, 14]. A cylinder made of superconducting material operating at a temperature below its critical temperature will prevent magnetic flux from entering or leaving the central space. Hence, not only will this material shield the nuclear material, it will also trap the flux within the space as its temperature cools below the critical temperature of the superconductor.

One method of detecting the Larmor precession of the nuclear material placed inside the superconducting tube is to use a superconducting quantum interference device (SQUID) magneto-meter [14]. In fact, these devices were used during the early development of cryogenic nuclear magnetic resonance gyroscopes to detect the free induction decay of a sample substance in a superconducting cylinder. The sample substance used was He^3 which, being gaseous at cryogenic temperatures, has the advantage of long relaxation times. This technique was unsuccessful because of the very poor signal to noise characteristics.

The cryogenic nuclear magnetic resonance gyroscope, as shown in Figure 4.23 [15], has many attractions, particularly the solid state nature of the construction and the extremely low drift which is possible, as low as arc seconds per annum. This accuracy and the anticipated size of a few litres suggest the most likely application to be ship's strapdown navigation systems.

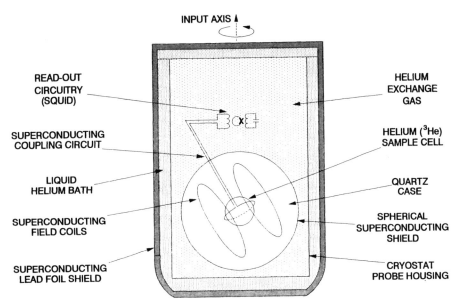

INPUT AXIS

READ-OUT
CIRCUITRY
(SQUID)

SUPERCONDUCTING
COUPLING CIRCUIT

LIQUID
HELIUM BATH

SUPERCONDUCTING
FIELD COILS

SUPERCONDUCTING
LEAD FOIL SHIELD

HELIUM
EXCHANGE
GAS

HELIUM (^{3}He)
SAMPLE CELL

QUARTZ
CASE

SPHERICAL
SUPERCONDUCTING
SHIELD

CRYOSTAT
PROBE HOUSING

Figure 4.23   Cryogenic nuclear magnetic resonance gyroscope

In the 1970s, the NMR gyroscope and the ring laser gyroscope (discussed in Chapter 5) were looked upon as rival technologies for higher accuracy navigation applications. Although the manufacture of a viable sensor based on NMR technology appeared feasible, its development was overshadowed by massive investment in ring laser technology throughout the 1970s and 1980s, the latter being considered the more promising technology at that time. As far as the authors are aware, there has been little in the way of research effort and resources directed towards the NMR sensor in recent years and, as a result, the full potential of such devices has never been realised.

4.5.2 SARDIN

Another example of a rotation sensor which exploits super-conductivity has been proposed by Brady [16]. This sensor is known by the acronym SARDIN (superconducting absolute rate determining instrument). The device is basically a superconducting cylindrical capacitor, the behaviour of which is governed by the principle that a closed superconducting ring always keeps the amount of flux linking it at a constant value. It does this by generating super-currents which flow in the ring for an indefinite length of time encountering no resistance to their motion.

Consider two concentric cylinders of superconducting material, one cylinder being raised to a potential V relative to the other. If the assembly is rotated around its central axis with angular velocity Ω, then the moving charges constitute currents which give rise to a net magnetic flux in the region between the two cylinders. As the magnetic flux inside a loop of superconductor cannot change, this field must be backed-off by further supercurrents (I) induced in each cylinder. A rigorous mathematical treatment of this effect [17] gives the governing equation:

$$I = \frac{C_0 V r \Omega}{\varepsilon} \tag{4.12}$$

where C_0 = the capacitance per unit length of the assembly
$\quad\quad\; r$ = the mean radius
$\quad\quad\; \varepsilon$ = the permittivity of the dielectric between the two cylinders

With a potential of 225 volts applied to such a device, having a mean radius of 2 cm, coupled to a SQUID magnetometer and an ammeter, Brady reported an angular rate sensitivity of 1 rad/s.

Following initial attempts to produce a superconducting rate sensor of this type in the early 1980s, development was not pursued because of the difficulties encountered in detecting the very small output signals generated. However, the development of higher temperature superconductors coupled with enhancements in signal processing techniques could possibly make this a viable technique in the future.

4.6 Electrostatically suspended gyroscope

Work to develop this class of gyroscope, also known as the ESG, started during the 1950s in the United States. The aim was to suspend a rotating sphere, known as a gyroscopic ball, by means of an electric field in an evacuated cavity. This form of suspension eliminates the undesirable characteristics of conventional gyroscopes such as bearings, flotation (in a liquid) or suspension.

The principle of operation is that a precision sphere, usually made of beryllium, is initially spun to very high speed whilst being suspended electrostatically in a hard vacuum. This sphere is then allowed to coast, its spin axis maintaining a fixed direction in inertial space provided that the sphere and suspension do not have any imperfections which can introduce torques on the sphere. Any angular movement of the case of the gyroscope in space can be determined optically or electrically by reference to this spin axis. Navigation may be achieved by determining the apparent change of this reference axis with respect to a local reference direction such as the local vertical. A schematic diagram of the ESG is shown in Figure 4.24.

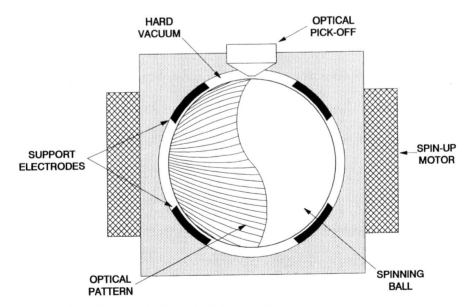

Figure 4.24   Electrostatically supported gyroscope

The development of the ESG was very successful, producing very accurate sensors capable of achieving drifts of the order of 0.0001°/hour. The sensors were developed primarily for use on stable platforms, giving navigation accuracies of the order of 0.1 nautical miles per hour. However, they are capable of being used in a strapdown application, particularly in a benign environment.

The major difficulties with this sensor have been associated with:

- manufacturing the inertial element, i.e. the ball, to sufficient accuracy to eliminate mass unbalance;
- producing a read-out device that does not disturb the motion of the sphere;
- producing a design that avoids grounding the ball, or dropping the ball, as it is sometimes referred to;
- avoiding self destruction if the power supply fails.

Errors in the production of the spherical inertial element lead to oscillations, nutations and whirls when the sphere is rotated at the very high speed which is necessary, typically of the order of 150 000 rpm. These undesired excursions are not very well damped because of the high vacuum inside the sensor. Additionally, the sensor requires special techniques to provide shock and vibration protection as the gyroscope does not have any inherent mechanism to damp out linear disturbances of the sphere. The addition of the anti-shock and anti-vibration mounts add to the size of the sensor.

The ESG is a very high accuracy sensor and has been used for many decades in specialised applications in aircraft, ships and submarines. This sensor has also been suggested for detecting acceleration. Since the support electrodes provide the only forces for accelerating the sphere to keep it moving with its case, measurement of these forces provides a measure of acceleration.

Unfortunately, despite being a very simple concept, the design is complex and the gyroscope is large and expensive. However, it is one of the most accurate gyroscopes ever to be designed and produced.

A similar sensor was produced by Rockwell in the USA known as the gas bearing free rotor gyroscope. This sensor was based on a spherical bearing, like a ball and socket arrangement, using a self acting gas lubricated bearing to support the ball. However, this sensor differs from the ESG as it only has limited angular freedom and is not really suitable for strapdown applications.

4.7 Other devices for sensing angular motion

There are a number of other devices that are either used or could be developed to sense angular motion. They are considered separately as they do not fit easily in any of the above classes of gyroscope, and are generally not the prime angular motion sensor used in a strapdown navigation system. Included in this category are fluidic sensors and fluxgate magnetometers, both of which offer well established technology suitable for various applications. The sensors tend to be small, reliable and rugged. Fluidic devices are often used for providing short term references, such as in stabilisation tasks, whereas magnetometers are generally used for long-term aiding of systems.

4.7.1 Fluidic (flueric) sensors

This term has several meanings, such as sensors which make use of the flow of fluid for either the propulsion of the rotor or its support. Another meaning, and of more direct relevance to the immediate consideration here, is the use of a fluid for the sensing of angular motion. Sometimes, this class of angular motion sensor is known by the term flueric, the term fluidic being reserved to describe those sensors which use a fluid either for support or for powering the rotor. The text will concentrate on the use of a fluid for the sensing of angular motion, i.e. flueric sensors.

Development of flueric sensors started in the 1960s. They appear to offer an interesting alternative to the electromechanical instruments, but despite much effort, it has not proved possible to make this type of sensor suitable directly for inertial navigation applications. The major difficulty is achieving adequate stability, resolution and insensitivity to environmental effects, particularly temperature changes. However, this class of sensor has found many applications, including flight control, stabilisation and in flight safety systems.

One form of sensor is a flueric gyroscope. In this device there is a spherical cavity with porous walls and a rotating mass or swirl of gas within the cavity. When the sensor case is rotated, the direction of the swirl of gas remains fixed and the displacement can be detected by monitoring pressure changes. Figure 4.25 depicts the form of this sensor.

Another form of this class of instrument has a continuous laminar flow of gas from an orifice which impinges on a pair of hot wire detectors. When angular motion is applied about axes orthogonal to the gas flow, the gas jet appears to be deflected laterally relative to the case. This results in a differential cooling of the hot wires, with a consequential change in resistance which is detected using a bridge circuit. The output signal is proportional to the applied angular rate. This form of sensor is shown schematically in Figure 4.26. It tends to be sensitive to temperature gradients, acceleration, vibration and shock.

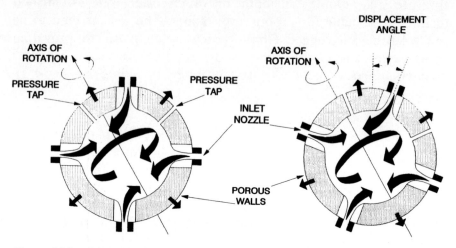

Figure 4.25   Schematic diagram of a fleuric attitude sensor

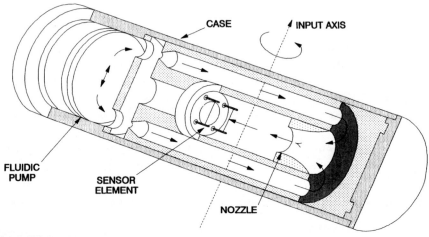

Figure 4.26   Gas jet sensor

These sensors may either use fluid bled from an engine, such as a jet engine's efflux, or may be pumped in a closed cycle.

Such sensors provide an inexpensive, short term reference with turn rate measurement accuracies of around 1% of the applied rate. Generally, they also show significant sensitivity to most environmental features, particularly temperature changes. However, some designs are very rugged and reliable and are capable of operating over a wide temperature range.

4.7.2 Fluxgate magnetometers

It was during the 1950s that an airborne magnetometer was developed at the Royal Aircraft Establishment, Farnborough (now part of the Defence Research Agency) for attitude measurement. Since that time there have been many developments of this type of attitude sensor [18, 19], usually based on fluxgate elements. A magnetometer has three such magnetic sensing elements, mounted mutually orthogonal to each other. These axes are usually arranged to be aligned with the principal axes of the vehicle, a configuration which enables the attitude of the vehicle to be determined with respect to the Earth's magnetic vector. The magnetometer alone cannot give an unequivocal measurement of a vehicle's attitude. Measurements made with such a sensor define the angle between the Earth's magnetic field and a particular axis of the vehicle. However, this axis can lie anywhere on the surface of a cone of semi-angle equal to that angle about the magnetic vector. Hence, an additional measurement is required to determine attitude with respect to another reference such as the gravity vector.

The basic operation of a fluxgate magnetometer is similar to the operation of an electrical transformer. However, the magnitude of the excitation signal is chosen to drive the core, linking the excitation and pick-up coils, into saturation on alternate peaks of the excitation signal. The high permeability of the core magnifies the effect of the changing magnetic field (H) generated from the changing current in the excitation coil. A simple fluxgate magnetometer element is shown in Figure 4.27.

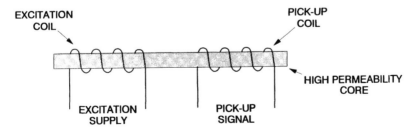

Figure 4.27   A simple fluxgate magnetometer

The magnetic induction (*B*) in the core increases as the applied magnetic field from the excitation coil increases up to a maximum value (B_{max}), determined by the core material. As the magnetic field increases beyond the saturating value (H_{sat}) the induction remains constant. This is illustrated in the *B-H* curves shown in Figure 4.28. This effect is reversible, and so as the magnetic field decreases below the saturation value the induction decreases until it reaches its negative saturation. ($-B_{max}$).

A voltage signal is only induced in the pick-up coil whilst the magnetic induction in the core is changing. Consequently, there is no signal in the pick-up coil during periods of induction saturation. The figure also shows the induction waveforms which result when a square wave signal is applied to the excitation coil with and without an external magnetic field acting along the core. In addition, the figure includes the resulting voltage waveform across the pick-off coil and the driving magnetic field profile.

In the absence of an external magnetic field, the positive and negative excursions in the voltage appearing across the pick-off coil are of equal magnitude, as shown in the left hand side of Figure 4.28, and there is no net output from the device. In the presence of an external magnetic field, this field acts either to aid or to impede the field generated by the alternating current in the excitation coil. As a result, the core is saturated more rapidly and remains in saturation slightly longer for one cycle of the excitation, the cycle in which the excitation and external fields are acting in the same direction. The converse is true when these fields are in opposition. As shown in the right hand side of the figure, the induction waveform is no longer symmetrical and a modified voltage waveform appears across the pick-off coil. The amount of advance and delay in the voltage waveform is proportional to the strength of the external magnetic field.

The change in the voltage waveform characteristic is not easy to measure in the simple scheme described above. A preferred arrangement, which enables a signal proportional to the size of the external magnetic field to be extracted with relative ease, is shown in Figure 4.29.

In this device, a pair of cores is used with the excitation coils wound in series opposition. In this case, the magnetic induction in the two cores cancels out in the absence of an external magnetic field. However, when an external field is present, the magnetic induction waveforms for the two cores are modified in the manner described above. The resulting voltage waveforms for each of the cores are shown in the lower part of the diagram, Figure 4.29. The pick-up coil

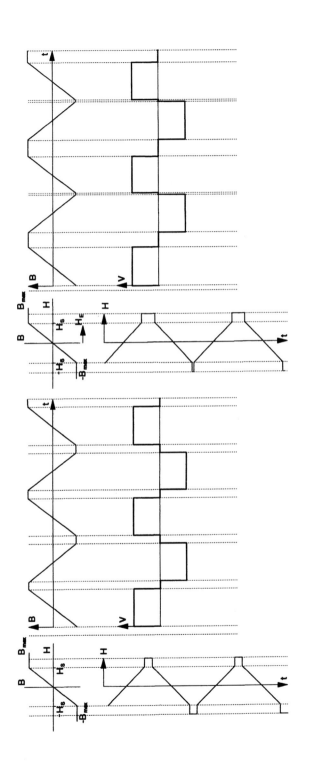

Figure 4.28   Magnetometer waveforms

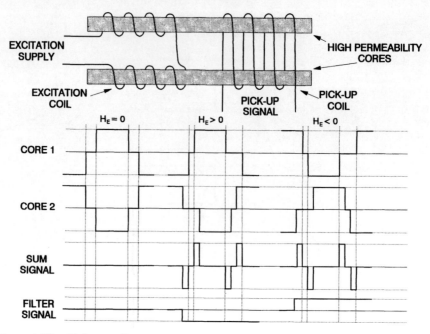

Figure 4.29   Twin-core fluxgate magnetometer and waveforms

is wound so that it sums these two contributions giving rise to the sum channel waveform shown in the Figure. This signal can be applied to a low pass filter and a steady output produced indicating the magnitude of the external magnetic field.

Such analysis for the two-core configuration is dependent on the use of perfectly matched cores and windings. This fundamental requirement for accurate measurement of the magnetic field is very difficult to achieve in a practical device. A common solution to this problem is to use a toroidal core over wound with an excitation coil. The excitation coil is energised with an alternating current signal having a frequency of 1 kHz to 1.5 kHz. The pick-up coil is a single coil wound over the toroid. A second coil can be wound over the toroid at right angles to the pick-up coil, and this can be used as a nulling device. This magnetometer arrangement is shown in Figure 4.30.

When operated in a closed loop mode, commonly known as the closed loop second harmonic mode, a current is passed through the second over wound coil to null the effects of the detected magnetic flux. It is the amplitude of this current which is used to give a measurement of ambient magnetic field strength. The output from this coil is filtered to select only the second harmonic frequency, hence avoiding saturation of the amplifiers by unwanted frequencies.

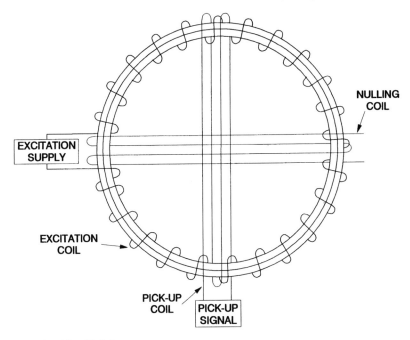

Figure 4.30   *Toroidal fluxgate magnetometer*

The major problem with this form of instrument is that it has poor angular resolution capability, typically about 0.1°, and sensitivity to magnetic anomalies. However, this latter sensitivity can be valuable provided that the magnetic anomaly in the region of operation is stable and well charted. A more serious limitation is often posed by the effects of the structure of the vehicle in which the sensor in mounted. The vehicle is likely to have ferromagnetic materials present as well as changing fields associated with power supplies and other instrumentation. Hence, such extraneous magnetic fields are likely to determine the precision with which the attitude of the vehicle can be determined, with respect to the magnetic field vector of the Earth. Therefore, these sensors are usually mounted at the extremities of a vehicle, as far away as possible from the source of extraneous fields. In some applications, it is possible to route wiring so as to cancel induced fields or else calibrate the sensor in its host vehicle.

A further potential problem arises in a strapdown configuration in a rolling airframe travelling at latitudes where the dip angle of the Earth's magnetic field is large. The roll motion will produce the effect of an alternating field on the magnetometer. An error will result unless the frequency response of the magnetometer is adequate to follow this rate of change of field intensity. In some circumstances it may be necessary to incorporate another sensor to monitor the rapid roll rate in order to compensate the output of the magnetometer.

Generally, fluxgate magnetometers are cheap, small, rugged and reliable, and are capable of working over a very wide temperature range. They are commonly used in low cost systems to provide an attitude reference in pilotless aircraft (UAV) for example, or in more sophisticated reference systems to provide a long term attitude reference. In this case, the data supplied by the magnetometer supplement the estimates made by other sensors, such as gyroscopes, enabling the attitude error to be bounded.

The dimensions of the high permeability core can be modified to give the desired response characteristics. For example, shorter length cores give greater linearity, whereas long thin cores provide a higher output for a given drive power, hence greater sensitivity.

4.7.3 The transmission line gyroscope

The use of the inertia of an electromagnetic field as the rotation sensing element has been demonstrated many times, in applications such as the ring laser gyroscope and the fibre gyroscope, and is discussed in the following chapter. Consideration has also been given by some researchers to the use of non-propagating fields such as the electrostatic gyroscope.

Analysis has also been undertaken by Forder[20] of general relativistic effects which may be observed in a closed ring of parallel wire or coaxial transmission line, appropriately energised, with either static or propagating electromagnetic fields. Three distinct effects can be identified, *viz.* electrostatic, magnetostatic and electromagnetic, allowing this type of device to be operated as an angular motion sensor in a number of different modes. It is predicted that the changes in line voltage and current detected in a circular loop of transmission line of radius R and line impedance Z rotated at an angular rate W are given by:

$$\Delta V = -(ZI)\ \Omega\ R/c$$

or
$$\Delta V = -(V/Z)\ \Omega\ R/c \tag{4.13}$$

where V = charging potential
       I = line current
      c = speed of light

Clearly, one of the major difficulties with the implementation of such devices is the detection of the very small changes in potential and current predicted above. However, practical forms of such sensors based on the above predictions may become feasible and practical in the future as various technologies advance.

References

1 CASE, W.: 'The gyroscope: an elementary discussion of a child's toy', *Am. J. Phys.*, 1977, **45** (11)
2 COUSINS, F.W.: 'The anatomy of the gyroscope', in HOLLINGTON, J.H. (Ed.): Agardograph 313, Parts I, II and III (AGARD, 1990)
3 SAVET, P.H.: (Ed.): *'Gyroscopes: theory and design'* (McGraw-Hill, 1961)
4 EDWARDS, C.S.: 'Inertial measurement units—building blocks into the 1990s'. Proceedings of DGON symposium on *Gyro Technology*, Stuttgart, Germany, 1986
5 PHILPOT, J.StL. and MITCHELL, J.H.: British Patent 599826, 1948
6 CRAIG, R.J.G.: 'Theory of operation of an elastically supported tuned gyroscope', *IEEE Trans.Aerosp. Electron. Syst.*, 1972, **8** (3) 280–297
7 BRYAN, G.H.: 'On the beats in the vibrations of a revolving cylinder or bell,' *Proceedings Cambridge Philosophical Society* , 1890,**7** (101)
8 JOHNSON, B. and LONGDEN, I.M.: 'Vibrating structure gyroscopes and their applications'. Proceedings of DGON symposium on *Gyro Technology*, Stuttgart, Germany, 1994
9 ELWELl, J.: 'Progress on micromechanical inertial instruments'. Proceedings of DGON symposium on *Gyro Technology*, Stuttgart, Germany, 1991
10 RUSHWORTH, F.A. and TUNSTALL, D.P.: *'Nuclear magnetic resonance'* (Gordon and Breach, 1978)
11 KARWACKI, F.A.: 'Nuclear magnetic resonance gyro development', *Navig. J. Instit. Navig.*, 1980, **27** (1)
12 KANEGSBURG, E.: 'A nuclear magnetic resonance (NMR) gyroscope with optical magnetometer detection', *SPIE 157, Laser Inertial Rotation Sensors*, 1978, **73**
13 ILLINGWORTH: *'Dictionary of physics'* (Penguin, 1990)
14 POTTS, S.P. and PRESTON, J.: 'A cryogenic nuclear magnetic resonance gyroscope', *Navig. J. Instit. Navig.*, 1981, **34** (1), pp. 19–37
15 SHAW, G.L. and TABER, M.A.: 'The ^{3}He gyro for an all cryogenic inertial measurement unit'. Proceedings of DGON symposium on *Gyro Technology*, Stuttgart, Germany, 1983
16 BRADY, R.M.: 'A superconduction gyroscope with no moving parts', *IEEE Trans.*, 1981, **MAG-17**, pp. 861–2
17 POTTS, S.P. and CREWE, P.P.: ASWE Memorandum XTN 82010, 1981
18 NOBLE, R.: 'Fluxgate magnetometry', *Electron. Wirel. World*, 1991, **97** (1666), p. 726
19 HINE, A.: *'Magnetic compasses and magnetometers'* (Adam Hilger, 1968)
20 FORDER, P.W.: 'General relativistic electromagnetic effects in transmission line gyroscopes', *Class. Quantum Gravit.*, 1986, **3**, p. 1125

Chapter 5
Gyroscope technology 2

5.1 Optical sensors

5.1.1 Introduction

This term is applied to those classes of gyroscope which use the properties of electromagnetic radiation to sense rotation. Such devices often use the visible wavelengths, but it is also possible to operate in the near infrared. Some mechanical gyroscopes use optical angle pick-off sensors, but for this discussion they are not classed as optical gyroscopes. Optical gyroscopes use an interferometer or interferometric methods to sense angular motion. In effect, it is possible to consider the electromagnetic radiation as the inertial element of these sensors.

It was during the late nineteenth century that Michelson pioneered work with optical interferometers, although his goal was not to produce an optical gyroscope. In 1913, the Sagnac effect was reported [1, 2] and this is the fundamental principle on which optical gyroscopes are based. When light travels in opposite directions (clockwise and anti-clockwise) around an enclosed ring, differences arise in the apparent optical length of the two paths when the ring is rotated about an axis orthogonal to the plane containing the ring. In 1925, this concept was applied by Michelson and Gale [3] in Chicago using a ring gyroscope with a perimeter of over one mile. By sending ordinary light through evacuated water pipes, they were able to detect the shift produced by the rotation of the Earth.

Further impetus to produce an optical sensor resulted from the demonstration of a laser by Maiman in 1960 [4, 5, 6]. These devices produce a well collimated and highly monochromatic source of electromagnetic energy between the ultra violet and far infra-red part of the spectrum, the wavelength being determined by the laser gain medium. In 1963, the first ring laser was demonstrated by workers at the Sperry Gyroscope company [7].. This marked the beginning of the development of the ring laser gyroscope. About a decade later the fibre optic gyroscope was first demonstrated [8]..

Clearly, the history of the development of optical gyroscopes is far more recent than the history of the mechanical sensors. Further impetus for the development of these sensors, besides the development of the laser, was the interest in the application of strapdown technology and the desire to capitalise on the benefits anticipated from the use of solid-state inertial sensors. One of the main difficulties in the application of strapdown technology from the point of view of gyroscope performance was the lack of adequate dynamic range of the mechanical sensors for the more accurate applications. Initial estimates of performance suggested that the ring laser gyroscope could provide the solution.

The spectrum of performance of optical gyroscopes ranges from the very accurate with a bias of less than 0.001°/hour, usually ring lasers, to tens of degrees per hour, often from the simpler fibre optic gyroscopes. Hence the range covered by optical devices is very similar to that covered by the mechanical gyroscopes. Generally, all the types of optical gyroscope are suitable for various strapdown applications, depending of course on the demanded accuracy of the system.

It appears that the application of optics to the sensing of angular rate can offer a number of advantages over the use of the well established mechanical technology. Some of the advantages often cited are listed below:

1 wide dynamic range;
2 instant start-up;
3 digital output;
4 output independent of some environmental conditions (acceleration, vibration or shock);
5 high rate capability;
6 easy self-test;
7 system design flexibility;
8 extended running life.

5.1.2 Fundamental principles

Optical gyroscopes rely upon the detection of an effective path length difference between two counter-propagating beams of light in a closed path. The mathematical development given here shows how this path difference arises in the presence of an applied turn rate about an axis perpendicular to the plane containing the light path.

Consider a perfect stationary circular interferometer with the light constrained to travel around the circumference of a circle of radius R, as shown in Figure 5.1 below. Light enters the ring at the point X, where there is a beam splitter which directs two beams of light in opposite directions around the complete ring, these beams recombining later at the beam splitter. The transit time (t), the time for the light to make one complete pass around the ring whilst the ring is stationary, is identical for both beams and is given by:

$$t = \frac{2\pi R}{c} \tag{5.1}$$

where c is the velocity of light, and is considered to be invariant.

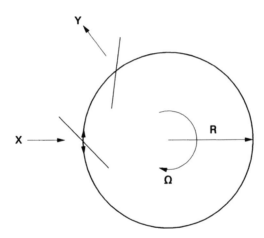

Figure 5.1   Circular rotating (Sagnac) interferometer

However, when the interferometer is rotated with angular velocity Ω, the time for each light beam to pass around the circumference is modified. This is because of the motion of the beam splitter during the time taken for the light to pass around the ring. As shown in the Figure, the beam splitter will have moved to position Y. Therefore, light travelling in a clockwise direction will have to travel further than the distance travelled when stationary. The converse is also true for the anti-clockwise beam. More generally, with respect to inertial space, light travelling with the direction of rotation must travel further than when the interferometer is stationary. Light travelling against the direction of rotation will have its path length reduced when compared with the stationary condition. Hence, the single pass transit time for the two beams is given by the following equations:

$$\text{clockwise path,} \quad t_1 = \frac{2\pi R + \Delta L_+}{c}$$
$$\text{anti-clockwise path,} \quad t_2 = \frac{2\pi R - \Delta L_+}{c} \tag{5.2}$$

Now, $\Delta L_+ = R\Omega t_1$ and $\Delta L_- = R\Omega t_2$ are the increment and decrement in the path length respectively. As reported by Aronowitz [9], this can also be interpreted as the velocity of light being different for the two counter-propagating beams and the path length being invariant.

From the above equations, the difference in transit time, Δt, is given by:

$$\Delta t = t_1 - t_2 = 2\pi R \left\{ \frac{1}{c - \Omega R} - \frac{1}{c + \Omega R} \right\} \tag{5.3}$$

To first order approximation, this becomes:

$$\Delta t = \frac{4\pi R^2 \Omega}{c^2} \tag{5.4}$$

The optical path length difference $\Delta L = c\Delta t$, and may therefore be expressed as:

$$\Delta L = \frac{4\pi R^2 \Omega}{c} \tag{5.5}$$

The area (A) enclosed by the path length is πR^2. Hence, the above equation may be rewritten as follows:

$$\Delta L = \frac{4A\Omega}{c} \tag{5.6}$$

Aronowitz [9] gives a more rigorous equation for the difference in closed path transit time for counter-propagating light beams on a rotating frame. This is based on the loss of synchronisation between a clock travelling on a rotating reference frame compared with one on a stationary reference frame. The conclusion from this study is that the optical path difference, $4A\Omega/c$, is independent of the position of the axis of rotation. As noted by Aronowitz, measurement of the optical path difference enables an observer, located on a rotating reference frame, to measure the so-called absolute rotation of his/her reference frame.

The various optical sensors described below rely on generating a path difference in an interferometer, the major differences being how the light is generated and how the path difference is observed.

5.1.3 Ring laser gyroscope

5.1.3.1 Introductory remarks

As stated above, serious development started in the early 1960s. The first ring laser gyroscopes were large and somewhat delicate. Substantial investment has lead to the production of very compact devices which produce extremely low bias, of 0.001°/hour or better. Typical path lengths for the accurate sensors are about 300 mm. Very small laser gyroscopes have also been produced with a path length of about 50 mm. Generally, these small sensors have a bias in the region of 5°/hour. Currently, the more accurate gyroscopes are used in strapdown navigation systems in both commercial aircraft as well as military fixed wing and rotating wing aircraft. Deployment in guided 'long range' weapons is probably imminent.

5.1.3.2 Principle of operation

Operation of a ring laser gyroscope relies on the fact that an optical frequency oscillator can be assembled as a laser using three or more mirrors to form a continuous light path. Typically, three mirrors are used to form a triangular shaped light path. If a light beam is generated at any point in this path, it can travel around the closed path, being reflected in turn from each mirror, to arrive back at its starting point. Sustained optical oscillation occurs when the returned beam is in phase with the out-going beam. Two such travelling wave laser beams are formed independently, one moving in a clockwise direction and the other in an anti-clockwise direction.

When the sensor is stationary in inertial space, both beams have the same optical frequency. However, when the sensor is rotated about the axis perpendicular to the plane containing the light beams, changes occur in the optical path lengths of the two beams. The frequency of each beam changes to maintain the resonant condition required for laser action such that the frequency of the beam with the longer path length decreases whilst the frequency of the other beam increases. This path difference is very small, no more than 1 nm, thus a source with high spectral purity and stability, such as a helium neon gas laser, is required to make the laser gyroscope concept feasible.

Maintenance of laser action requires a constant phase at a given mirror surface after every round trip in order to maintain the resonant condition.

Hence, if L_a is the anti-clockwise path length and L_c the clockwise path length, then the resonant condition is given by:

$$L_a = p\lambda_a$$
$$Lc = p\lambda_c$$

(5.7)

where p is the mode number, typically of the order of a million, and λ_a and λ_c are the two wavelengths of laser energy. When this interferometer is rotated at a rate Ω, these path lengths differ and are given by:

$$L_a = p\lambda_a = L + \frac{2A\Omega}{c}$$
$$Lc = p\lambda_c = L - \frac{2A\Omega}{c}$$

(5.8)

where L is the perimeter length and the path difference $\Delta L = 4A\Omega/c$.

Now if v_a and v_c are the optical frequencies of the two beams, $v_a\lambda_a = v_c\lambda_c = c$. Substituting for wavelength in the above equations, we have:

$$v_a = \frac{cp}{L_a} \quad \text{and} \quad v_c = \frac{cp}{L_c}$$

(5.9)

Hence, small changes in path length result in small changes in frequency, Δv, given by the relation:

$$\frac{\Delta v}{v} = \frac{\Delta L}{L}$$

(5.10)

Substituting for ΔL in this equation, this beat frequency can be expressed as:

$$\Delta v = \frac{4A\Omega}{cL}v = \frac{4A\Omega}{L\lambda}$$

where

$$\lambda = \frac{\lambda_a + \lambda_c}{2}$$

(5.11)

and

$$v = \frac{v_a + v_c}{2}$$

It follows from eqn. 5.11 that the turn rate (W) may be determined from the frequency difference (Dn) generated in its presence. The scale factor of the sensor is directly proportional to the area (A) enclosed by the optical path. Changes in A result from variations in the cavity length. Use of active laser gain control and cavity path length control usually contain these excursions to a few parts per million, or less, for most designs.

Substitution of typical values into eqn. 5.11 shows the beat frequency to be from a few hertz up to megahertz. This beat frequency can be detected, even for very slow rotation rates. As noted by Aronowitz, thermal and mechanical instabilities can cause frequency variations in the individual beams that are far greater than the rotational beat frequency, as can be deduced from eqns. 5.9, 5.10 and 5.11. Successful operation of this type of sensor is achieved since both beams occupy the same laser cavity and therefore are subject to identical perturbations.

In order to detect the rotational motion, a small amount of light from each beam is allowed to escape through one of the mirrors, known as the output mirror, and the two beams are combined using a prism to form an interference pattern on a set of photodiodes. The frequency difference between the two beams causes the interference fringes to move across the detectors at a rate equal to the difference in frequency between the two beams, which is also proportional to the rotational motion.

Hence, the movement of a single fringe past the detector corresponds to an incremental rotation, $\Delta\theta$, where:

$$\Delta\theta = \frac{\lambda L}{4A} \tag{5.13}$$

This equation can be used to determine the sensitivity of a ring laser gyroscope. For a device having an equilateral triangular path of total length L, the area is given by:

$$A = \frac{1}{2}(L/3)^2 \sin 60° \tag{5.14}$$

Substituting in the preceding eqn., it can be shown that the sensitivity is inversely proportional to the path length, viz:

$$\Delta\theta = \frac{3\sqrt{3}\lambda}{L} \qquad\qquad (5.15)$$

For instance, considering a helium-neon laser in which the optical wavelength is 0.6328 micrometers (632.8 nm), the sensitivity of a 30 cm path length device is 2.25 arc seconds per fringe changing to 5.6 arc seconds for a 12 cm path length device.

5.1.3.3 The lock-in phenomenon

At very low rotation rates, the two laser beams in the cavity cease to oscillate at different frequencies and assume the same frequency, or lock together. The interference pattern does not change so there is no output signal. This phenomenon of frequency synchronisation is illustrated in Figure 5.2 and is known as the lock-in condition, or simply lock-in. It is analogous to the mutual coupling common in electronic oscillators working in close proximity at similar frequencies. In the optical case, it is caused by the radiation of one laser beam being scattered into the other beam causing the host mode to change frequency towards that of the back scattered energy, with the consequence of both beams being shifted to the same frequency. There are many sources of back scattering, but careful design and the use of very high quality mirrors allows the effect to be minimised and

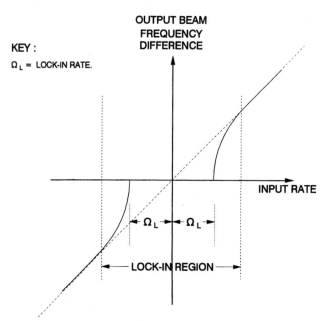

Figure 5.2   Laser gyroscope input/output characteristic

the lock-in condition is restricted to a very narrow zone close to the zero rotation rate.

Alleviation of lock-in

One of the most common methods used to alleviate the lock-in problem is the use of mechanical oscillation. Mechanical dithering consists of applying angular vibrations to the entire cavity at high frequency but at low amplitude and through small angles, thereby avoiding low frequency outputs. Through the use of a so-called large dither product (dither frequency multiplied by the amplitude), having a high frequency motion but small displacement, very little time is spent by the sensor in the lock-in region, hence greater accuracy is achieved through missing fewer pulses.

The dither frequency has a random frequency component superimposed on it which randomises slightly the motion of the cavity. The result of this randomisation is that the motion has a randomised rate noise rather than a mean bias which would be produced by a sinusoidal motion of the cavity block [10]. This motion produces a random walk in angle which appears on the output of the sensor.

The use of mechanical dither causes an increase in size, weight and complexity. It is necessary to subtract the dither motion from the gyroscope's output and this may be accomplished either optically or electronically. Any difference between the actual and compensated output is termed dither spillover which leads to a scale factor error.

Another technique that is currently being applied is called magnetic mirror biasing. This electro-optical technique uses a non-reciprocal magneto-optical effect (the transverse Kerr effect [11]). One of the highly reflective mirrors has a magnetic coating on its top surface. The magnetic coating, when saturated by an applied magnetic field, causes a difference in phase delay between the two counter-propagating laser beams, biasing the frequencies away from the lock-in zone. In order to prevent any drifts in bias voltage being interpreted as a rotation rate, it is necessary to switch between two bias points so that any drifts average to zero. A potential disadvantage with magnetic mirrors is the introduction of higher cavity losses which may exclude it from high accuracy applications. However, it is a genuine solid-state sensor [12] which is smaller and less complex than the mechanically dithered ring laser gyroscope.

Multi-oscillator concepts have been demonstrated [13–15] where more than a single pair of beams propagates in the same cavity, usually four beams in a square configuration. Independent lasing of left and right polarised modes are propagated in each direction in the cavity, giving a total of four modes. Avoiding the phenomenon of lock-in still applies, so it is necessary to bias the modes away from this zone. The reciprocal splitting between the right hand and left hand circularly polarised modes can be several hundred megahertz, achieved by using a quartz retarder plate, or a non-planar cavity configuration. The real difficulty is to achieve adequate biasing of the direction dependent (non-reciprocal) modes, i.e. the clockwise and anti-clockwise right hand circularly polarised modes, and similarly with the opposite handed modes.

Three methods have shown promise: use of a Faraday rotation element in the laser cavity; a mirror with a magneto-optic coating that uses either the polar or transverse Kerr effect; application of a Zeeman field to the discharge to induce a frequency change. Sometimes the ring laser gyroscope that has a Faraday cell in its cavity is called the differential laser gyroscope or DILAG; the other name is the four frequency gyroscope. One distinct advantage that this form of gyroscope has over the conventional ring laser gyroscope is an enhanced scale factor, giving a sensitivity that is twice that of the equivalent sized conventional laser gyroscope.

5.1.3.4 Detailed description of the sensor

The more commonly used ring laser gyroscope configuration which uses three mirrors is illustrated in Figure 5.3. Successful configurations using four mirrors have also been produced.

The major components of the gyroscope, as shown in Figure 5.21 are:

1 The laser block, which is formed in a low expansion ceramic glass such as Zerodur or Cervit. Contained within it is the lasing medium, which is usually a mixture of helium and two isotopes of neon to enable the two laser modes to propagate without competition.
2 The optical components, usually just the mirrors and photodetectors, but an optical biasing element may also be included in the laser cavity. There are two types of mirror; the partially reflecting (partially transmitting) output mirrors, as mentioned above, and the mirror with a very high reflection coefficient. All of the mirrors are multi-layer dielectric stacks of alternate layers of materials with a different refractive index, deposited on extremely high quality polished substrates to give very low back scatter. One of the mirrors is attached to a piezoelectric device so that it can be moved in and out to maintain a constant

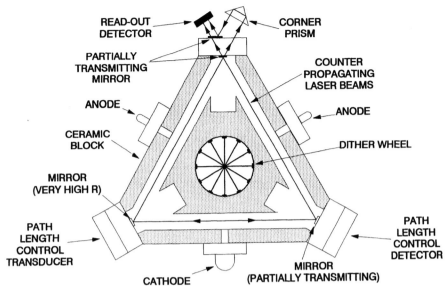

Figure 5.3   Schematic diagram of a ring laser gyroscope

path length (at the resonant condition) as the temperature changes. The other mirrors are firmly bonded directly to the laser block. A variety of techniques are used including optical contacting, using soft metal seals such as gold and indium, and an alternative bonding technique known as frit sealing.

3 The non-optical components, the usual configuration is to have one cathode and two anodes which produce a discharge when a high voltage is applied to these electrodes. This discharge then provides the source of excitation for the laser action. The usual laser wavelengths that are used are either the red line at 632.8 nm or the 1.152 μm line in the infrared part of the electromagnetic spectrum.

4 The biasing mechanism required to overcome the lock-in phenomenon described in detail above. A bias can be applied by various techniques such as mechanical dither, magnetic mirror or the use of optical elements within the laser cavity.

A photograph of a mechanically dithered ring laser gyroscope is shown in Figure 5.4.

The primary disadvantages of this technology are the precision engineering that is required to make and polish the faces of the laser block and the high technology required to produce the mirrors. These tend to make the cost of the sensor quite high, although

Figure 5.4   Mechanically dithered ring laser gyroscope
Courtesy of British Aerospace plc

techniques for reducing this are being sought. A further anticipated problem is the potential for helium to leak out of the cavity through one of the many seals. Radio frequency pumping of the laser cavity has been demonstrated and is a method of reducing the number of components fitted into the block and hence reducing the number of orifices and seals through which this gas can leak.

Mirror assessment prior to assembly of the sensor is crucial to the performance and yield achieved in the production of these devices. It is usual to evaluate scatter, loss, surface quality and flatness. The two mirror parameters that are most closely related to sensor performance are the scatter and loss. Deterioration of the mirrors is minimised by operating the gyroscope at the lowest possible internal laser intensity consistent with reliable performance.

5.1.3.5 Sources of error

There are three types of error which are characteristic of a ring laser gyroscope:

(i)   The lock-in phenomenon considered in detail above;
(ii) null shift, where the input/output characteristic does not pass through the origin so that the sensor records some counts from the detector even when stationary;
(iii) scale factor changes resulting from mode pulling effects.

A null shift error arises when one of the laser beams experiences some difference in its optical path when compared with the other laser beam. Hence, the use of a split discharge and the balancing of the discharge currents in the two discharges in order to make the laser cavity as isotropic or reciprocal as possible. Similarly, the sensor is usually shielded from stray magnetic fields in order to minimise any unwanted magneto-optic effects, particularly in the mirrors.

Mode pulling effects, considered by Lamb [16], give rise to dispersion effects, normal or anomalous. Any changes in the dispersive effects of the laser medium can give rise to instabilities and continuous changes to the scale factor. All of these errors are considered in detail by Aronowitz [9].

The stability of the sensing axis is also a key parameter which influences system performance. This is defined by the plane containing the laser beams, which can move owing to disturbances in the laser block and movements of the beam induced by non-parallel motion of the cavity path length control mirror.

The output of a ring laser gyroscope ($\tilde{\omega}_x$) may be expressed mathematically in terms of the input rate (ω_x) and the rates about the axes which lie in the lasing plane (ω_y and ω_z) as:

$$\tilde{\omega}_x = (1 + S_x)\omega_x + M_y\omega_y + M_z\omega_z + B_x + n_x \tag{5.16}$$

where S_x = scale factor error
    M_y, M_z = misalignments of the gyroscope lasing plane
        with respect to the nominal input axis
    B_x = fixed bias
    n_x = random bias error

The random bias term includes the random walk error referred to earlier which gives rise to a root mean square magnitude of angular output which grows with the square root of time. Although present to some extent in mechanical gyroscopes, the effect is generally an order of magnitude larger in optical sensors. In a mechanically dithered ring laser gyroscope, this error is largely caused by the random phase angle error introduced as the input rate passes through the lock-in region. An additional noise term gives rise to a bounded error and is the result of scale factor errors in the mechanism used to eliminate lock-in.

5.1.3.6 Typical performance characteristics

With careful design, this form of gyroscope does not exhibit any significant acceleration or vibration sensitivity. The typical range of performance that can be achieved from these devices is as follows:

g-independent drift (bias)	<0.001 - 10 °/hour
g-dependent bias	usually insignificant for most applications
g^2-dependent bias	usually insignificant for most applications
scale factor errors	few parts per million to 0.01%
	(of maximum rotation rate)
bandwidth	>200 Hz (can be made very large)
maximum input rate	several thousand degrees per second

Hence, the key parameters are the bias repeatability, random noise, scale factor repeatability and sensing axis stability.

5.1.4 Three-axis RLG configuration

Various schemes have been proposed and implemented to produce a single sensor with three sensitive axes using ring laser technology. Such devices are commonly called triads. These configurations are generally based on the use of three mutually orthogonal square laser cavities within a single cubic block. This arrangement enables each mirror to be shared by two of the laser cavities so that only six mirrors are required for this device [18, 19]. Similarly, the cathode is shared between discharges. The use of mechanical dither, applied about a body diagonal of the laser block, enables a bias to be applied simultaneously to each of the individual sensors within the monolith, and hence alleviate the lock-in problem for each of the axes.

Use of such a configuration can be very attractive for a number of applications as it provides great stability between the axes. The cost and complexity can also be reduced compared with the use of three independent sensors by using one dither mechanism, one discharge circuit and reducing the number of mirrors required through sharing.

The major disadvantage of such a system is the care needed and difficulty in machining the laser block to the necessary accuracy and avoiding damage during production, i.e. achieving a high yield. Additionally, a single fault could mean that all three axes of angular motion information is lost.

A schematic diagram of a triad is shown in Figure 5.5.

5.1.5 Fibre optic gyroscope

5.1.5.1 Introductory remarks

Work at the US Naval Research Laboratories in the late 1960s suggested that multiple circulations of a Sagnac interferometer may

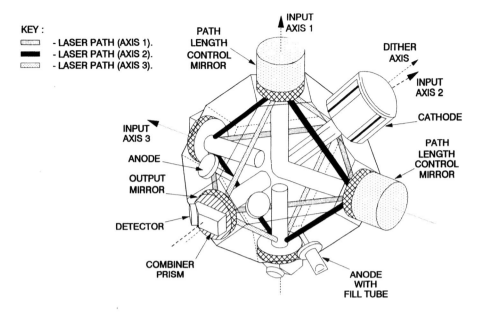

KEY :
▨▨ - LASER PATH (AXIS 1).
▰▰ - LASER PATH (AXIS 2).
▨▨ - LASER PATH (AXIS 3).

PATH LENGTH CONTROL MIRROR

INPUT AXIS 1

DITHER AXIS

INPUT AXIS 2

CATHODE

INPUT AXIS 3

PATH LENGTH CONTROL MIRROR

ANODE

OUTPUT MIRROR

DETECTOR

COMBINER PRISM

ANODE WITH FILL TUBE

Figure 5.5   Schematic diagram of a triad

give sufficient sensitivity to enable angular rotation to be detected and measured. By the mid 1970s, research progressed significantly on the use of passive interferometric techniques to sense angular motion, by applying optical fibre technology to form the light path [8]. This approach was seen as offering a far cheaper alternative to the ring laser technique, as the need to machine and polish surfaces to fractions of an optical wavelength would not be required. However, it was recognised that this approach was unlikely to produce a sensor with true high performance inertial performance characteristics, i.e. bias values in the region of 0.01°/hour or better.

In contrast with the ring laser technology, the fibre optic gyroscope senses angular motion by detecting the phase difference between the two beams passing around the light path in opposite directions. The gyroscope can be constructed as a genuine solid-state sensor, even in a closed loop mode, by the use of integrated optical components (chips). Use of this technology means that this type of inertial instrument can be very compact. However, extreme care and good design is required with the necessary fibre connections to avoid failure in harsh environments. Currently this is the subject of various research projects in many parts of the world [20].

These sensors have found some applications, particularly in the robotics and automobile industries. Aerospace applications are imminent, especially for stabilisation and sub-inertial navigation.

5.1.5.2 Principle of operation

Operation of the fibre optic gyroscope is dependent on the formation of a Sagnac interferometer [21]. In its simplest form, light from a broad-band source is split into two beams that propagate in opposite directions around an optical fibre coil. These two beams are combined at a second beam splitter to form an interference pattern where the resultant intensity is observed using a photodetector. When the interferometer is stationary, the path length of the two counter-rotating beams is identical so that there is no phase difference resulting in maximum amplitude. However, when the fibre coil is rotated about an axis normal to itself, the light travelling in the same direction as the rotation travels slightly further than the light travelling in the opposing direction. The consequent phase difference results in a change in amplitude of the interference pattern formed when the two beams are recombined.

For a rotating fibre gyroscope with a single turn of fibre, the phase difference ($\Delta\Phi$) between the counter-propagating beams of light may be expressed in terms of the path difference (ΔL) generated when the device rotates as:

$$\Delta\Phi = 2\pi\frac{\Delta L}{\lambda} \tag{5.17}$$

Substituting for ΔL from eqn. 5.6 gives :

$$\Delta\Phi = \frac{8\pi A\Omega}{c\lambda}$$

where A is the area enclosed by the fibre coil
Ω is the applied rotation rate
c is the velocity of light

For a coil of N turns this becomes:

$$\Delta\Phi = \frac{8\pi AN\Omega}{c\lambda} \tag{5.18}$$

This may be expressed in terms of the length of the fibre ($L = 2\pi RN$) as:

$$\Delta\Phi = \frac{4\pi RL\Omega}{c\lambda} \tag{5.19}$$

Consider a coil of radius 40 mm containing a 100 m length of fibre. If the optical wavelength is 850 nm, the phase differences which occur for rotation rates of (a) 15°/hour and (b) 500°/second are as follows:

(a) $\Delta\Phi = 0.0008°$
(b) $\Delta\Phi = 98.6°$

Clearly, if a sensor is going to be capable of detecting Earth's rate or comparable rotations, a high level of dimensional stability will be necessary. Hence, light travelling one way around the fibre coil must travel exactly the same path as the light which travels in the opposite direction, i.e. reciprocity must be maintained.

Comparing this equation with eqn. 5.11 for the ring laser gyroscope, the difference in sensitivity between the two sensors is obvious owing to the occurrence of the velocity of light in the denominator of the above equation. Hence, it is necessary to measure minute phase shifts to achieve high performance, which is a non-trivial task.

5.1.5.3 Detailed description of the sensor

The fundamental optical components of a fibre optical gyroscope, as illustrated schematically in Figure 5.6, are:

1 A light source, usually a broad-band source with a coherence length chosen to minimise the scattering effects within the fibre.
2 Couplers to link energy into and out of the fibre. It is usual to use 3 dB couplers so that they act as beam splitters.
3 The fibre coil, the angular motion sensing element. As a small single coil is unlikely to provide sufficient sensitivity, multiple turns are used. Depending on the desired sensitivity, high birefringence monomode or polarisation maintaining fibre may be used.
4 The detector, a photodiode used to detect the changes in the fringe pattern.

The non-optical components include the former on which the fibre coil is wound and the electronic components.

The fibre optic gyroscope can be operated in either an open loop mode or a closed loop mode [21–23]. When it is used in the simple open loop configuration, it is particularly sensitive to any non-reciprocal effects, consequently reducing the sensitivity of the device.

Open loop operation

A scheme was devised which ensured that both of the returning waves had propagated along the identical path, but in opposite directions. This was achieved by projecting polarised light into the interferometer through a single mode waveguide, such as a monomode optical fibre, and observing the returning interference wave which had been

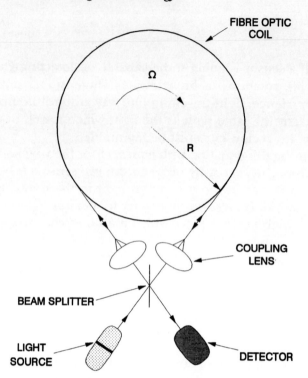

Figure 5.6   Open loop fibre optic gyroscope

filtered by the same waveguide prior to detection. This arrangement is known as the reciprocal or minimum configuration gyroscope, and is shown schematically in Figure 5.7.

The returning light beams from the fibre coil are combined at the second beam splitter and emerge from the so-called reciprocal port. These two beams are exactly in phase when the fibre coil is at rest, but the resultant intensity varies sinsoidally with the rate of angular rotation of the coil. The major disadvantage of this form of fibre gyroscope is the lack of sensitivity at small applied input rates, owing to the cosinusoidal shape of the fringe pattern, as shown in Figure 5.7.

It is possible to modify the fringe pattern to enhance the sensitivity at low rotation rates by incorporating a phase modulator at one end of the fibre coil. This modulator acts like a delay line. It is operated asymmetrically to give a phase dither of $\pm\pi/2$, which appears at the detector at twice the modulation frequency. Consequently, the gyroscope (detector) output is now biased to give its greatest sensitivity at and around small rotation rates. However, the response is still sinusoidal. This is illustrated in Figure 5.8 which also shows a schematic diagram of a phase-biased fibre gyroscope.

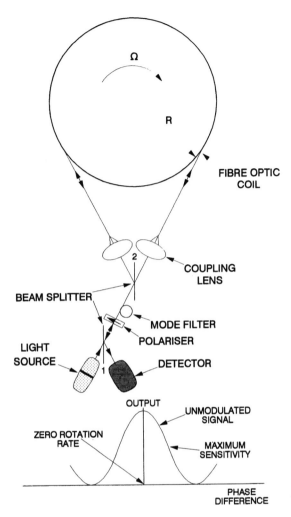

Figure 5.7 Reciprocal configuration fibre gyroscope and detector response

The phase modulator can be made by winding a few turns of optical fibre around a piezoelectric cylinder. A square wave signal may be applied to this cylinder to make it change shape and consequently to modulate the optical path length of the fibre coil.

Closed loop operation

There are many applications that require good accuracy over a wide angular rate range, possibly up to hundreds of degrees per second, and not merely at or close to zero rate. It is generally desirable that the

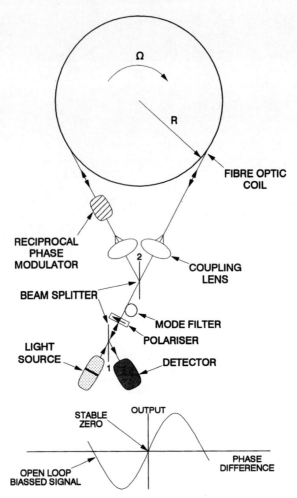

Figure 5.8   Phase-biased fibre gyroscope and detector response

scale factor linking the perceived angular motion to the actual motion should have good stability, be linear and be independent of the returning optical power. This can be achieved using a closed loop signal processing approach. Two techniques, or architectures, have been demonstrated, namely phase nulling and frequency nulling.

In the case of phase nulling, a second phase modulator is added to the other end of the fibre coil, as shown in Figure 5.9. It is operated at twice the frequency of the dither (biasing) modulator and is used to back-off or null the effects of the Sagnac induced phase shifts caused by the angular motion of the fibre coil. The open loop signal is used as an error signal to generate an additional phase difference ($\Delta\phi_n$)

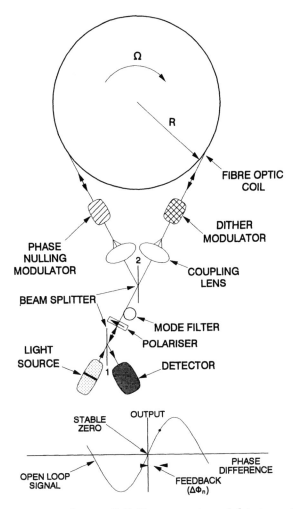

Figure 5.9   Closed loop (phase-nulled) fibre gyroscope and detector response

that has an opposite sign to the rotation induced phase difference. Consequently, the total phase difference is arranged to be at or very close to the zero value, so that the system is operated about the point of greatest sensitivity. The value of the additional feedback ($\Delta\phi_n$) is used as a measurement of angular rate. It has a linear response with good stability, since it is independent of the power of the returning optical signal and the gain of the detection system. However, the accuracy of the scale factor does depend on the stability of the source wavelength and the geometric stability of the sensing coil.

An alternative technique for achieving closed loop operation is through the use of a frequency shift generated by an acousto-optical modulator, or Bragg cell, placed at one end of the sensor coil. This frequency shift is used to produce a differential phase shift to null that caused by the Sagnac effect [20]. By varying the voltage applied to the Bragg cell, the frequency shift it induces can be varied. Hence, the voltage that needs to be applied to the Bragg cell to null the detector's output is directly proportional to the applied angular motion.

The frequency shift given to the light by the Bragg cell is chosen so that the sensor is operated in the region where it is most sensitive to low rotation rates, as described above. This frequency shift produced by the Bragg cell is dithered about the centre frequency using a square wave modulator and the intensities of the light beams incident on the detector are monitored. When the sensor coil is rotated about its input axis, a small phase shift is introduced between the two beams giving rise to a mismatch between the intensities of the two beams on the detector. Consequently, the output signal from the detector will be modulated at the dither frequency. A phase sensitive detection system is used to deduce the amount by which the centre frequency of the Bragg cell needs to be altered in order to return the intensities to their original matched state, thereby nulling the sensor. This change in the voltage applied to the Bragg cell is directly proportional to the applied angular motion.

The principal drawback of this architecture is the generation of a suitable bandwidth of frequencies in the modulator. This can be achieved by using two acousto-optical modulators at opposite ends of the sensor coil and dithering their frequencies about a centre frequency. An alternative approach is to use two modulators in opposition at one end of the fibre coil. With both arrangements, great care is required to achieve satisfactory mechanical stability of the whole assembly.

One technique that is currently used is the so-called phase ramp, serrodyne modulation [21], which relies on the fact that a frequency can be considered to be a time derivative of phase. In practice, a sawtooth waveform is used to modulate the applied phase shift, with a very fast flyback at the reset positions. An alternative method that alleviates the flyback problem is the digital phase ramp. In this case, phase steps are generated with a duration equal to the group delay difference in time between the long and short paths which connect the phase modulator and the beam splitter. These phase steps and the resets can be synchronised with a square wave biasing modulation. The use of digital logic enables this technique to be implemented easily.

One of the current developments of the fibre optic gyroscope is the demonstration of the so-called integrated fibre optic gyroscope. In this device, all the bulk optical, or fibre, components are replaced with components fused into a lithium niobate substrate [21]; the substrate is used to produce the beam splitter or couplers, optical waveguides and the necessary modulation to the light required to measure the rotation rate accurately. Fibre leads are used to connect the source and detector to this so-called integrated optics chip. This form of gyroscope has the potential to be compact, rugged and have a long shelf-life whilst retaining the advantages of optical sensors listed earlier in Section 5.1. Currently, these sensors are about 70 to 80 mm in diameter, a size which is a compromise between producing a compact sensor and avoiding excessive biases induced by the strain in the fibre when bent into a small radius of curvature. An exploded view of a packaged fibre gyroscope is shown in Figure 5.10.

Use of a reciprocal path in optical sensors like the fibre optic gyroscope should be free from errors associated with the environment, such as temperature, acceleration and vibration. However, regrettably, these devices do exhibit some sensitivity to these effects. Fortunately, through careful design, particularly the winding of the sensor coil, these sensitivities can be minimised. These effects are considered below.

5.1.5.4 Sources of error

Changes in ambient temperature can cause a bias or drift to be observed because of a multitude of effects within the sensor, temperature gradients within the sensor being a particular problem. As the ambient temperature changes, the source wavelength changes and the sensitivity of the sensor is inversely proportional to the source wavelength. Temperature changes result in variation of the refractive index of the optical fibre leading to changes in the modulation. If possible, the expansion coefficient of the fibre and the coil former should be well matched, or else differential stresses are induced by thermal expansion which result in measurement errors. Thermal changes also alter the size of the coil resulting in changes in the scale factor of the gyroscope.

A bias occurs when there is a time dependent thermal gradient along the optical fibre as a result of a temperature gradient across the coil. This results in a non-reciprocity occurring when corresponding wave fronts of the counter-rotating beams cross the same region at different times. This is known as the Shupe effect [24]. Anti-Shupe

Figure 5.10 *Fibre optic gyroscope*
Courtesy of British Aerospace plc

windings have been devised so that the parts of the optical fibre that are equal distances from the centre of the coil are adjacent.

When an acceleration is applied to a coil it can result in the distortion of the coil, producing a change in the scale factor of the gyroscope. Distortions also change the birefringence of the fibre and hence, the bias of the sensor. Additionally, distortions can lead to changes in the direction of the sensitive axis and, in the case of a three-axis configuration, changes in the relative orientation or alignment of the three input axes.

Application of vibratory motion to a coil of fibre, or length of fibre, can lead to a distortion of the coil and the fibre, depending on the amplitude of the input motion. As above, this leads to errors in the angular motion measurements made by the sensor.

The presence of stray magnetic fields also can have an adverse effect owing to interaction with the non-optical components, and magnetic fields can produce changes in the state of polarisation of the light in the optical fibre through the Faraday effect. These sensitivities lead to a bias in the output signal from the device. The use of magnetic shielding can minimise this effect.

The bias and drift in the output signal are a consequence mainly of birefringence in the optical fibre and propagation of cladding modes, as well as polarisation modulation of the optical signals. The problem is essentially one of many modes existing within the fibre, each with a different phase and with a lack of coherence, so fading occurs. One method of overcoming spurious coherent effects, including scattering within the fibre, is to use a low coherence source, i.e. one with a short coherence length. A superluminescent diode fulfils this criterion. Use of this source also minimises the bias generated by the Kerr electro-optic effect, caused by changes in the refractive index of the optical medium through variations in the power of the two counter-propagating beams.

5.1.5.5 Typical performance characteristics

Typical performance parameters for fibre optic gyroscopes are given below:

g-independent bias	<0.5–50°/hour
g-dependent bias	~1°/hour/g
g^2-dependent bias	~0.1°/hour/g^2
	(g-dependent biases can be made negligible with good design but often show some sensitivity)
scale factor errors	0.05–0.5%
bandwidth	>100 Hz
maximum input rate	>1000°/second

It is anticipated that these sensitivities to the environment will be solved in the near future. It is worth noting that these problems are far less severe than those which occurred with the mechanical gyroscopes 30 years earlier!

5.1.6 Fibre optic ring resonator gyroscope

This is yet another implementation of the Sagnac effect and could be considered to be a development of the fibre optic gyroscope. Work to develop this concept started at various institutions in the early 1980s, much of the pioneering work being led by Dr Ezikiel at the Massachusetts Institute of Technology.

  This type of gyroscope uses a re-circulating ring resonant cavity to enhance the phase differences induced by the Sagnac effect when the sensor is rotated. A ring cavity is formed by joining the ends of a short optical fibre cable, typically 10 m or less, to a coupler, as shown in Figure 5.11. The light enters and leaves the ring cavity though this coupler.

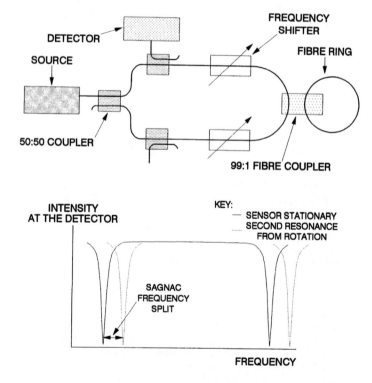

Figure 5.11   Fibre optic ring resonator gyroscope and intensity waveform

The principle of operation of the ring cavity is very similar to that of a Fabry-Perot interferometer [21], but with multiple interference between the re-circulating waves and the light being introduced to the cavity. Resonance can be achieved in the ring either by stretching the fibre or by sweeping the frequency of the light, thus resulting in constructive interference occurring in the ring and hence maximum light intensity.

At resonance, there is a dip in the light intensity reflected at the output port resulting from the majority of the light entering the fibre ring and then being lost by scattering after many re-circulations in the ring cavity. Figure 5.11 shows the form of the intensity waveform at the detector. The width of the resonance is set by the finesse of the resonator [25]. This parameter is inversely proportional to the coupling ratio into the ring. Hence, it is necessary to have a high coupling ratio. The quality of the cavity, and its sensitivity, is defined by the finesse, which is the ratio of the free spectral range to the linewidth of the resonance. The resonant condition is observed by detecting the energy reflected by the coupler linking the fibre ring to the source and detector.

When the sensor is stationary, the resonant frequencies of the two counter-propagating beams are identical. When the ring is rotated about its sensitive axis, the resonant frequencies of the two counter-propagating beams differ. This results in a shift in the positions of the resonances and, hence, in the light intensity from the output port.

The eqn. 5.19, for the fibre optic gyroscope gives the phase difference for the two beams. Because the two counter-propagating beams experience different loop paths, there is a different resonance frequency (v_d) generated in each direction. It is given in terms of the radius of the coil by substitution for $A(= \pi R^2)$ and $L(= 2\pi R)$ in eqn. 5.11. Hence:

$$v_d = \frac{2R\Omega}{\lambda} \tag{5.21}$$

where R is the radius of the coil
        Ω is the applied rate
        λ is the wavelength

Clearly, the number of turns of fibre does not influence its sensitivity, as this is dependent on the radius of the loop for a given wavelength and rotation rate. Hence, given practical considerations, the length of optical fibre required is very much less, typically of the order of 10 m.

This sensor is used in a feedback mode so that the frequency of the two counter-propagating beams is shifted when the sensor is rotated, thus keeping the two beams at resonance simultaneously even under rotation. Sinusoidal phase modulation with frequency modulation can be applied to the two counter-propagating beams. The difference between the two modulations required to maintain resonance in the two beams is proportional to the applied rotation rate.

The fundamental requirement for the light energy used in this sensor is that it has a narrow bandwidth and a high coherence. Hence, a light source is needed that is very stable in order to maintain the narrow width resonance in the ring. Currently, the lasers that fulfil this requirement are expensive, but with the development of quantum well lasers the price should reduce. Careful design is required to avoid incoherent back scattering from one beam to the other and the consequent onset of the familiar lock-in problem. It is usual to control the polarisation of the light before it enters the ring to avoid intensity fading.

As in the case of the fibre optic gyroscope, the bulk optical components can be replaced with integrated optical materials using guided waves. Hence this, together with the very short length of fibre, typically 10 m or less, can lead to a very compact sensor.

Currently, this sensor is being developed by a number of companies but definite performance and error data are not available. However, it is expected to have a similar performance to the fibre gyroscope; data for that sensor are given in Section 5.1.5.5. Generally, the ring resonator gyroscope is considered to be more sensitive than the fibre gyroscope, possibly by a factor of three [26[.

Ring resonator rotation rate sensors—non-linear mode

Currently, the measurement of rotation rate by the exploitation of non-linear effects, induced in the energy in the ring resonator, is being investigated. These non-linear effects occur when there is only a small loss, of the order of 5%, in the ring, leading to about a 20 fold increase in power. The non-linear interactions are usually divided into two groups, intensity dependent refractive index effects and scattering phenomena. In the case of the fibre optic gyroscope, these non-linear effects are considered detrimental as they degrade performance.

The non-linear refractive index effect makes use of the Kerr Effect [11]. The electric field of the optical energy propagating through a medium produces a refractive index change. Now, when a ring resonator, operating in a resonant condition, is rotated, the intensity in one direction decreases whilst the intensity in the other direction increases. Additionally, the refractive index in each direction also changes, amplifying the intensity change. This amplification can be varied by changing both the parameters of the resonant ring and the Kerr material, which, in turn, improves the sensitivity of the resonator.

The propagation of a high power optical beam through a dense medium produces phonons, energy is transferred to these phonons resulting in a change in frequency of the beam. This results in a scattered optical wave with a modified frequency. The two major types of non-linear scattering are called Brillouin and Raman [27], and are threshold effects; Brillouin occurs at lower power levels. Brillouin scattering has a characteristic linewidth which requires a source with a coherence length shorter than the life time of a phonon. Consequently, a coherent source produces Brillouin scattering in preference to Raman.

It has been suggested [28] that a ring laser, exhibiting Brillouin scattering, pumped from both directions will produce a frequency difference between the two components of the Brillouin scattered energy which is directly proportional to the rotation rate. This sensor should produce a signal similar to that produced by a ring laser gyroscope, i.e. a rotation rate dependent frequency. However, there should not be any lock-in effects, as the two scatter mechanisms should be independent.

A sensor has also been proposed based on the Raman scattering phenomenon using high power pulses from a mode locked laser [29]. It appears that the major drawback will be the size of the sensor. Currently there are no reports indicating that any of these sensors has been demonstrated.

5.1.7 Ring resonator gyroscope

This sensor is very similar to the fibre optic ring resonator, but the fibre ring is replaced with an optical waveguide which can be etched into a suitable substrate [30]. The principle of operation is very similar to that described for the fibre optic ring resonator. Typical rings are about 50 mm in diameter. Hence, as the problems associated with scattering and coupling are overcome, the prospect of a gyroscope on a chip can become a reality, along with all the advantages of optical sensors.

This type of sensor has been developed by Northrop in the USA [10] and is known as the micro-optic gyroscope (MOG). Rugged sensors with a diameter of about 25 mm have been produced with performance in the 1 to 100°/hour category.

5.2 Summary of gyroscope technology

There are many types or classes of sensor that can be used to sense or detect angular motion. Many of these devices have been considered in the foregoing text, particularly those that are used currently, or could be applied in the future, in strapdown applications. These instruments range from the conventional mechanical gyroscopes, using a rotating mass, to the unconventional, using atomic spin.

A great deal of effort has been, and still is being, expended to develop the so-called novel technology, the aim being to produce a gyroscope on a chip. New technology used in industry, such as robotics, is helping to sustain this effort. However, this is also being applied to the conventional technology and is helping to keep the mechanical gyroscope competitive.

The range of accuracy that can be achieved from the spectrum of rotation sensing devices spans many orders of magnitude. Some sensors have a bias of less than 0.0001°/hour, whilst others are in the 1°/second class or worse. Most sensors show some unwanted sensitivity to the environment in which they operate. The goal of much of the research is to reduce these sensitivities or improve the ruggedness of the instruments as many, particularly some high precision devices, are quite sensitive to vibratory motion.

A summary of typical performance characteristics for a range of sensors suitable for strapdown application is given in the following table[1].

Table 5.1 Performance characteristics for a range of gyroscopes

Characteristic	RIG	DTG	Flex gyro MHD	DART/ gyro	Vibratory	RLG	FOG
g-independent bias (°/hour)	0.05–10	0.05–10	1–50	360–1800	360–1800	0.001–10	0.5—50
g-dependent bias (°/hour/g)	1–10	1–10	1–10	180	36-180	0	<1
anisoelastic bias (°/hour/g²)	1–2	0.1–0.5	0.05–0.25	18–4	18	0	<0.1
scale factor non-linearity (%)	0.01–0.1	0.01–0.1	0.01–0.1	0.5–1	0.2–0.3	5–100	0.05–0.5
bandwidth (Hz)	60	100	100	100/80	>100	500	>200
maximum input rate (°/second)	>400	1000	>500	800/400	>1000	>1000	>1000
shock resistance	moderate	moderate	moderate	moderate	>25000g	good	good

[1]These are typical values applicable over the range of parameters stated. In many cases, the values given could be improved. However, it is not normally possible to have all the best case values in an single unit, particularly for the conventional sensors. These values are only for general indicative purposes.

In general, a significant amount of precision engineering and high technology is required to produce a device that is functional. As the accuracy required from the sensor increases, so does the precision and the size needed to fulfil the requirement, although this is not universally true. In some of the recent research programmes, an effort has been made to alleviate the need for ultra high precision for the high accuracy instruments. Usually, however, this has led to the need to apply very high technology, such as superconductivity, which imposes its own demands such as cooling to cryogenic temperatures.

References

1 SAGNAC, G.: 'L'ether lumineux demontre par l'effet du vent relatif d'ether dans un interferometre en rotation uniforme', *C. R. Acad. Sci.*, 1913 (157), p. 708
2 SAGNAC, G.: 'Sur la preuve de la realite de l'ether lumineux par l'experience de l'interferograph tournant', *C. R. Acad. Sci.*,1913 (157), p. 1410
3 MICHELSON, A.H. and GALE, G.H.: 'The effect of the Earth's rotation on the velocity of light', *Astrophys. J.*, 1925 (61), p. 140
4 MAIMAN, T.H.: 'Stimulated optical radiation in ruby', *Nature*, (187), p. 493
5 MAIMAN, T.H.: 'Stimulated optical emission in fluorescent solids', *Phys. Rev.*, 1961 (123), p. 1145
6 MAIMAN, T.H.: 'Optical and Microwave experiments in ruby', *Phys. Rev. Lett.*, (1960), **4**, p. 564
7 MACEK, W.M. and DAVIS, D.T.M.: 'Rotation rate sensing with travelling wave ring lasers', *Appl. Phys. Lett.*,1963 (2), p. 67
8 VALI, V. and SHORTHILL, R.W.: 'Fibre ring interferometer' *Appl. Opt.*, 1976 (15), p. 1099
9 ARONOWITZ, F.: in ROSS, M. (Ed.): '*Laser Applications vol 1—the laser gyro*' (Academic Press, 1971) Academic Press, 1971
10 LAWRENCE, A.W.: '*Modern inertial technology—navigation, guidance and control*' (Springer-Verlag, 1993)
11 YOUNG, M.: '*Optics and Lasers*' (Springer-Verlag, 1992)
12 PERLMUTTER, M.S., BRESMAN, J.M. and PERKINS, H.A.: 'A low cost tactical ring laser gyroscope with no moving parts' Proceedings DGON symposium on *Gyro technology*, Stuttgart, Germany, 1990
13 BRECHMAN, J., COOK, N. and LYSOBURY, D.: 'Differential laser gyro development', *Navig. J. Inst. of Navig.*, 1977, **24** (2), p. 153
14 DICKINSON, M.R. and KING, T.A.: 'Polarisation frequency splitting in non-planar ring laser resonators', *J. Mod. Opt.*, **34** (8), p. 1045
15 KNIPE, C., ANDREWS, D.A. and KING, T.A.: 'Mode interactions in four-frequency degeneracy-lifted ring lasers', *J. Mod. Opt.*, **35** (3)
16 LAMB, W.E. JR.: 'Theory of an optical maser', *Phys. Rev. A, At. Mol. Opt. Phys.*,1964 (134), p. 1429
17 SIMMS, G.J.: 'Ring laser gyroscopes' US Patent 4 407 583 1983
18 SIMMS, G.J.: UK Patent GB 2076213B4, 1980
19 SIMMS, G.J.: 'A triaxial laser gyroscope', *Proc. Inst. Mech. Eng. C., J. Mech. Eng. Sci.*, 1987, **C56/87**
20 SMITH, R.B.: 'Fibre optic gyroscopes 1991—a bibliography of published literature'. Proceedings of SPIE 15th anniversary conference on *Fibre optic gyros*., 1991, p. 464
21 LEFEVRE, H.: '*The fibre-optic gyroscope*' (Artec House, 1993)
22 BURNS, W.: '*Optical fibre rotation sensing*' (Academic Press 1994)
23 CULSHAW, B. and GILES, I.P.: 'Fibre optic gyroscopes', *J. Phys. E; Sci. Instrum.*,

1983 (6), p. 5

24 SHUPE, D.N.: 'Thermally induced non-reciprocity in the fibre-optic inter-ferometer', *App. Opt.*, 1980, **19** (5), p. 654

25 MALVERN, A.R.: 'Optical fibre ring resonator gyroscope', Proceedings of IMechE symposium on *Technology of inertial sensors and systems.* 1991

26 GILES, I.P.: 'Optical fibre ring resonator gyroscopes' Proceedings of Royal Aeronautical Society symposium on *Lasers and fibre optic gyros*, 1987

27 MILLS, D.L.: '*Non-linear optics*' (Springer-Verlag, 1991)

28 THOMAS, P.J., VAN DRIEL, H.H. and STEGEMAN, G.I.A.: 'Possibility of using an optical fibre Brillouin ring laser for inertial sensing', *Appl. Opt.*, 1985 (19), p. 12

29 NAKAZAWA, M.: 'Synchronously pumped fibre Raman gyroscope', *Opt. Lett.*, 1985 (10), p. 4

30 BERNARD, W. and ENGLERT, R. *et al:* 'Waveguide ring resonators in glass for optical gyros'. Proceedings of DGON symposium on *Gyro technology*, Stuttgart, Germany, 1986

Chapter 6
Accelerometer and multi-sensor technology

6.1 Introduction

As described in Chapter 1, inertial navigation relies on the measurement of acceleration which can be integrated successively to provide estimates of changes in velocity and position. Measurements of acceleration are used in preference to direct measurements of velocity or position because velocity and position measurements require an external reference whilst acceleration can be measured internally.

The form of construction of devices which may be used to sense acceleration may be classified as either mechanical or solid-state. The technology of mechanical sensors is well established [1, 2] and devices capable of sensing acceleration over a wide accuracy range, from 50 milli-g down to a few micro-gs and to a similar level of resolution, are currently available. There have been significant advances in the development of solid state sensors in recent years, particularly with silicon technology.

The concept of using a single instrument to measure acceleration and angular motion has been the subject of research for a number of decades and during the 1980s was developed by a number of institutions and companies. This device has become known as the multi-sensor and has tended to be based on either vibratory technology or gyroscopic mass unbalance technology. Evaluation of this technology has generally shown it to be capable of providing estimates of linear acceleration and angular motion compatible with sub-inertial[1] navigation applications.

---

[1]The term sub-inertial is sometimes used when describing system performance for short duration navigation systems. Typically, sub-inertial systems use gyroscopes and accelerometers with measurement biases of the order of 1°/hour and 1 milli-g (1σ) respectively.

6.2 The measurement of translational motion

The translational acceleration of a rigid body, resulting from the forces acting upon it, is described by Newton's Second Law of motion. A force *F* acting on a body of mass *m* causes the body to accelerate with respect to inertial space. This acceleration (*a*) is given by:

$$F = ma \tag{6.1}$$

Although it is not practical to determine the acceleration of a vehicle by measuring the total force acting upon it, it is possible to measure the force acting on a small mass contained within the vehicle which is constrained to move with the vehicle. The small mass, known as a proof or seismic mass, forms part of an instrument called an accelerometer. In its simplest form, the accelerometer contains a proof mass connected via a spring to the case of the instrument as shown in Figure 6.1.

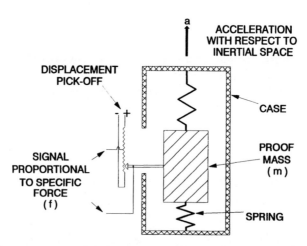

Figure 6.1   A simple accelerometer

When the case of the instrument is subjected to an acceleration along its sensitive axis, as indicted in the Figure, the proof mass tends to resist the change in movement owing to its own inertia. As a result, the mass is displaced with respect to the body. Under steady state conditions, the force acting on the mass will be balanced by the tension in the spring, the net extension of the spring providing a measure of the applied force, which is proportional to the acceleration.

The total force (*F*) acting on a mass (*m*) in space may be represented by the equation:

$$F = ma = mf + mg \qquad (6.2)$$

where f is the acceleration produced by forces other than the gravitational field. In the case of a unit mass, $F = a = f + g$. The acceleration (a) may be expressed as the total force per unit mass. An accelerometer is insensitive to the gravitational acceleration (g) and thus provides an output proportional to the non-gravitational force per unit mass (f) to which the sensor is subjected along its sensitive axis. As described in Chapter 2, this is referred to as the specific force exerted on the sensor.

Taking the case of an accelerometer which is falling freely within a gravitational field, the case and the proof mass will fall together resulting in no net extension of the spring. Hence, the output of the instrument will remain at zero. In this situation, the acceleration of the instrument with respect to an inertially fixed set of axes becomes $a = g$ and the specific force is zero in accordance with the above equation. Conversely, in the situation where the instrument is held stationary, $a = 0$ and the accelerometer will measure the force acting to stop it from falling. Following from eqn. 6.2, this force, $mf = -mg$, is the specific force required to offset the effect of gravitational attraction. It is clear, therefore, that knowledge of the gravitational field is essential to enable the measurement provided by the accelerometer to be related to the inertial acceleration.

Many mechanical devices commonly used in present-day inertial navigation systems for the measurement of specific force operate in a manner analogous to the simple spring and mass accelerometer described above.

In order to carry out the full navigation function, information is required about the translational motion along three axes, as described in Chapter 3. Commonly, three single-axis accelerometers are used to provide independent measurements of specific force, although multi-axis instruments can be used. It is common practice to mount the three accelerometers with their sensitive axes mutually orthogonal, although such a configuration is not essential, as will be discussed later in Chapter 8.

The various principles of operation and performance of current accelerometer technology are reviewed in the sections which follow, covering both the mechanical and solid-state instruments. Later in this Chapter, multi-sensors and angular accelerometers are reviewed in a similar way. Linear accelerometers may also be used to measure rotational motion [3]. However, owing to the need for very accurate measurements, as well as precise sequencing and timing of the measurements, this technique is rarely used.

6.3 Mechanical sensors

6.3.1 Introduction

This is the broad division of sensors primarily described in the Section 6.2 as mass-spring type devices. These sensors have been developed over many decades. Different construction techniques have been identified for use in different environments. Compact and reliable devices giving high accuracy and wide dynamic range have been produced in large quantities. The most precise force feedback instruments are capable of measuring specific force very accurately, typically with resolutions of micro-g, or better. This class of mechanical sensors is used in both inertial and sub-inertial applications.

6.3.2 Principles of operation

As in the case of gyroscopes, accelerometers may be operated in either open or closed loop configurations. The basic principle of construction of an open loop device is as follows. A proof mass is suspended in a case and confined to a zero position by means of a spring. Additionally, damping is applied to give this mass and spring system a realistic response corresponding to a proper dynamic transfer function. When the accelerations are applied to the case of the sensor, the proof mass is deflected with respect to its zero or null position and the resultant spring force provides the necessary acceleration of the proof mass to move it with the case. For a single axis sensor, the displacement of the proof mass, with respect to its null position within the case, is proportional to the specific force applied along its input, or sensitive, axis.

A more accurate version of this type of sensor is obtained by nulling the displacement of the pendulum, since null positions can be measured more accurately than displacements. With a closed loop accelerometer, the spring is replaced by an electromagnetic device which produces a force on the proof mass to maintain it at its null position. Usually, a pair of coils is mounted on the proof mass within a strong magnetic field. When a deflection is sensed, an electric current is passed through the coils in order to produce a force to return the proof mass to its null position. The magnitude of the current in the coils is proportional to the specific force sensed along the input axis. The force feedback type is far more accurate than the open loop devices and is currently the type most commonly used in inertial navigation systems.

6.3.3 Sensor errors

All accelerometers are subject to errors which limit the accuracy to which the applied specific force can be measured. The major sources of error which arise in mechanical accelerometers are itemised below. Further details relating to specific types of accelerometer will be given later in this Chapter where the physical effects which give rise to each type of error are discussed more fully.

fixed bias, is a bias or displacement from zero on the measurement of specific force which is present when the applied acceleration is zero. The size of the bias is independent of any motion to which the accelerometer may be subjected and is usually expressed in units of milli-g or micro-g depending on the precision of the device involved;

scale factor errors, errors in the ratio of a change in the output signal to a change in the input acceleration which is to be measured. Scale factor errors may be expressed as percentages of the measured full scale quantity or simply as a ratio; parts per million (ppm) being commonly used. Scale factor non-linearity refers to the systematic deviations from the least squares straight line, or other fitted function, which relates the output signal to the applied acceleration;

cross-coupling errors, erroneous accelerometer outputs resulting from accelerometer sensitivity to accelerations applied normal to the input axis. Such errors arise as a result of manufacturing imperfections which give rise to non-orthogonality of the sensor axes. Cross-coupling is often expressed as a percentage of the applied acceleration;

vibro-pendulous errors, dynamic cross-coupling in pendulous accelerometers arises owing to angular displacement of the pendulum which gives rise to a rectified output when subjected to vibratory motion. This type of error can arise in any pendulous accelerometer depending on the phasing between the vibration and the pendulum displacement. The magnitude of the resulting error is maximised when the vibration acts in a plane normal to the pivot axis, at 45 degrees to the sensitive axis and when the pendulum displacement is in phase with the vibration. This error may be expressed in units of g/g^2.

As in the case of gyroscopic sensors, repeatable errors, temperature dependent errors, switch-on to switch-on variations and in-run errors arise in sensors of this type. Even with careful calibration, the residual errors caused by the unpredictable error components will always be present, restricting the accuracy of inertial system performance.

6.3.4 Force feedback pendulous accelerometer

6.3.4.1 Detailed description of sensor

These devices are also known as restrained pendulum accelerometers. The main components of such a sensor are:

1 A pendulum, which has a proof mass attached to it or as an integral part of it.
2 A suspension mechanism or hinge element. This flexible member attaches the pendulum to the case and is usually either a flexible hinge or a pivot type arrangement.
3 A pick-off device to sense motion of the pendulum. It may use optical, inductive or capacitive techniques. The optical system may be very simple, a detector measuring the change in transmittance of a light beam through a slit in the pendulum. The inductive system involves measuring the differential current in coils fixed to the case interacting with a plate on the pendulum, which affects the mutual inductance of the coils. This system measures the relative position of the pendulum between the pick-off coils and not the null position. In the case of a capacitive system, movement of the pendulum causes a change in capacitance between the faces of the pendulum and two electrodes in close proximity to the pendulum. This change is sensed using a bridge circuit.
4 A force rebalance mechanism to oppose any detected movement of the pendulum. This component usually takes the form of two identical poles of two magnets arranged centrally about the proof mass and a pair of coils mounted symmetrically on the pendulum. A current flowing in the coils generates an electromagnetic restoring force. This component is often referred to as the torquer.
5 The various components are usually hermetically sealed in a case. The case may be filled with a low viscosity oil to give resistance to shock and vibratory forces in both its quiescent and active states. Alternatively, the case may be filled with a dry gas such as air.

Such a sensor is shown schematically in Figure 6.2.

Displacement of the pendulum, which occurs in the presence of an applied acceleration, is sensed by the pick-off. In the most simple devices, this displacement provides a direct measure of the applied acceleration. Generally however, a device of this type operates with an electronic rebalance loop to feed the pick-off signal back to the

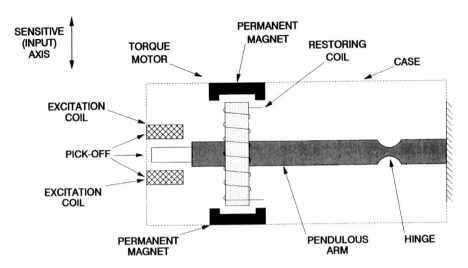

Figure 6.2   Force feedback pendulous accelerometer

torquer. The electromagnetic force, produced by the torquer, acts to offset any displacement of the pendulum and maintain the pick-off output at zero. The current in the torquer coil is proportional to the applied acceleration. Operating the sensor in this mode means that the hinge is not under any bending stress.

6.3.4.2 Sources of error

Accelerometers of this type are capable of very high performance with good linearity, small biases and with a dynamic range in the region of 10^4 to 10^5. This is a dimensionless quantity obtained by dividing the maximum acceleration which the sensor can measure by its resolution. The dominant sources of error are as follows:

fixed bias which arises as a result of residual spring torques and null shift in the electrical pick-off device used;

scale factor error, principally caused by temperature effects and non-ideal behaviour of components;

cross-axis coupling which gives rise to a measurement bias when the sensor is under g loading in the direction of the hinge axis or the pendulum axis, the latter being essentially a hinge interaction effect;

vibro-pendulous error which can give rise to a measurement bias under certain conditions when the sensor is subject to vibration along the sensitive and pendulum axes simultaneously;

random bias caused by instabilities within the sensor assembly.

Further errors occur in the measurements provided by pendulous accelerometers, such as those resulting from hysteresis effects, non-repeatability of bias and higher order scale factor errors. Changes in the characteristics of the permanent magnets may also change the scale factor by a process known as ageing. This may be corrected by periodic recalibration.

The measurement provided by such sensor (a_x) may be expressed in terms of the applied acceleration acting along its sensitive axis (a_x) and the accelerations acting along the pendulum and hinge axes, a_y and a_z respectively, by the equation:

$$\tilde{a}_x = (1 + S_x)a_x + M_y a_y + M_z a_z + B_f + B_v a_x a_y + n_x \qquad (6.3)$$

where S_x = scale factor error, usually expressed
                in polynomial form to include non-linear effects
     M_y, M_z = cross-axis coupling factors
     B_f = measurement bias
     B_v = vibro-pendulous error coefficient
     n_x = random bias

6.3.4.3 Typical performance characteristics

Typical performance figures for the moderate accuracy sensors are as follows:

input range	up to ±100 g
scale factor stability	~0.1%
scale factor non-linearity	~0.05% of full scale
fixed bias	0.0001 g–0.01 g
bias repeatability	0.001 g–0.03 g
bias temperature coefficient	~0.001 g/°C
hysteresis	<0.002 g
threshold	~0.00001 g
bandwidth	up to 400 Hz

Most of these figures are improved significantly with the very high accuracy accelerometers. Biases as low as a few micro-g can be achieved with very high precision sensors, whereas those likely to experience high accelerations in very dynamic environments usually have a bias of a few milli-g.

6.3.5 Pendulous accelerometer hinge elements

The hinge element of a pendulous accelerometer is the component that enables the proof mass to move in one plane normal to the hinge axis. It must be stiff normal to the hinge line to maintain the mechanical stability of the hinge relative to the case under conditions of dynamic loading. However, it must be flexible about the hinge line and must minimise unpredictable spring restraint torques which cannot be distinguished from applied accelerations. The hinge must not be overstressed by either shock acceleration or vibratory motion. It must also return to its null position exactly when the proof mass is displaced, in order to give the sensor good bias stability. Hinge elements exist that enable the proof mass to move in two orthogonal directions. These are essentially a complex combination of two single axis elements, as described in Section 6.3.6.

The two basic forms of hinge elements are flexures and pivots, there being several variations of each type.

6.3.5.1 Flexure hinges

The materials used to form the hinge are selected for their low mechanical hysteresis in order to minimise unpredictable spring torque errors. Hysteresis effects are minimised by choosing the hinge dimensions so that hinge stresses under dynamic forces, and pendulum movement, are well below the yield stress for the hinge material. A material that is commonly used is the alloy beryllium-copper since, because of the high ratio of its yield stress to its Young's modulus [4], it is capable of sustaining a large deflection without exceeding its yield stress. Fuzed quartz is another very suitable material. Some designs have both the pendulum and the hinge etched from a quartz substrate.

The main advantages of flexure hinges are that they exhibit very low static friction so offer almost infinite resolution and very low threshold. However, these hinges have a significant temperature dependent bias that requires calibration and compensation for the most accurate applications. Additionally, these hinges can be susceptible to damage from shock accelerations and also demand very tight tolerancing, typically in the region of a micrometre, if the desired flexure compliance is to be attained.

6.3.5.2 Jewelled-pivot hinges

This form of hinge supports the pendulum between a pair of spring loaded synthetic jewel assemblies. The spring loading provides three-dimensional shock protection. These hinges have very small temperature dependent bias characteristics. However, stiction, under very quiescent conditions, can limit the resolution and wear of the pivots in very harsh vibratory environments can be a problem. This can be partially alleviated by the use of very hard materials on the bearing surface such as silicon carbide.

There are many applications, particularly those requiring a low maximum acceleration capability (20 g or less), where jewel and pivot hinges can offer a cheaper and more sensitive instrument. However, with higher accelerations and high vibratory environments, flexure hinges tend to provide enhanced performance.

6.3.6 Two-axis force feedback accelerometer

This form of instrument has many applications including some of the most demanding such as ship's inertial navigation systems. It has a pendulum which has freedom to swing about two orthogonal axes and, like the single axis device described above, it is restrained to its null position by electrically energised coils working in a permanent magnetic field.

Clearly, it is necessary to have a hinge that constrains the pendulum to deflect about these two orthogonal axes, albeit by very small angles. Typically, the pendulum is attached rigidly to a plate at its top end, which is attached by two weak leaf springs to another plate. This second plate is attached to the case by means of a second pair of similar hinges, which are mounted at 90° to the first pair of hinges. Motion of the pendulum is often damped by filling the case with silicone fluid.

The principle of operation is identical to that described above for the single-axis sensor, and its performance is similar to that which can be obtained using the higher grade single-axis devices.

6.3.7 Open loop accelerometers

A common form of this sensor is the mass spring device of the type described in Section 6.2. Generally, these instruments are less stable and less accurate than the closed loop accelerometers. Undesirable characteristics inherent in open loop accelerometers are sensitivity to supply voltage variations, non-linearity of the displacement caused by the applied acceleration and high thermal coefficients of bias and scale factor. Consequently, they are generally inappropriate for most inertial navigation applications and the mechanical variant will not be discussed further. Currently, however, an optical open loop pendulous fibre optic accelerometer is being developed and the principle of operation will be discussed below.

6.3.7.1 Optical fibre accelerometer

The fundamental principle of operation of this sensor is identical to the mechanical device. The major difference essentially lies in the form of the pick-off mechanism and pendulous mechanisms which allow accelerations about two axes to be sensed. Optical fibres have excellent mechanical strength and elastic modulus characteristics, and additionally have negligible thermal expansion over the normal operating temperature range of inertial sensors, but need to be selected carefully to have isoelastic properties.

In this sensor, the pendulum is a length of fibre cable with a proof mass attached, together with a micro-lens at the bottom of the fibre and light from a solid state laser coupled into the top. When an acceleration is applied to the case along any axis normal to the fibre, the bottom is deflected. Its displacement is sensed and measured by means of the laser light passing through the optical fibre and being focused on to a two dimensional photo-sensitive array. A suitable array is a charge coupled imaging device (CCID) which can provide both *x* an *y* co-ordinates of the displacement. A schematic diagram of a fibre optic pendulous accelerometer is shown in Figure 6.3

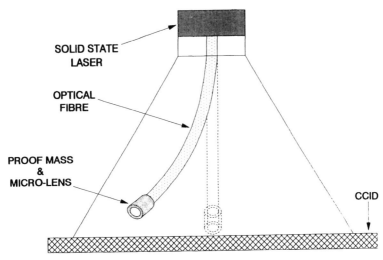

Figure 6.3   Pendulous fibre optic accelerometer

Several parameters determine the performance range of this sensor:

- size of proof mass;
- diameter of the optical fibre;

- length of optical fibre;
- height of the suspension point above the photo-sensitive detectors;
- size of the photo-sensitive array.

Currently, the accuracy is limited by the pixel density of the photo–sensitive array. Performance data are not currently available.

6.4 Solid state accelerometers

During recent years, there has been intensive research effort to investigate various phenomena which could be used to produce a solid state accelerometer. Various devices have been demonstrated, with surface acoustic wave, silicon and quartz devices being most successful. These sensors are small, rugged, reliable and offer the characteristics needed for strapdown applications.

Many of the devices discussed below have been the subject of research studies into the concepts only and have not been developed for particular applications as far as the authors are aware. In these cases, performance figures are not given.

6.4.1 Vibratory devices

These are open loop devices which use quartz crystal technology. A common configuration uses a pair of quartz crystal beams mounted symmetrically back-to-back, each supporting a proof mass pendulum. A schematic representation of such a device is shown in Figure 6.4.

Each beam is made to vibrate at its own resonant frequency. In the absence of any acceleration along the axis sensitive to acceleration, both beams vibrate at the same resonant frequency. However, when an acceleration is applied along the sensitive axis, one beam experiences compression whilst the other is stretched, or under tension, owing to the inertial reaction of the proof mass. The result is that the beam in compression experiences a decrease in frequency, whereas the beam in tension has an increase in frequency. The difference in frequency is measured and this is directly proportional to the applied acceleration.

Some of the errors often associated with this type of technology can be minimised by careful design. The symmetrical arrangement of the beams produces a cancellation of several errors that exist if only one beam is used. Error effects that are usually alleviated, or even eliminated, by this design include variations in nominal beam frequency owing to temperature changes and ageing of the quartz,

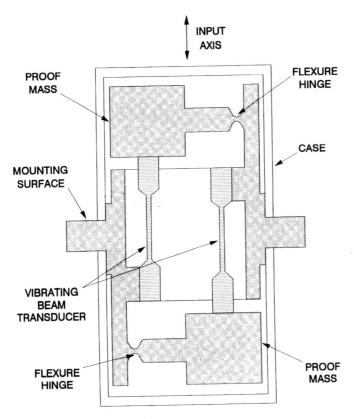

Figure 6.4   Vibrating beam accelerometer

asymmetrical scale factor non-linearities, anisoinertia errors and vibro-pendulous effects.

Typical performance data are shown below:

input range	±200 g
scale factor stability error	~100 ppm
scale factor non-linearity	~0.05% of full scale
bias	~0.1–1 milli-g
threshold	<10 micro-g
bandwidth	>100 Hz

6.4.2 Surface acoustic wave accelerometer

This sensor is an open loop instrument with a surface acoustic wave resonator electrode pattern on the surface of a piezoelectric quartz cantilever beam [5, 6]. This beam is rigidly fixed at one end to the case of the structure but is free to move at the other end, where a proof mass is rigidly attached, as shown in Figure 6.5. A surface acoustic wave train [7] is generated by use of the positive feedback between a pair of the metal electrode interdigital arrays, its wavelength being determined by the separation of the metal electrodes, often called fingers.

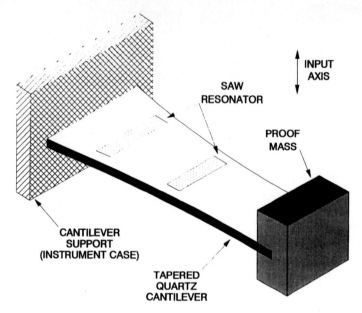

Figure 6.5   Surface acoustic wave accelerometer

When an acceleration is applied normal to the plane containing the beam, the inertial reaction of the assembly causes the beam to bend. When the surface of the beam is subjected to an applied strain, as occurs when the beam bends, the frequency of the surface acoustic wave changes in proportion to the applied strain. Comparison of this change with the reference frequency provides a direct measure of the acceleration applied along the sensitive axis.

The effects of temperature and other effects of a temporal nature can be minimised by generating the reference frequency from a second oscillator on the same beam. Lock-in type effects are prevented by ensuring that this reference signal is at a slightly different frequency from that used as the sensitive frequency.

Typical performance data are shown on the next page.

input range	±100 g
scale factor stability error	0.1–0.5%
scale factor non-linearity	<0.1%
bias	<0.5 milli-g
threshold	1–10 micro-g
bandwidth	~400 Hz

6.4.3 Silicon sensors

Over the last decade or so, there have been research studies directed at fabricating accelerometers from silicon [8, 9]. As a material, silicon has many advantages over other materials [10]; it is inexpensive, very elastic, non-magnetic, it has a high strength to weight ratio and possesses excellent electrical properties allowing component formation from diffusion or surface deposition. Additionally, it can be electrically or chemically etched to very precise tolerances, of the order of micrometres.

In one concept, micro-machining techniques were used to form cantilevered beams of silicon dioxide over shallow cavities etched in silicon. The end of the cantilever beam was gold plated to provide the proof mass and hence increase the sensitivity of the instrument. The cantilever was metal plated along its top surface to form one plate of a capacitor, the silicon substrate forming the other plate of the capacitor, as illustrated in Figure 6.6. This form of accelerometer can be operated in either an open loop mode or as a closed loop device. In the open loop mode, the capacitance between a pair of metal plates changes with the deflection of the cantilever, i.e. the applied acceleration. In the closed loop mode, as shown in Figure 6.6, a pair of electrodes are used to null any deflections of the cantilever. Use of the closed loop mode increases its sensitivity. Although such devices tend not to be very accurate, they are very small and quite rugged.

A monolithic accelerometer was developed in the United States during the early 1980s. A cylindrical proof mass was supported by single crystal silicon diaphragm discs which were hinged on a Cervit frame, as shown in Figure 6.7. This instrument was operated open loop, using a differential capacitive pick-off on each end to detect motion of the proof mass when subjected to an applied acceleration. The materials were chosen to provide a thermally stable path. The major problem areas with this instrument have centred around

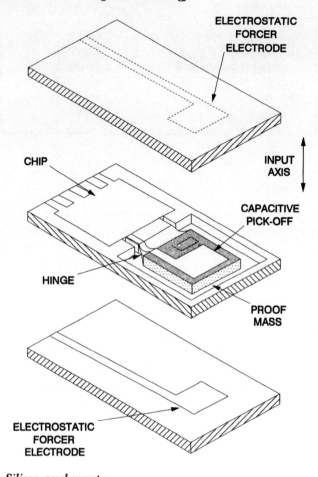

Figure 6.6   Silicon accelerometer

difficulties with machining the materials, achieving adequate scale factor linearity and bonding the components together. Currently, performance data are not available for this sensor.

Another form of silicon accelerometer that is currently under development has frequency sensitive resonant tie bars integrally attached to a silicon seismic mass. These tie bars are maintained at mechanical resonance, typically vibrating at frequencies between 40 kHz and 100 kHz depending on the configuration. When an acceleration is applied along the sensitive axis, movement of the seismic mass induces a strain in the tie bars resulting in a change in frequency of the order of tens of hertz for each applied unit g. This change in frequency is reasonably detectable. A conceptual diagram of this sensor is shown in Figure 6.8.

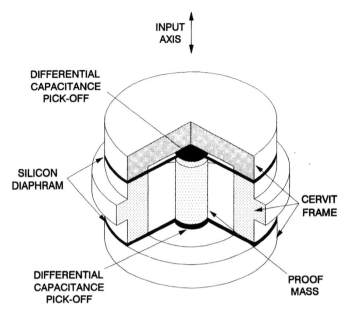

Figure 6.7   Monolithic accelerometer

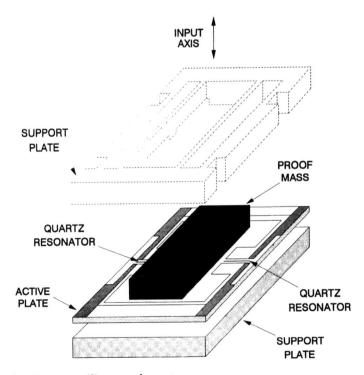

Figure 6.8   Resonant silicon accelerometer

Typical performance parameters are:

input range	±100 g
scale factor stability	0.5–2%
scale factor non-linearity	<0.1–0.4%
bias	<25 milli-g
(with compensation)	
threshold	1–10 micro-g
bandwidth	~400 Hz

Work in the UK has investigated a thermal excitation method as an alternative to the use of piezoelectric transducers for excitation of the proof mass. This thermal excitation technique is achieved by depositing a form of bimetallic strip on the tie bars, which is used in place of the piezoelectric transducer.

A bimetallic element is formed on a tie bar by the deposition of a resistor on the top surface of a tie bar. Application of a potential difference to this resistive load produces localised heating on the top surface of the tie bar. Consequently, there is an expansion of the hot surface with respect to the cooler surface which causes the tie bar to bend. If an alternating potential is applied to this resistive load, then the localised heating will be periodic and the top surface of the tie bar will expand and contract with respect to the lower surface, depending on the heating cycle of the resistive material. The frequency of the applied current is chosen to be synchronous with one of the natural resonant frequencies of the tie bars. As a result of this periodic bending of the tie bars, the proof mass is forced to oscillate, as described above, for the piezoelectric excitation technique.

A second resistor is located on each of the driving tie bars and is used as a detector to sense the oscillation frequency. This is then used as the feedback signal to modify the frequency of the applied alternating current. The drive and control electronics can also be formed in the silicon material. Quality factors in excess of 1000 have been demonstrated with such designs.

Variation in the heating effect produced by the resistive material on the tie bars is achieved by applying a suitable bias in combination with the alternating drive current. Consequently, the variation in the polarity of this applied potential allows the heating effect of the resistive material to be modulated at the frequency of this applied potential.

The main motivation for the development of this excitation technique was that an all silicon sensor could be developed. Several techniques exist for the deposition of the resistive heating elements on to the tie bars. Examples include direct diffusion doping or poly-silicon deposition. Similar techniques can be used to form the detector.

6.4.4 Fibre optic accelerometer

The use of fibre optical elements is very attractive for many applications as the fibre optical waveguide is immune to electromagnetic interference. One form of fibre optic sensor has already been described above, as it is very similar in operation to the pendulous accelerometers, the fibre optics merely providing an alternative form of read-out. Other forms of fibre optic accelerometer rely on some physical change in a component which can be sensed using electromagnetic radiation.

Ensuring that these changes are linear functions of acceleration in known directions remains a difficult development problem, although the use of fibre technologies gives a very sensitive position read-out.

6.4.4.1 Mach-Zehnder interferometric accelerometer

A Mach-Zehnder interferometer [11] uses either one or two optical fibres attached to an inertial mass as its sensitive element. When an acceleration is applied along the axis of the optical fibre, this will produce a small change in length which is proportional to the applied acceleration. The change in length can be detected by interferometric techniques similar to those described for the fibre optic gyroscope. The use of two optical fibres allows each fibre to form an arm of the interferometer and the use of nulling techniques enables greater sensitivity to be achieved, along with compensation for temperature changes in the fibres. Additionally, it is necessary to constrain the proof mass to move only along the sensitive axis of the instrument.

Schematic illustrations of the sensitive elements of two possible configurations are shown in Figure 6.9.

A very sensitive sensing element for accelerometers can be produced by winding a fibre optic coil around a compliant former, such as a rubber cylinder. When an acceleration is applied to the sensing element it changes dimensions and hence produces a phase change in the interferometer which is proportional to the applied acceleration. The sensitivity of the device is proportional to the number of optical fibre turns on the cylinder. Maximum sensitivity can be achieved by operating the device in a feedback mode, as shown in

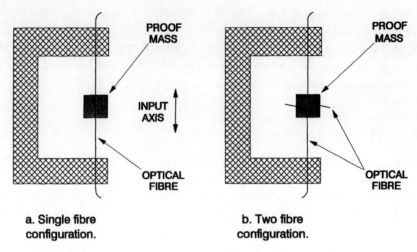

a. Single fibre
configuration.

b. Two fibre
configuration.

Figure 6.9   Sensitive elements of a Mach-Zehnder interferometric accelerometer

Figure 6.10. The intensities of the two light beams in the interferometer are detected separately and compared in a differential amplifier. The output signal from this component can then be used to drive a piezoelectric device to null the phase change introduced by the distortion of the sensing element. The output of the differential amplifier is proportional to the applied acceleration. Again, it is necessary to constrain the movement of the element to be only along the sensitive axis of the device. Other technological concerns are the longer term stability of the compliant component and the effect of the different thermal expansion coefficients.

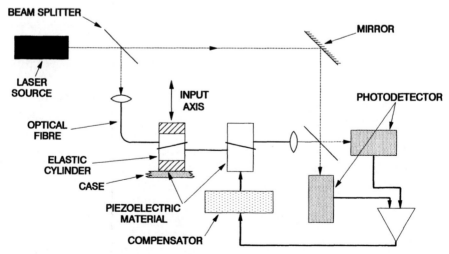

Figure 6.10   Interference accelerometer

6.4.4.2 Vibrating fibre optic accelerometer

A short length of single mode optical fibre is fastened and tensioned between two pivot points in a rigid structure. This structure is vibrated so that the optical fibre oscillates at its fundamental frequency. In the absence of any applied acceleration, the displacements are symmetrical and the maximum stretch occurs at the maximum displacement with relaxation as it passes the centre line. Light passing through this optical fibre is phase modulated at $2f$, and at higher order even harmonics of f, where f is the fundamental frequency. However, when the sensitive element is subjected to an acceleration parallel to the plane containing the oscillation, the displacement of the fibre will now be asymmetrical.

Light passing through the optical fibre will now be phase modulated at f and at the odd harmonics of f. The first and odd harmonic phase modulation has an amplitude proportional to the applied acceleration, and its phase relative to the drive signal will depend on the sense of the applied acceleration. Again, fibre optic interferometric techniques are used to sense the phase changes. Care is necessary in the choice of fundamental frequency and in the design in order to reduce the effects of orthogonal acceleration sensitivities and environmental vibratory motion. The displacement of the fibre is shown schematically under the conditions of no acceleration and applied acceleration in Figure 6.11.

It may be possible to produce an amplitude modulation system by using lossy multi-mode fibre, which is optimised for micro-bending losses, in the scheme described above. In this case, light which is guided along the vibrating optical fibre, is coupled into the cladding

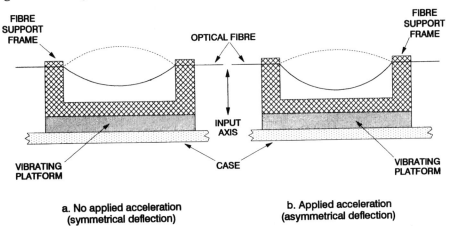

a. No applied acceleration
(symmetrical deflection)

b. Applied acceleration
(asymmetrical deflection)

Figure 6.11   Oscillating modes of a vibrating fibre accelerometer

surrounding the optical core at the support points. This occurs as a consequence of the bending of the fibre decreasing the barrier between the core and cladding modes. Such a system would not need to use interferometry to determine the magnitude of the applied acceleration. This is because this technique converts the device from a phase modulator to an amplitude modulator.

6.4.4.3 Photo-elastic fibre optic accelerometers

The sensitive element in this device is a birefringent material [11]. Suitably polarised light is coupled into the sensitive element using multi-mode optical fibre. When an acceleration is applied to the photoelastic material the transmission of light through it changes and the change is proportional to the applied acceleration. Research is continuing with this form of sensor. A schematic diagram of an engineering concept is shown in Figure 6.12.

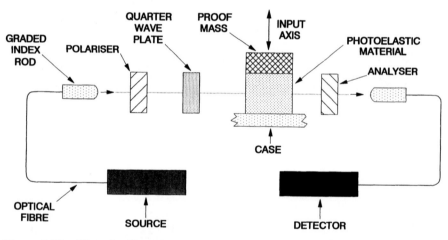

Figure 6.12   Photoelastic accelerometer

6.4.4.4 Bragg grating fibre accelerometer

Recent work in the USA and in Europe, at the Microelectronic Centre of Denmark, has demonstrated an accelerometer containing a Bragg grating in an optical waveguide. The centre wavelength of a Bragg grating is determined by the characteristics of the grating but can be changed by changes in the temperature, strain and pressure applied to the grating [12, 13]. Thus, when an optical waveguide containing a Bragg grating is distorted by an acceleration applied along the waveguide, the wavelength of light transmitted along the waveguide changes. This change in wavelength is proportional to the applied acceleration and is small, but can be detected using a fibre interferometer [14].

The effect of the applied acceleration can be enhanced if the fibre waveguide is rigidly bonded to a proof mass. The general layout of this sensor is shown in the schematic diagram in Figure 6.13a. Care is necessary to ensure that the proof mass and fibre move in the direction of the applied acceleration and do not deflect when cross-accelerations are applied. This is accomplished using guides as shown in Figure 6.13b. Clearly, care is also necessary to ensure that movement of the proof mass is not impeded by these guides. Performance

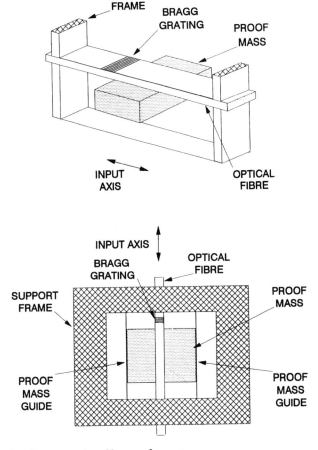

Figure 6.13   *Bragg grating fibre accelerometer*
    a Schematic layout
    b Section view

data are not available, but initial measurements suggest sensitivities of these devices to be in the micro-g regime.

6.4.4.5 Combined fibre optic sensors

The use of similar materials such as solid state lasers, photodetectors, optical fibres and common techniques in the fabrication of the sensors, suggests that there is plenty of scope for producing integrated devices, enabling both angular rate and linear acceleration to be sensed in a single device. The operation of the individual aspects of each sensor has already been dealt with in each appropriate section and will not be repeated here. The major problems are associated with the integration of the individual components and the sharing of components. Additionally, it is necessary to isolate particular processes, such as modulation frequencies, in order that effects can be identified uniquely.

6.4.5 Other acceleration sensors

Many physical effects have been exploited over the last half century or more in an attempt to measure acceleration. For completeness, two other interesting concepts known to the authors are discussed below. Generally, these programmes are not active, but some could become active if there is a significant change in a relevant technology.

6.4.5.1 Solid state ferroelectric accelerometer

Attempts have been made to use the piezo-optic and dielectric properties of ferromagnetic materials. It was hoped that the magnitude of the applied acceleration could be measured as a function of the strain or pressure induced in a thin fibre of this material. However, technological limitations in the past prevented the feasibility of the device being demonstrated.

6.4.5.2 Solion electrolytic accelerometer

This is a solid state ion device, making use of a shift in ions in a solution resulting from the application of an acceleration. This motion causes a resultant change in the potential in the electrolyte and this potential change was found to be proportional to the applied acceleration, with good linearity. However, the electrolyte is, by its very nature, thermally sensitive. This device was originally developed as part of the German missile programme during World War II.

6.5 Multi-sensors

6.5.1 Introduction

Previous discussions in this book have indicated that elements that are vibrating change their frequency of resonance when rotated or accelerated. Alternatively, cantilevered piezoelectric materials can be used as transducers by measuring the change in electrical charge across a crystal when it is deflected by an applied force. These principles have been applied by mounting several such elements at particular orientations with respect to each other. This enables one sensor to produce information about both the applied acceleration along and the rotation rate about an axis. These instruments are often called multi-sensors.

Multi-sensors are not confined to bending, cantilever or vibrating beam technologies. It will be recalled that a mechanical gyroscope, with a mass unbalance in its rotor support, will drift when subjected to an applied acceleration about an appropriate axis. This phenomenon can be applied using a cluster of three two-axis sensors, suitably oriented with appropriate known mass unbalance, to produce an inertial measurement unit which will provide information on both linear acceleration and rotation sensed about three reference axes.

The use of multi-sensors offers the distinct advantage of reducing the number of inertial instruments required to measure the rotation and linear motion of a vehicle. Only three instruments are required for some sensor types to give full inertial data in three axes. However, the information is generally mixed in each axis and needs to be separated at some particular frequency, usually the spin frequency of the assembly. There can also be problems in achieving satisfactory or compatible performance from both accelerometer and gyroscopic channels of a multi-sensor for some applications. An additional problem area can also be cross-coupling between the different channels, although careful design can minimise this effect.

6.5.2 Rotating devices

Research began on these devices in the United States and in Britain in the late 1970s. Such devices operate by detecting the change of dynamic input, or force, which is applied to a piezoelectric transducer. Such a device is mounted on a cantilever as shown in Figure 6.14 and can be attached rigidly to a rotating element. This transducer produces an alternating electric signal proportional to the applied input.

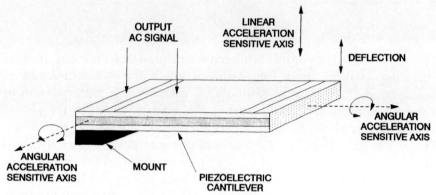

Figure 6.14   piezoelectric accelerometer

The two principle components of a rotating multi-sensor are:

(i)   a rotating assembly;
(ii) piezoelectric transducers.

Additionally, it is necessary to have a set of slip rings to transfer the electrical signals from the transducers to the electrical connecting pins on the case.

Piezoelectric accelerometers do not have a very low detection threshold sensitivity, nor do they have good day to day stability characteristics. These deficiencies can be reduced if the sensor is designed to operate at only a set frequency, and then synchronous demodulation used to remove d.c. uncertainties.

Typically, there are four piezoelectric transducers mounted rigidly as cantilevers to the rotating assembly. The transducers are mounted in pairs, each pair being orthogonal to each other and orthogonal to the spin axis of the rotating member as shown in Figure 6.15. The construction and mounting is such that one set of transducers generates signals which are proportional to angular rate, whilst the other set produces signals proportional to linear acceleration.

Each transducer is made of layers of piezoelectric ceramic. These are ferroelectric materials, which are non-symmetric crystals and have a built in electric dipole. In its polarised form, any stress applied to the structure results in a variation of the dipole moments causing a voltage to appear across the electrodes. Thus the material can convert mechanical energy to electrical signals and vice versa.

When a stationary cantilevered beam is oriented so that an applied acceleration deflects the beam, then a steady voltage results. However,

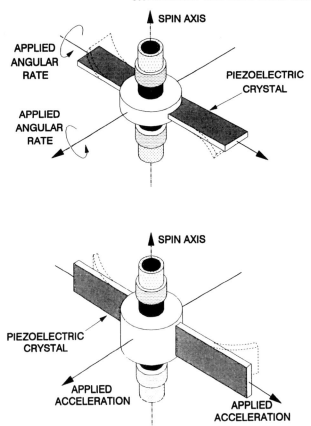

Figure 6.15 Principle of operation of a rotating multisensor

if the beam is now rotated about an axis so that when it has rotated by 180° the direction of the deflection is in the opposite direction, then the signal will have the reverse sense. In the case of continuous rotation of the beam, the output signal voltage is a sinusoid, with a frequency equal to the rotation frequency, the peak voltage being proportional to the applied acceleration.

As noted above, there are usually two identical cantilevered beams mounted on the rotating assembly, with their flexing axes colinear with the spin axis. In this orientation, these transducers sense linear acceleration in the plane perpendicular to the spin axis. These rotating beams produce a suppressed carrier modulated spin frequency signal, with a peak amplitude proportional to the amplitude of the applied acceleration. The amplitude is a maximum when these transducers sense the total applied acceleration, and becomes a minimum when rotated through 90°.

The principle of detection of angular rate is based on the gyroscopic behaviour of an elastically restrained body which is rotated about an axis. Usually, the cantilevered transducers are mounted 180° apart on the rotating assembly for common mode rejection and also signal enhancement. As mounted, these transducer elements act both as the inertial members and the restraining springs. When an angular rotation rate is applied to the spinning assembly, about an axis orthogonal to the spin axis, the angular momentum of the rotating transducers generates a forcing function, bending the transducer. This forcing function is a suppressed carrier signal modulated at the spin frequency of the mounting assembly. A plot of the amplitude of the signal generated is a sinusoid, its magnitude being proportional to the magnitude of the applied rate. The phase of the output is such that the maximum signal occurs when the sensing transducer is colinear with the applied input. The minimum voltage occurs one quarter of a rotation away from the maximum signal.

Clearly, such an instrument can sense accelerations applied along two axes in the plane perpendicular to the spin axis. Similarly, it can sense angular rotation rates applied along two axes in the plane orthogonal to the spin axis. Therefore only two sensors are required to be mounted so that their spin axes are not parallel to enable three axes of angular rate data and three axes of linear acceleration data to be generated. The redundant information generated along and about the fourth axis is available for system checks.

These sensors are open loop devices and consequently tend to have poorer scale factor characteristics when compared with closed loop sensors, such as the floated rate integrating gyroscope. Additionally, the scale factor can change as the piezoelectric crystals age. However, these open loop devices do not consume extra power, and consequently do not liberate heat, when measuring high rates of rotation.

Although these sensors are capable of very accurate measurement of angular rotation rates and linear acceleration, to achieve the high accuracies required for inertial navigation purposes, careful calibration and characterisation are necessary and temperature compensation is usually vital. Currently, the very accurate instruments tend to be quite large; up to 150 mm long by 35 mm diameter. This is offset by the fact that this single sensor provides four of the six measurements required by a navigation system.

Careful choice of certain components is necessary in order to contain certain error sources. Use of low noise bearings minimises the noise coupled into the crystals producing a background signal. Noise generated by the slip rings is synchronous with the piezoelectric signals and consequently represents an acceleration or angular rate error. The measurement bandwidth of the sensor is dependent on the spin speed of the rotating assembly. Consequently, as it requires at least two readings per cycle to define a sinusoid, the rotating assembly must have a spin frequency at least twice that of the measurement bandwidth. For high bandwidth applications, this can lead to significant generation of bearing noise giving rise to potential saturation of the electronic measurement system.

Variations in temperature can also lead to changes in bearing generated noise as a result of variations in the bearing characteristics such as internal loading and viscosity changes to the lubricant. Additionally, variations in the temperature can alter the reference electronics used to resolve the acceleration and angular signals. This appears like a scale factor error.

Typical performance data are given below:

gyroscope:	
maximum input rate	300–400°/second
g-independent bias	1–10°/hour
g-dependent/mass unbalance bias	5–10°/hour/g
anisoelastic bias	0.1–0.2°/hour/g^2
scale factor stability error	0.1–0.5%
scale factor non-linearity	0.03–0.1%
bandwidth	60–100 Hz
accelerometer:	
input range	up to ±100 g
scale factor stability error	0.1–2%
scale factor non-linearity	0.03–0.1%
bias	1–10 milli-g
threshold	1–10 micro-g
bandwidth	>70 Hz

One concept which was the subject of a recent research programme was based on the use of rotating surface acoustic wave accelerometers mounted on a common shaft, instead of piezoelectric sensors. Angular motion and linear acceleration of the sensor's case are sensed by mounting pairs of surface acoustic wave accelerometers as cantilevers on a body that is rotated at a constant speed. When angular motion or linear acceleration is applied, the cantilevers are deflected owing to the various physical effects described earlier for the piezoelectric based sensor. The output signals are also generated using similar techniques.

The use of the surface acoustic wave elements offered several advantages as various effects, such as temperature induced biases, can be compensated. These effects can be compensated by the use of two surface acoustic wave oscillators on the same cantilever, as described in Section 6.4.2. Additionally, a digital output can be generated directly on the element and passed through the slip rings allowing the effect of noise on small signals to be eliminated. However, work on these sensors appears to be dormant.

6.5.3 Vibratory multi-sensor

Research into this form of sensor has been most active in the United States. It is a single axis device based on vibrating sensor technology enabling measurements to be made of both angular rotation rates and linear acceleration. The sensor measures continuous angular rates by the oscillating Coriolis acceleration induced on an accelerometer which is being vibrated, and forces subject to oscillatory linear velocity. Typically, the accelerometer in this device uses silicon solid state technology.

The principal components of this instrument are:

(i)   an accelerometer;
(ii) a vibrator (vibratory platform).

The principle of operation is essentially that of a pendulous accelerometer which is vibrated along its hinge, or pivot, axis. As a result of this vibration, an oscillatory linear velocity is imparted to the pendulum. Consequently, the accelerometer will sense a Coriolis acceleration at the frequency of vibration proportional to the angular motion applied about the axis of the pendulum. Additionally, the accelerometer will measure any linear acceleration applied along its input axis. The electrical signal generated by such a sensor will have a d.c. value proportional to the applied linear acceleration and an a.c. signal at the vibration excitation frequency. The latter signal can be demodulated to produce a signal which is proportional to the applied rotation rate about the axis of the pendulum.

In a practical device, an accelerometer using vibrating beam technology is excited by piezoelectric crystals. Typically, two accelerometers are vibrated 180° out of phase to give common mode rejection, thus preventing random inputs at the vibrating frequency from corrupting measurements. This type of sensor has many advantages through the elimination of rotating elements and

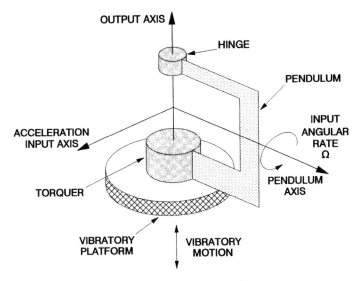

OUTPUT AXIS

HINGE

PENDULUM

INPUT
ANGULAR
RATE
Ω

ACCELERATION
INPUT AXIS

PENDULUM
AXIS

TORQUER

VIBRATORY
PLATFORM

VIBRATORY
MOTION

Figure 6.16   Principle of operation of a vibratory multisensor

bearings, but has the disadvantage of small signal to noise ratio for the angular rate measurement. Schemes have been devised for mounting three accelerometers at various orientations on a plate, enabling them to be vibrated, or dithered, about the body diagonal of the reference axis set. Using one common activation axis has several advantages, such as the elimination of cross-talk and aliasing between the sensors. It also allows some common electronic circuits to be used resulting in a very compact three-axis inertial measurement unit.

Anticipated performance parameters for this form of multi-sensor are as follows:

gyroscope:	
maximum input rate	±1000°/second
g-independent bias	5–10°/hour
scale factor stability error	~0.1%
scale factor non-linearity	~0.05%
bandwidth	>100 Hz
accelerometer:	
input range	up to ±200 g
scale factor stability error	~0.05%
scale factor non-linearity	<0.1%
threshold	~10 micro-g
bandwidth	>100 Hz

6.5.4 Mass unbalanced gyroscope

Angular momentum gyroscopes, such as the floated rate integrating gyroscope and the dynamically tuned gyroscope, are precision instruments requiring careful assembly to achieve the levels of performance normally required for most applications. During manufacture, particular care must be taken to ensure that the spinning rotor, or rate integrating gyroscope rotor/float combination, is balanced accurately. In the presence of a linear acceleration normal to the float axis in the case of the rate integrating gyroscope, or normal to the rotor spin axis in the dynamically tuned gyroscope, any mass unbalance will induce a torque, causing the rotor to precess, and so produce an erroneous rate measurement. However, by introducing a known amount of mass unbalance into gyroscopes of this type, it is possible to obtain a measure of the acceleration to which the instrument is subjected, in addition to the angular rates that it is sensing. Much of the pioneering work on this form of sensor was undertaken in Germany and the United States, but more recently in France [15].

This concept dates from the 1950s and, indeed, instruments have been manufactured based on this principle for many years. The Honeywell precision integrating gyroscopic accelerometer (PIGA) is one such device. This type of device was also developed in the UK by Ferranti, now part of GEC. The design of the PIGA was based on a single-axis floated gyroscope, in which the rotor was made pendulous with respect to the output axis of the instrument. Generally, this form of sensor was intended for use on stable platforms.

More recently, attention has focused on a development of the dynamically tuned gyroscope, which has the centre of suspension of its rotor displaced slightly with respect to its centre of gravity.

The displacement occurs along the motor drive shaft, in a manner which causes accelerations applied perpendicular to the drive shaft, i.e. parallel to the input axes, to produce torques which cause precession of the rotor. This is shown schematically in Figure 6.17. As with the conventional dynamically tuned gyroscope, the device operates in a torque rebalance mode. However, in this case the pick-off outputs are fed back to null the precession caused by both the input rates and the applied accelerations.

A perfect two-axis gyroscope operating in a torque rebalance mode has the steady state relationship between the input rates, ω_x and ω_y, and the applied torquer moments, M_x and M_y, given by:

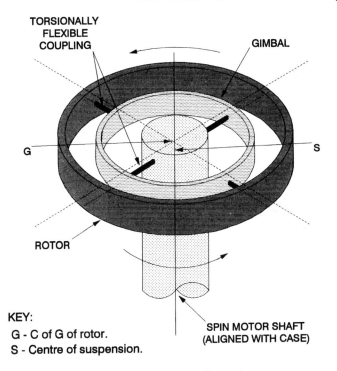

TORSIONALLY
FLEXIBLE
COUPLING

GIMBAL

G

S

ROTOR

KEY:
G - C of G of rotor.
S - Centre of suspension.

SPIN MOTOR SHAFT
(ALIGNED WITH CASE)

Figure 6.17   Mass unbalance gyroscope rotor configuration

$$\omega_x = -M_y/H$$
$$\omega_y = M_x/H \tag{6.4}$$

where H is the angular momentum of the rotor. If, in addition to the torquer moments, moments act as a result of the unbalance in the rotor suspension, the above equations take the form shown below, where the unbalance torques are proportional to the applied accelerations, a_x and a_y,

$$\omega_x = -(M_y + Ba_x)/H$$
$$\omega_y = (M_x - Ba_y)/H \tag{6.5}$$

The factor B is a function of the displacement between the rotor centre of gravity and its centre of suspension, and the inertia of the rotor about an axis perpendicular to its spin axis. Rearranging eqn. 6.4 and writing $B/H = b$, the measurements, m_1 and m_2, provided by a single sensor may be expressed as follows:

$$m_1 = -M_y/H = \omega_x + ba_x$$
$$m_2 = M_x/H = \omega_y + ba_y \tag{6.6}$$

Thus a single gyroscope can provide a weighted sum of the turn rate about, and the acceleration along, each input axis. The constant b is referred to as the mass unbalance coefficient and may be expressed in units of $°/s/g$. The choice of b depends on many factors including the required measurement range of the gyroscope and the motion of the vehicle in which it is to be installed.

By combining three mass unbalanced gyroscopes of this type in an inertial measurement unit, it is possible to obtain estimates of angular rates and linear accelerations in three mutually orthogonal directions provided:

- the mass unbalance coefficient is different for each gyroscope;
- the spin axes of the gyroscopes are not coplanar.

A conventional strapdown system using dynamically tuned gyroscopes would require two such gyroscopes and three accelerometers. It is postulated therefore, that the three accelerometers may be replaced by one additional gyroscope to form a mass unbalanced navigation system. Such a system has the advantage of using identical rebalance loop electronics for all sensors, and uses fully the information from each input axis. Potential disadvantages include reduced dynamic range compared with the conventional system, some additional computing complexity associated with the extraction of separate angular rate and linear acceleration estimates. Additionally, there is the possibility of additional dynamic cross-coupling between these quantities.

Analysis of the effects of bias, scale factor and cross-coupling errors in a mass unbalanced system, reveals that each error term produces inaccuracies in the estimates of both angular rate and linear acceleration, the latter being functions of the mass unbalance coefficient, b. For example, take the case of an orthogonal system in which the mass unbalance coefficients for two of the gyroscopes are equal, but of opposite sign, and zero for the third gyroscope. The general form of the rate and acceleration estimation errors, $\delta\omega$ and δa, in such a system is illustrated by the following matrix form:

$$
\begin{pmatrix} \dot{\delta\omega} \\ \cdots \\ \dot{\delta a} \end{pmatrix} = \begin{pmatrix} \mathbf{B}_f \\ \cdots \\ \mathbf{B}_f b \end{pmatrix} + \begin{pmatrix} \mathbf{S} & \cdot & b\mathbf{S} + \mathbf{B}_g \\ \cdots & \cdots & \cdots \\ \mathbf{S}/b & \cdot & \mathbf{S} + \mathbf{B}_g/b \end{pmatrix} \begin{pmatrix} \omega \\ \cdots \\ a \end{pmatrix} \tag{6.7}
$$

where **B**$_f$ = fixed bias
B$_g$ = g-dependent bias or uncertainty in the mass unbalance
coefficient
S = matrix containing scale factor errors and cross-coupling
terms
a = applied linear acceleration
ω = applied turn rate

Preliminary tests of this type of device carried out under laboratory conditions [15] suggest that measurements of turn rate and acceleration can be derived to an accuracy of significantly better than 100°/hour and less than 10 milli-g respectively, and that the scale factor error for such a device would be less than 10^{-3}. Errors in the acceleration measurements may be deduced, to a large extent, from the axial unbalance factor, which is assumed here to be in the region of 5°/hour/g. In addition to the usual bias and scale factor errors which arise when separate gyroscope and accelerometers are used, some additional cross-coupling between the rate and acceleration estimates arises in a mass unbalanced system.

6.6 Angular accelerometers

This form of inertial sensor provides a means for sensing angular motion. Traditionally, non-gyroscopic angular motion sensors have used a balanced mass suspended in bearings which generate a torque proportional to the applied angular acceleration. When the mass is constrained by a spring, the angular displacement is a measure of the angular acceleration.

There have been significant developments in the technology used in angular accelerometers. Small, compact, rugged and accurate sensors can now be produced, and have been applied to several applications. The devices may be operated in a closed or open loop mode, depending in configuration.

6.6.1 Liquid rotor angular accelerometer

Recently, progress in the development of this type of instrument in the United States has enabled it to evolve from a heavy, and often fragile, device to a small lightweight sensor. The modern devices have almost instantaneous readiness, reduced power consumption, enhanced ruggedness and increased sensitivity when compared with the older designs, and have eliminated rotating elements.

A schematic diagram of a liquid rotor angular accelerometer is shown in Figure 6.18.

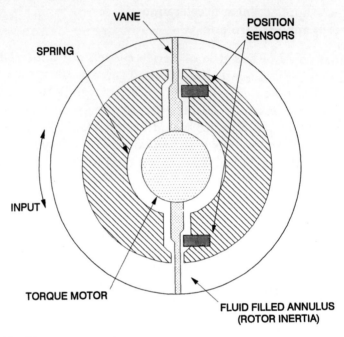

Figure 6.18   Fluid rotor angular accelerometer

The fluid filled sensor has an annular tube containing a liquid such as silicon oil, or a high density liquid of the type used in rate integrating gyroscopes. This liquid forms the seismic or proof mass in the sensor. The annular tube is blocked by a disc connected to a galvanometer movement, supported by jewel and pivot bearings. This arrangement forms a servoed torque generator.

Application of an angular acceleration about the axis of the annular tube would accelerate this tube leaving the inertial mass behind. However, the disc causes the fluid to move with the case, with a consequential reaction at the disc. This motion is sensed by the position sensing mechanism and provides feedback to the galvanometer torquer, which provides the torques necessary to accelerate the fluid with the case. The magnitude of this feedback signal is directly proportional to the angular acceleration acting about the input axis.

This form of system provides a good deal of flexibility as the electronic gain can be set to generate full scale deflections for any given displacement of the disc. Various other parameters, such as the cross-section, diameter and sensing area of the tube, as well as the density of the fluid in this tube, may be selected individually to provide the desired frequency response and sensitivity.

One common problem with this sensor is the effects of change in temperature and thermal gradients across the sensor which can produce non-linear responses and sensitivity to linear acceleration. Changes in ambient temperature can be corrected using a volume compensator. However, careful design and thermal screening are necessary to avoid thermal gradients across the instruments.

Typical performance parameters are given below:

input acceleration	up to 50 r/s^2
scale factor non-linearity	~0.1 % of full range
bias	~0.001 r/s^2
bias temperature coefficient	~0.0005 r/s^2/°C
threshold	~0.005 % of full range
bandwidth	up to 60 Hz

6.6.2 Gas rotor angular accelerometer

This design of angular accelerometer has some similarities with the instrument which uses a liquid rotor. However, in this case, a high density gas at a high pressure is contained in a single tube, its end being attached to a pressure sensor. Generally, this device is operated in an open loop mode.

When angular motion is applied about an axis perpendicular to the plane containing the gas filled tube, there is relative motion between the tube and the gas. This motion is sensed using the pressure sensor which forms a barrier across the tube and prevents free flow around it.

The pressure generated by the gas rotor is usually quite small, typically in the range 10–100 Pa. Hence the pressure sensor must have high sensitivity. Additionally, it must impart a stiffness to the system to produce the necessary dynamic characteristics of the instrument. A pressure sensor with an electrically conductive membrane positioned between two circular electrodes is one possible simple design that could be used to detect the gas motion. In this case, when a differential pressure is applied across the membrane, resulting from the motion of the gas, its displacement results in a differential capacitance change between the membrane and the electrodes. Alternative pressure sensors giving greater accuracy can also be used, although they are usually more expensive.

The tube containing the gas can be formed into any shape. However, a helix is the most common one used. The pressure generated is proportional to the mean radius of the helix, the number of turns, the density of the gas and, of course, the applied acceleration. Successful designs, with diameter of about 40 mm and a few tens of turns have been demonstrated, a constriction in the tube being necessary to provide damping of the motion of the gas.

A schematic diagram of a gas rotor angular accelerometer is shown in Figure 6.19.

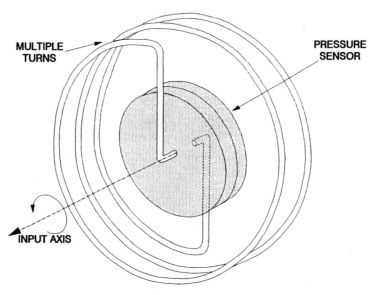

Figure 6.19   Gas rotor sensor

As in the case of the liquid rotor, temperature gradients across the sensor must be avoided. Additionally, careful screening of leads and components is necessary to avoid stray capacitances corrupting the output signals.

Inherently, this design is very robust and offers a long operational life and low cost. However, there does not appear to be any current development activity of this type of sensor, in contrast with the status of the liquid rotor devices.

6.7 Inclinometers

An inclinometer is a gravity reference device capable of sensing tilt. The instrument is basically a special implementation of a linear accelerometer with low maximum acceleration capability. The accelerometer output is usually processed to give a d.c. voltage directly proportional to the angle of tilt. Typical applications of the inclinometer are platform levelling for target acquisition systems and fire control systems, and in inertial component testing.

6.8 Summary of accelerometer and multi-sensor technology

Many different types and designs of inertial sensors can be used for sensing and measuring the magnitude of an accelerating force. The review has included mechanical sensors, using the classical pendulum principle, through to modern solid state devices. Generally, all these instruments are suitable for strapdown applications and in such an environment will give accuracies ranging from tens of micro-gravitational acceleration (micro-g) to fractions of g.

The mechanical accelerometers come in various forms, with a selection of materials and designs for the pendulum's hinge mechanism. These sensors may be fluid filled in order to improve the damping of the motion of the pendulum, and the pendulum may be constrained to very small displacements, through the use of force feedback techniques, in order to achieve high accuracy. Alternatively, the sensor may be operated in an open loop mode.

Solid state technology offers various techniques which may be applied to enable small, reliable and relatively inexpensive instruments to be produced. A variety of techniques has been reviewed, including the use of optical fibres, vibratory devices, surface acoustic wave devices and silicon materials. These sensors are generally operated in an open loop mode, but some designs are amenable to the use of closed loop techniques. In the case of the closed loop mode, the displacement of the proof mass is generally not returned to its null position; instead, the sensor operates by nulling an observed effect, such as a frequency change or a modified resonant condition.

A summary of typical performance characteristics[1] for a range of accelerometers is given in Table 6.1.

Table 6.1   Performance characteristics for a range of accelerometers

Characteristic	Accelerometer type				
	Force f/b pendulous	Vibrating fibre optic	Vibrating quartz	SAW	Silicon
Input range (g)	±100	±20	±200	±100	±100
Scale factor error stability (%)	0.1	0.001	0.01	0.1–0.5	0.5–2
Scale factor nonlinearity (% full scale)	0.05	0.05	0.05	<0.1	0.1–0.4
Fixed bias (mg)	0.1–10	1	0.1-1	<0.5	<25
Threshold (μg)	10	1	<10	1–10	1–10
Bandwidth (Hz)	400	100	400	400	400

It is noted that substantially higher performance can be achieved using force feedback devices. Precision devices capable of detecting accelerations as small as a few micro-g have been made with scale factor stabilities of 10^{-5}%. However, such instruments are not normally designed to measure accelerations of ±100 g.

Since the 1980s, there have been significant developments in the performance of so-called multi-sensors, enabling a simple instrument to sense both linear and angular motion along and about two axes. These sensors offer significant potential for some applications in the future.

Finally, there has been progress in the state of the art of the manufacture of angular accelerometers, which can offer an alternative to the use of gyroscopes for some applications. The use of a fluid ring rotor to sense the applied motion has enabled small, sensitive, rugged and reliable angular accelerometers to be produced.

---

[1]These are typical values applicable over the range of parameters stated. In many case, the values given could be improved. However, it is not normally possible to have all the best case values in an single unit. These values are only for general indicative purposes.

References

1 MCLAREN, I.: 'Open and closed loop accelerometers', *NATO AGARDograph*, 1974 (160)
2 SMITHSON, T.G.: 'A review of the mechanical design and development of a high performance accelerometer'. *Proc. Inst.Mech.Eng. Sci.*, Conference on mechanical technology of inertial sensors, 1987, **C49/87**
3 SCHULER, A.R., GRAMMATIKOS, A. and FEGBY, K.A.: 'Measuring rotational motion with linear accelerometers', *IEEE Trans. Aerosp. Electron. Syst.*, **3**, p. 465
4 NEWMAN, F.H. and SEARLE, V.H.L.: *'The general properties of matter'* (Edward Arnold, 1963)
5 PARKER, D.F. and MAUGIN, G.A. (Eds.): *'Recent developments in surface acoustic waves'* (Springer Verlag, 1988)
6 ACHENBACH, J.D.: *'Wave propagation in electric solids'* (North-Holland, 1973)
7 'Component performance and system applications of surface acoustic wave devices'. Proceedings of IEE International Specialist Seminar, 1973 (Conference publication 109)
8 BARTH, P.W.: 'Silicon sensors meet integrated circuits', *IEE Spect.*, 1981 (9) p. 33
9 MIDDLEHOEK, S. and AUDET, S.A.: *'Silicon sensors'* (Academic Press, 1989)
10 PETERSEN, K.E.: 'Silicon as a mechanical material', *Proc. IEEE*, 1982, **70** (5), pp. 420–457
11 YOUNG, M.: *'Optics and Lasers'* (Springer Verlag, 1992)
12 LUCKOSZ, W.: 'Integrated optical nanomechanical devices as modulators, switches, and as tunable frequency filters, and as acoustical sensors' SPIE, 1973, Integrated Optics and Microstructures, 1992
13 MOREY, W.W.: 'Distributed fibre grating sensors' Proceedings OFS 90, pp. 285–289, 1990
14 KERSEY, A.D., BERKOFF, T.A. and MORSEY, W.W.: 'Fibre-grating based strain sensor with phase sensitive detection' Proceedings 1st European Conference on smart structures and materials, pp. 61–62, 1992
15 MICHELIN, J.L. and MASSON, P.: 'Strapdown inertial systems for tactical missiles using mass unbalanced two-axis rate gyros' AGARD Lecture Series No. 133, 1984

Chapter 7

Testing, calibration and compensation

7.1 Introduction

The evaluation testing of inertial sensors is required to establish their suitability for a given application, i.e. to ensure that they satisfy all the performance requirements of that application.

Inertial sensors and systems are designed and manufactured for a very wide range of applications which includes the provision of extremely precise navigation, in ships and submarines for example, and measurements for flight control of short time-of-flight missiles. The performance required for such diverse applications spans eight or nine orders of magnitude. Similarly, the environment in which the sensors and system are required to operate varies widely, from the potentially more benign maritime applications to the very high dynamic forces experienced by highly agile surface to air and air to air missiles travelling at supersonic or hypersonic speeds.

Testing and calibration methods need to reflect the type of application and also, but very importantly, the environment in which the sensors and systems are to operate. It is crucial to establish that the sensors not only survive and operate reliably whilst being subjected to the vibrations, shocks and accelerations induced by the host vehicle, but also have sufficient endurance and resistance to survive the testing and calibration procedures.

It is possible to represent the behaviour or performance of an inertial sensor by the use of a mathematical expression as described in Chapters 4, 5 and 6. One purpose of testing an inertial sensor is to evaluate the coefficients of these equations, the various error terms, so that the performance of a sensor can be predicted for particular circumstances. Having established the performance figures, or characterised the sensor, any systematic errors may be compensated for, thus enhancing its accuracy. Other purposes of testing are to enable the output signals to be calibrated and to understand the behaviour of the device in various situations and environments.

Although often neglected, the testing and calibration of sensors and systems has great importance with very significant consequences, in terms of cost and performance of the host vehicle, if the testing is either inadequate or too demanding. As a consequence of the various requirements and demands of a test programme, there are many different approaches that can be followed.

There are typically three distinct categories of testing undertaken on sensors: qualification, acceptance and reliability tests. The qualification tests tend to be the most extensive and stringent tests to which a sensor is subjected. These tests usually precede the production of a sensor by a manufacturer and are intended to show that a particular design will meet the requirement of a customer with adequate margins for production tolerances. They are likely to include all the investigations discussed in the sections that follow. Acceptance tests are undertaken on sensors during production in order to check selected parameters and to establish data for the calibration of the sensors. Tests may be undertaken on each sensor produced or on a sample batch selected from a production run, the number tested being determined by some statistically based rules. Reliability testing usually involves a sample of sensors selected at random from a production batch and run under normal operating conditions in order to establish the mean time between failure.

7.2 Testing philosophy

Depending on the sensor or system to be investigated, and the form of evaluation required, use may be made of either static or dynamic test methods. In the case of a static test, the device is kept fixed and the response to some natural effect or phenomenon observed. For example, the specific force due to the Earth's gravity could be observed with an accelerometer in various orientations. When dynamic testing is undertaken, the sensor under test is moved and the response of the device to that disturbance is monitored and compared with the stimulus.

A three stage process may be followed to characterise the performance of a sensor or system:

(i) coarse checking or evaluation using very simple tests, such as a single stationary position test on a bench, to establish that the response is compatible with the designer's or manufacturer's predictions;

(ii) static testing and/or calibration to derive performance parameters of the device from multi-position tests as defined in Sections 7.5/6;

(iii) dynamic testing where the device under test is subjected to motion such as an angular rotation or linear movement with acceleration. This form of testing requires specialised test equipment such as a rate table or vibrating table.

Inertial sensors and systems are subjected to different testing and calibration schemes throughout their development. At the prototype or initial research and development phase, the testing strategy will be designed to estimate the boundaries of performance of the device, establishing what is it good for, without breaking it, as prototype devices are usually very expensive and in short supply! Depending on the type of sensor, the testing may well be on a bench rather than in a sophisticated test laboratory. Tests will normally be arranged so that only one of the vast range of environmental stimuli is changed during any one series of tests, to enable the response of the sensor to be understood and characterised. Throughout this and subsequent stages of testing, it is crucial that an accurate log is kept to record details of all tests and the results obtained.

Many projects have a so-called integrated test plan. The idea is to have a structured plan to determine the performance and reliability of a system using the minimum amount of testing, as this is often time consuming and expensive. Consequently, in the structured test plan, the data gathered from the qualification and acceptance evaluation tests can form an input to the integrated test plan, particularly if data are recorded on run-times and any variation in performance with time. All these data can help estimate the potential reliability and mean time between failures for this system.

As an inertial sensor or system progresses through its research and development phase, the testing becomes more intensive in order to investigate progressively more of its performance envelope. Generally, the tests are undertaken in a sophisticated test laboratory, specialised test equipment being necessary for the more accurate evaluation of sensors. Some units will be tested to destruction, but this will usually be in a very controlled and deliberate manner. It is important to evaluate as many sensors as possible in order to establish the confidence limits of the parameters measured in the tests.

When the inertial sensors are being produced in significant quantities, the manufacturer's testing is usually to establish that the sensors are conforming to the production specification, and also to enable the sensor to be calibrated. In the case of large production quantities where there may be a possible significant range to the performance of the individual sensors, the testing can be used to grade the sensor performance, and hence direct the sensors to the appropriate application. This technique has been applied with success by many manufacturers

The philosophy of acceptance testing is usually to establish that a sensor, or class of sensors, but more generally, an inertial system is compatible with the host vehicle. These tests will establish that the device, in whatever form, will operate satisfactorily within the vehicle and not jeopardise its integrity or safety. It will also be established that the device will fit its specified location and that it will achieve the required accuracy. Generally, the device will undergo very specialised testing to establish the desired compliance with the objectives of the project application.

Various standards exist [1–9] to ensure the use of common terminology for users and manufacturers. Consideration is also given to recommended procedures for testing.

7.3 Test equipment

Over the years, during the development of inertial sensors and systems, many different testing methods and procedures have been perfected [2–9]. Many, if not most, require very specialised, accurate and costly equipment, housed in laboratories that are often built specifically for this special requirement. Special foundations are normally required so that the test equipment can be isolated from shocks, vibrations and other perturbations induced by the local environment. The temperature of the environment, as a rule, is also carefully controlled, although environmental cabinets are generally used for thermal cycle testing. The application of the digital computer to the field of testing, both for the control of equipment and the analysis of data collected during the testing, has advanced the testing procedures significantly during the last two decades.

It is important not only to match the testing schedule and procedures to the requirements of the application, but also to ensure that the test equipment is of sufficient accuracy and precision to be compatible with the desired test accuracies. This enables a given and known stimulus to be applied to the inertial sensors or systems and their response observed. Generally, the test equipment should be capable of measuring a given quantity to an accuracy significantly in excess of that required by the sensor—a factor of between five and ten is typical. Similarly, the data collection system and the algorithms manipulating those data must be compatible with the anticipated accuracy of either the sensor or its application. Additionally, the test equipment should be calibrated regularly where necessary, and investigated to ensure that it is providing the desired stimulus or disturbance to the device under test.

An example is a vibratory platform which should provide oscillatory motion along an axis which should be, for example, normal to the horizontal plane. As the equipment degrades with use, it is possible for the axis of motion to move about a cone. Thus any gyroscope under test may detect this angular motion, depending on its accuracy and alignment, and therefore produce an output signal reflecting its detection of this motion. If the investigator was not aware of the defect in the test equipment, then the gyroscope would be reported to have a particular vibration sensitivity, which of course had been induced by a defect in the test equipment. Similarly, this defect may be masking or compensating a real sensitivity in that gyroscope.

7.4 Data logging equipment

The electrical signals produced by inertial sensors or inertial systems can be in various forms, such as direct current or alternating current, and may be in continuous or pulsed form. The form of the signals is dependent on the types of sensors, their pick-offs and the nature of any rebalance loops used. Depending on the signal being monitored and the accuracy being required, a chart recorder or some form of digital meter may be used. Chart recorders are often used to show trends in sensor performance.

The other technique, most common for accurate work, is to use either a digital voltmeter or a digital ammeter to monitor the signals. The signals are often integrated over a period of time, its length depending on the testing, and then logged in a suitable form, usually through a data bus to a computer. For many years small computers have played a most important role in the control and conduct of tests, as well as the collection of data, their subsequent manipulation and ultimate presentation.

The rules concerning the accuracy of the data monitoring and recording equipment are similar to those for the test equipment as outlined in the previous section. It is vital that the resolution of the whole of the chain of data recording equipment exceeds that given by the sensors being tested, preferably by about an order of magnitude or more. Similarly, the dynamic range and stability of the monitoring and recording equipment, including any data buses, should exceed those of the instruments to be tested. Care has to be taken to ensure that various transient effects are not masked by inadequate data logging equipment or during the subsequent data manipulation.

7.5 Gyroscope testing

The gyroscopes to be tested are usually mounted in a test fixture, often a cube with very accurately machined faces for achieving very precise mounting in the test equipment. This enables the sensor to be transferred between the various pieces of test equipment used in a test programme and maintaining its mounting accuracy precisely; it also allows various designs of sensor to be tested on the same equipment. For the more accurate sensors, i.e. those with a bias in the region of 0.01°/hour, mounting accuracies of the order of ten arc seconds are required.

Prior to undertaking a series of tests for evaluating the performance of a gyroscope, it is usual to undertake some preliminary investigations. Such checks include measurement of electrical resistance and insulation strength, polarity, time for the rotor to reach its operating speed, time to stop rotating and power consumption.

The IEEE has published a number of documents [2–6] defining procedures for testing various types of gyroscope. The following text concentrates on general test techniques for strapdown gyroscopes, examples of test equipment and methods of analysis enabling the performance of an instrument to be evaluated.

7.5.1 Stability tests—multi-position tests

The purpose of stability testing is normally to evaluate the run-to-run, or switch-on to switch-on, drift/bias and in-run drift of a gyroscope. The gyroscope is placed on a block in a series of fixed orientation with respect to geographic axes and the local gravity vector of the Earth. Alternatively, the gyroscope may be positioned accurately in the required orientation using a Graseby table, a stabilised gimbal system with controlled and instrumented gimbal angles, as shown Figure 7.1.

For more accurate work the table, or gyroscope under test, is mounted on a plinth of granite that has its own foundations separate from, and vibrationally isolated from, the laboratory.

Figure 7.1   Graseby table
Courtesy of DRA, Farnborough

The gyroscope is operated within a fixed temperature range and is positioned in one of a set of eight (possibly up to 12) standard orientations as indicated in Table 7.1. The gyroscope is switched on and after a predetermined time, to allow thermal transients to settle, the signals from the gyroscope are recorded. This test is repeated a number of times, the number and duration being dependent on the required accuracy of the test, determined by the usual statistical rules for confidence levels [10–12], with a predetermined time for cool down between runs.

For sub-inertial quality sensors a set of runs in one test in this series may last for up to one hour; for higher quality sensors, the duration may be many hours. On completion of the series of runs with the gyroscope in a particular axial setting, a new orientation is selected and the test procedure repeated. From the different data sets, the effects of various systematic errors, and the effects of the rotation of the Earth, can be removed. The data are analysed to find the mean drift rate and its variance, or scatter, about this value, for each test run.

The run-to-run stability of the gyroscope is evaluated from the scatter in the mean level of gyroscope output for each run recorded during these tests when the gravity vector was not coaxial with the input axis of the gyroscope.

The in-run stability of the gyroscope drift rate is deduced from the average scatter of the measured drift in the output of the gyroscope about its mean value during a run, calculated for each test in a series. This value can be averaged over the whole series of tests, but care must be taken to avoid the inclusion of any data that have any anomalies without establishing the reasons for the anomalies. Figure 7.2 shows the form of the raw data collected over a period in a single run.

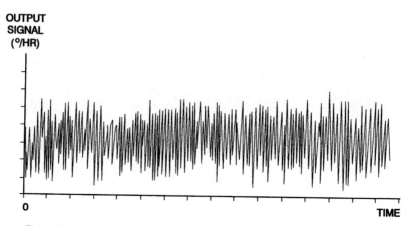

Figure 7.2   Raw data collected during stability testing

Table 7.1 illustrates typical axial settings for a dual-axis gyroscope used during stability tests, it also shows the components of gravity along each vector axis and the rotation of the Earth about each axis of the gyroscope.

Table 7.1 Gyroscope axial settings for stability tests

n, s, e, w are horizontal directions north, south, east and west
Ω = Earth's rate, 15.041°/hour
L = local angle of latitude
$\Omega \cos L$ = horizontal component of Earth's rate
$\Omega \sin L$ = vertical component of Earth's rate

Position	Direction of axis			Acceleration of axis			Component of Earth's rotation along axis	
	spin	input 1	input 2	spin	input 1	input 2	input 1	input 2
1	up	n	w	+1	0	0	$\Omega \cos L$	0
2	up	w	s	+1	0	0	0	$-\Omega \cos L$
3	up	s	e	+1	0	0	$-\Omega \cos L$	0
4	up	e	n	+1	0	0	0	$\Omega \cos L$
5	n	e	down	0	0	-1	0	$-\Omega \sin L$
6	n	up	e	0	+1	0	$\Omega \sin L$	0
7	n	w	up	0	0	+1	0	$\Omega \sin L$
8	n	down	w	0	-1	0	$-\Omega \sin L$	0

The Earth's gravitational force is taken to be an apparent acceleration acting vertically upwards, and this direction is positive in the convention used here. Data from these tests are used to solve a set of simultaneous equations expressing the gyroscopic drift or bias associated with each external stimulus in each position.

For example, from this series of tests it is possible to establish the acceleration dependence of the gyroscopic bias (g-dependent bias) by comparing the mean levels of signals for the cases when the sensor is mounted with its input axis orthogonal to and coaxial with the gravity vector. This is illustrated in the following analysis.

Under static conditions, assuming allowance has been made for Earth's rate component, the output of the gyroscope (ω_o) may be expressed as:

$$\omega_o = B_f + B_{gx}a_x + B_{gy}a_y + B_{gz}a_z \qquad (7.1)$$

where B_f is the g independent bias and B_{gx}, B_{gy}, B_{gz} are the g-dependent biases induced by accelerations a_x, a_y and a_z acting along the x, y and z axes of the sensor respectively. For the conventional rate integrating gyroscope, these correspond to the spin, input and output axes of the gyroscope.

If measurements are taken with the gyroscope positioned with its x axis coincident with the gravity vector pointing up and down, then the measurements obtained, m_1 and m_2, may be expressed as follows:

for input axis up: $m_1 = B_f + B_{gx} g$

for input axis down: $m_2 = B_f - B_{gx} g$ (7.2)

The coefficients B_f and B_{gx} may be calculated from the sum and difference of these two measurements. Similarly, the g dependent bias coefficients may be determined by taking sets of measurements with the y and z axes of the gyroscope aligned with the gravity vector.

7.5.2 Rate transfer tests

The purpose of these tests is to investigate the various characteristics of the scale factor of the gyroscope which relate the output signal to the input motion, and the maximum and minimum angular rotation rates which the gyroscope can measure or capture. A schematic representation of this characteristic is shown in Figure 7.3. This diagram also shows how the scale factor, resolution, dead band and threshold are defined.

The usual characteristics of the scale factor that are evaluated in these tests are:

- its mean value and scatter about this value;
- the change in mean value as the angular rotation rate of the rate table changes, i.e. its linearity;
- the variation of the mean value of the scale factor as the ambient temperature changes;
- any hysteresis in the response of gyroscope.

In the case of two-axis gyroscopes, such as the dynamically tuned gyroscope, this form of testing can also assess the orthogonality of the two input, or sensitive, axes. These tests are undertaken using a precision turntable called a rate table, as shown in Figure 7.4.

The equipment has a flat circular plate on which the gyroscope can be mounted. The plate can be rotated about an axis normal to the surface at very precise angular velocities, which are monitored and controlled very accurately. The angular velocity can often be varied from a fraction of one degree per hour to many hundreds of degrees per second. Manufacturers currently offer rate tables with maximum angular rates of about 3000°/s. The equipment is usually mounted with its axis of rotation vertical on a granite plinth to provide isolation and stability, and those operating at very high rotation rates are often

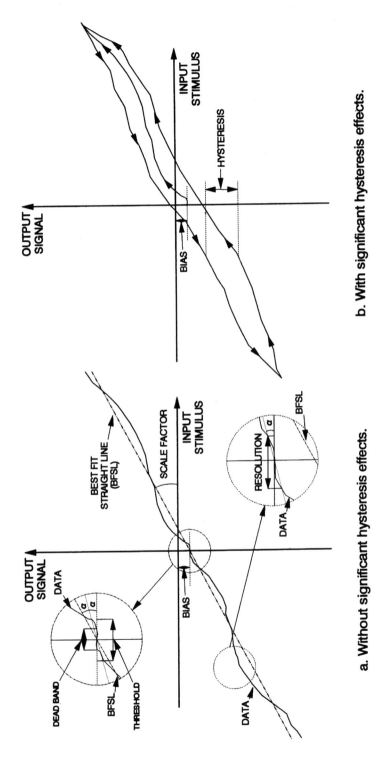

b. With significant hysteresis effects.

a. Without significant hysteresis effects.

Figure 7.3 Scale factor characteristic of a gyroscope

Figure 7.4   Rate table with a controlled environmental chamber
Courtesy DRA, Farnborough

caged for safety reasons. An environmental chamber may enclose the
test table for thermal evaluations, see Section 7.5.3.

During rate transfer tests, the gyroscope is mounted securely on the
turntable with its sensitive axis (or one of its sensitive input axes in the
case of a twin-axis sensor) parallel to, although not necessarily coaxial
with, the axis of rotation of the rate table. The rate table can be used
in many ways, but the basic principle is to compare the angular rate or

displacement measured by the gyroscope with that given by the rate table. Care is necessary to ensure that the rate table does not overshoot its intended angular rotation rate for any given measurement. Examples of the way in which the rate table may be used are:

1 During a typical test schedule, the rotation rate of the rate table is stepped through a series of angular rates starting at zero, recording data at each stage. The rotation speed is kept constant for a set period at each step and the sensor outputs allowed to stabilise, before recording the output signals. The applied angular rate is varied in incremental steps between the maximum and minimum desired rotation rates as shown in Figure 7.5. At each step, the signals from the gyroscope are recorded when the sensor is in equilibrium.

By cyclical variation of the applied rate over the expected measurement range of the sensor, this form of test allows hysteresis effects in the scale factor to be observed, sometimes called the 'bow-tie' or 'butterfly' effect owing to the shape of the plot of the error in the indicated rate from the gyroscope against the actual turntable rotation rate, as shown in the example plot in Figure 7.6.

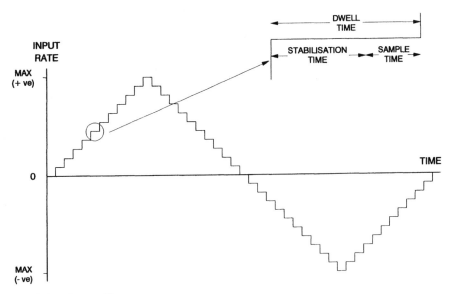

Figure 7.5   Rate table step sequence

2 This is essentially a variation of method 1 in which the rotation rate is stepped rapidly from rest to the maximum rotation rate in one direction, then slowed down in steps until at rest and stepped to the

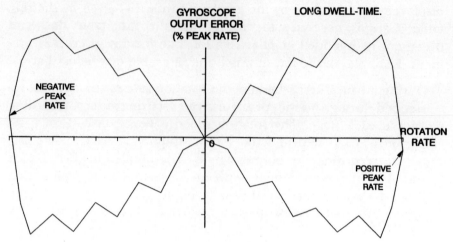

Figure 7.6   Sample data from a rate transfer test

maximum rotation rate in the opposite direction, and then brought to rest in steps. In this case, the dwell at each step is a few hundred milliseconds to allow the turntable to stabilise and allow the output signals from the gyroscope to be recorded. Collecting data at the correct moment is crucial for this form of test, it being co-ordinated by the controlling computer. This form of rate transfer test is often used when assessing gyroscopes for tactical missile applications as it can be completed in times that are comparable with the time of flight of the missile. This so-called rapid rate transfer test can be completed, in some circumstances, in as little as ten seconds. Typical data from such a test are shown in Figure 7.7.

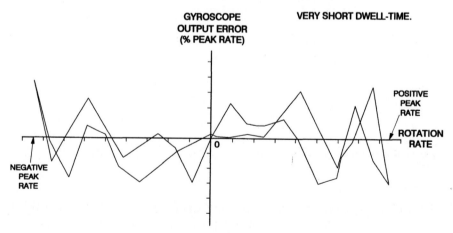

Figure 7.7   Sample data from a rapid rate transfer test

3 The rate table is set to a constant angular rate and data 3 The rate table is set to a constant angular rate and data collected for a given number of complete rotations which are monitored by optical means to give high precision read-out and control. This method is used for precision gyroscopes, such as the ring laser gyroscope, which are often as precise as the normal rate table control systems.

Orthogonality of the axes of a two-axis sensor may be investigated by recording data during the above tests from the second input axis which is nominally at right angles to the axis of apparent rotation. When the data from this axis have been corrected for systematic errors, the residual rate is a result of the input axes not being orthogonal.

Data from the rate transfer tests are normally analysed by comparing the output signal from the gyroscope with the corresponding turn rate of the table, usually measured by a tacho-generator. This process is repeated for all the data collected in a test sequence and a straight line is constructed through the data using a least squares fit procedure. The gradient of this line is the scale factor of the gyroscope. In order to take account of any non-linear trends, a curve may be fitted to the data. This is represented mathematically by a polynomial expression, the coefficients of which define the scale factor non-linearity of the gyroscope. Typical plots of the processed input to output characteristic are shown in Figures 7.6 and 7.7. These curves show deviations from the theoretical linear input/output plot. Although somewhat idealised, they illustrate the influence that time has on the recorded output signals from a gyroscope. With a long dwell-time, there is a sufficiently long period for various phenomena to reach equilibrium, particularly thermal effects. Further, it is usually possible to observe any hysteresis in the torquer scale factor with the longer dwell-time, whereas this is less evident for the very short measurement period.

Gyroscopes may be calibrated using the data provided by the rate transfer tests. The output signal(s) from the sensor are compared with the accurately known applied rotation rate and the scale factor defined, for instance, as so many milli-volts per degree per second of rotation rate.

7.5.3 Thermal tests

Variations in the performance of a gyroscope with changes in the temperature within the case of the gyroscope can be observed by enclosing the turntable in a climatic chamber, as shown in Figure 7.4.

The temperature within the chamber can be varied from sub-zero temperatures, typically down to about −55°C, to temperatures in the region of 75 or 80°C. Cooling is normally achieved by using carbon dioxide, so care must be exercised in venting the used gas. A schematic diagram of a typical experimental arrangement is shown in Figure 7.8.

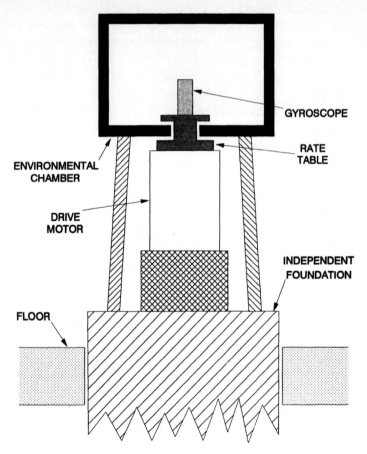

Figure 7.8   Illustration of thermal test equipment

There are various thermal tests that can be undertaken such as allowing the temperature of the gyroscope to stabilise, i.e. a soak test, or allowing a controlled increase or decrease over a given period, i.e. a thermal ramp test. The rate tests described above are repeated at the various temperatures and the output signal from the gyroscope recorded. Using this technique, the scale factor can be evaluated at various temperatures throughout the operating range of the sensor. A typical plot is shown in Figure 7.9. Any correlation with temperature variation can be defined by a mathematical expression which may be stored in a computer and used to provide on-line compensation for temperature variation, provided a thermal sensor is supplied with the gyroscope. Care must be taken when testing mechanical gyroscopes to identify so-called 'tombstone' effects. These are large biases caused by stiction occurring within narrow temperature bands.

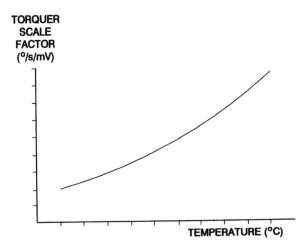

Figure 7.9   Typical variation of torquer scale factor with temperature

Variations in other performance parameters of the gyroscope with temperature can be evaluated similarly by enclosing the gyroscope in a climatic chamber for a specific test, such as a multi-position test, and recording the response of the gyroscope. Care has to be taken so that the imposition of thermal variation does not affect the performance of the test equipment itself.

7.5.4 Oscillating rate table tests

The purpose of these tests is to determine the frequency response characteristics of a gyroscope and its associated electronic control circuits to oscillatory rotation applied to the input axes of the sensor. Normally, both the bandwidth and the natural frequency of response of the sensor are evaluated in this test. The test equipment is very similar to the rate table already described for the rate transfer tests. In this case, the turntable, again mounted on a suitable plinth to provide stability, applies an oscillatory angular motion at various preset frequencies and so requires low inertia about the rotation axis. A photograph of an oscillating rate table is shown in Figure 7.10.

Figure 7.10   Oscillating rate table
Courtesy of DRA, Farnborough

The gyroscope to be evaluated is mounted and fixed to the turntable with its sensitive axis parallel to the axis of rotation of the turntable. A given maximum rotation rate is selected along with a maximum frequency of oscillation. The frequency of oscillation is increased in pre-determined steps up to the maximum value. The response of the gyroscope is recorded at each frequency step in the series, up to the maximum value, and then as the frequency is reduced to the starting condition. The tests are generally repeated for various input rate maxima.

Results from these tests are usually plotted as gain and phase graphs by comparing the amplitude and the phase of the signal generated by the gyroscope with the actual amplitude and phase of the disturbance applied by the turntable. A typical response is shown in Figure 7.11 from which the bandwidth and damping factor can be deduced.

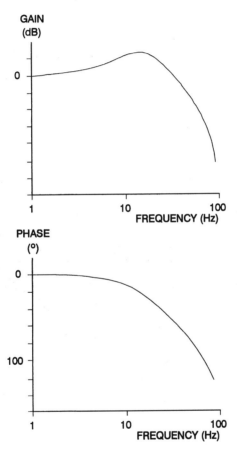

Figure 7.11   Typical gyroscope frequency response

Such information is needed to determine the speed of response of the sensor. This is particularly crucial in applications where the sensor is to be used to provide feedback control, as in a missile autopilot for example.

7.5.5 Magnetic sensitivity tests

The purpose of these tests is to examine and quantify any influence that external magnetic fields may have on the drift characteristics of a sensor. The form of the testing is identical to the multi-position testing described earlier, but with the addition of a pair of Helmholtz coils, as shown in Figure 7.12, which may be positioned to apply a magnetic field along each of the principal axes of the sensor under test.

Figure 7.12   Helmholtz coils
    Courtesy of DRA, Farnborough

Initially, data from the gyroscope are recorded for a given period, with the sensor aligned to a particular geographic orientation and with the coils unpowered. This test is then repeated with the appropriate current passing through the coils to generate the desired magnetic field strength. Generally, a whole series of measurements is made for different orientations of the magnetic field with respect to the axes of the gyroscope in which the magnetic field strength is incremented in suitable steps up to a maximum value. It is usual to keep the gyroscope running during these tests to eliminate any switch-on to switch-on effects, and similarly the gyroscope is allowed to stabilise before the first measurements in the series are recorded.

The recorded data are analysed to establish the mean value of the bias for each series of tests. Comparison is made between the mean value in the presence and absence of a magnetic field for each orientation of the magnetic field to establish the magnitude, if any, of the dependence of the bias on the strength and orientation of an applied magnetic field. Typical sensitivity of a conventional gyroscope is illustrated in Figure 7.13. It is important that the material chosen for the gyroscope's mounting block does not change the magnetic field applied to the sensor.

7.5.6 Centrifuge tests

A centrifuge provides a means of applying large steady or fluctuating accelerations to a gyroscope. A photograph of a centrifuge is shown in Figure 7.14. The purpose of the centrifuge testing is to investigate the response of the gyroscope to large accelerations and to establish its ability to withstand large continuous or fluctuating accelerations whilst the sensor is either operational or in a quiescent condition.

These tests are normally one of the last in a series of tests applied to an inertial sensor, as great care has to be taken to prevent permanent damage to the sensor. This is particularly true when the extremes of the gyroscope's performance envelope are being investigated. Care must be taken to align the sensitive axis, or axes, so that the maximum rotation rate of the gyroscope is not exceeded and allowance has to be made for the applied input rate which is sensed by the gyroscope. Alternatively, the sensor can be mounted on a counter-rotating table to null the effects of the rotation of the centrifuge. In this case, the applied acceleration is sinusoidal at the table rotation frequency. The magnitude of the acceleration applied along each axis of the gyroscope can either be controlled by the rotation speed or the distance of the sensor from the centre of rotation, the radius arm. The addition of an accelerometer adjacent to the gyroscope under test provides an accurate measure of the acceleration acting on that sensor.

To investigate the acceleration sensitivity of a gyroscope, the applied acceleration is increased in steps until a given maximum level is

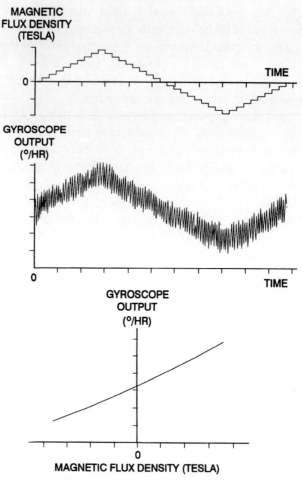

Figure 7.13   Illustration of magnetic sensitivity

reached. This process is repeated with the gyroscope mounted in different orientations in order to investigate its sensitivity to acceleration along different axes. Comparison is then made between the mean values of the output signals produced by the gyroscope for each acceleration applied to it, and hence the acceleration sensitivity of the sensor can be calculated. This value can be compared with the value deduced from the multi-position tests described earlier, where the acceleration sensitivity is evaluated at a low value of acceleration.

In the case of quiescent testing, a particular parameter of the gyroscope, or a range of parameters, are evaluated prior to the test. The sensor is subjected to the desired steady state acceleration, or acceleration profile, as described above, but with the sensor in an

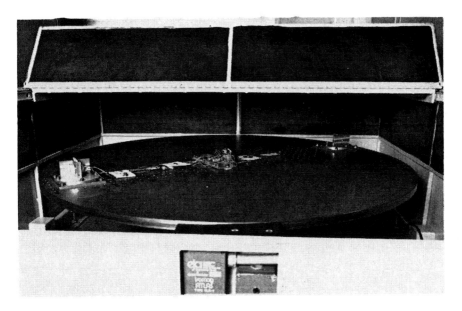

Figure 7.14   Photograph of a centrifuge used for sensor testing
Courtesy of DRA, Farnborough

unpowered state. The gyroscope is then re-evaluated to establish if any change in performance has taken place.

7.5.7 Shock tests

The purpose of this form of test is to measure the response of the gyroscope to an applied shock, and to establish the resilience of the sensor to such an applied acceleration over a very short duration, typically of the order of milliseconds. As in the case of the centrifuge tests, the gyroscope may be operational or quiescent during the test.

A shock may be applied to a gyroscope by either using a shock table or a vibrating table. When using a shock table, the sensor is mounted on a heavy metal table and this table is dropped over a given distance on to a suitably shaped piece of lead. In the latter case, a short duration single displacement is applied to the vibrating table on which the sensor is rigidly mounted [13].

In order to measure the response of a gyroscope in an operating condition to an applied shock, the sensor is suitably oriented on the table of the test apparatus and very firmly bolted to it. The output signals from the sensor are recorded for a given period before the shock is applied. If possible, the output signals are recorded during the application of the shock and then also for a period after the shock.

Comparison with the mean value of the gyroscopic bias before and after the application of the shock will indicate any transient or permanent change in the characteristic of the gyroscope.

When a gyroscope is tested in a quiescent state, it is normal to follow the procedure already described above for the quiescent centrifuge test. Again, a comparison is made of suitable characteristics of the tested sensor before and after the application of the shock acceleration.

7.5.8 Vibration tests

This is normally the last series of tests undertaken on a gyroscope owing to a potentially high risk of permanent damage to the sensor. Great care must be taken to ensure that resonances in the mounting fixture do not amplify the applied accelerations. It is strongly recommended that a thorough investigation of the modal behaviour of the structure used to mount the gyroscope on the vibrating table is undertaken before any investigations of random vibration are started.

The purposes of this form of testing are usually four-fold:

(i)   investigation of the frequency at which any resonant responses of the sensor occur and their magnitude;
(ii)   evaluation of the anisoelasticity or acceleration squared (g^2) bias dependency of the gyroscope;
(iii)   examination of the resilience and survivability of the sensor in a particular vibratory environment; the sensor may be either quiescent or operational, depending on the reason for the test;
(iv)   estimation of the change in noise characteristic of the output signals of a sensor experiencing a vibratory environment.

A vibration table provides a means of applying various forms of vibratory motion to a sensor. There are several different forms of such equipment. Usually, the vibrator has a table on which the sensor is mounted. The table is driven along a well defined axis by an electromagnet field with a current of the required frequency and wave shape. The table may oscillate vertically or it may be rotated through 90° and be attached to a slip-table and vibrated horizontally. A typical vibration table is shown in Figure 7.15.

Two forms of motion may be applied to the test table, either sinusoidal vibrations or random frequency vibrations. In the case of the former type, the displacement of the table is varied sinusoidally over a given frequency band without exceeding a pre-defined

Figure 7.15   *Photograph of a vibration table used for sensor testing*
      Courtesy of DRA, Farnborough

acceleration level. In the latter case, random vibration is applied according to a given power spectral density and frequency bandwidth.

Throughout testing with a vibrator, it may be necessary to use degaussing coils as some vibratory test equipment generates very large magnetic fields. These fields can alter the 'drift' performance of a gyroscope as indicated by the magnetic sensitivity testing of Section 7.5.5. Thus, care must be taken to ensure that genuine effects are being observed and analysed correctly, not merely artifacts introduced by the test equipment.

It is good practice to carry out an initial series of tests using a low level of acceleration with the gyroscope operating, in order to establish if any resonances arise within the sensor and the frequencies at which they occur, known as a resonance search. This enables the investigation of the sensor to be undertaken at frequencies well away from those that will excite any of these resonances. The sensor is suitably orientated and firmly clamped to the vibration table. During these tests, a small 'feedback' accelerometer is mounted on the test sensor or its mounting fixture to provide a measure of the accelerations applied to the sensor. The vibrating table is used in a feedback mode to ensure that a fairly even acceleration amplitude is applied to the sensor under test. A small peak acceleration, in the region of 1g, is chosen and a sinusoidal displacement is applied to the sensor by the vibrating table. The vibration frequency is changed slowly from an initial value of a few Hertz to an upper limit, normally in the region of 10 kHz. This is often called a sine sweep. During this sweep, the output signals from the gyroscope are monitored continuously to ensure that any resonances encountered do not destroy the sensor, and the frequency at which they occur is recorded. This test can be repeated for other orientations of the gyroscope to establish resonance free frequency zones.

When testing mechanical gyroscopes, it is usual to mount the sensor with two of its axes at 45° to the axis of table movement in order to maximise the effect of any anisoelastic biases and to allow them to be identified. In this case, the third axis of the gyroscope is perpendicular to this axis of motion and aligned with one of the cardinal geographic axes. The gyroscope is tested to establish the mean and standard deviation of its 'drift' in this selected orientation over a suitable period. Sinusoidal motion is applied to the sensor at a frequency of oscillation selected well away from any resonances detected during the preliminary sine sweep tests. A pre-selected maximum acceleration is applied and the output signals of the gyroscope are collected over the period of oscillation, typically a duration of many minutes. Such tests are repeated with different peak accelerations at a number of spot frequencies. The statistics of the gyroscope data collected during these tests are evaluated and a comparison made of the mean bias in the absence of vibration, to allow the acceleration squared dependency of the sensor to be evaluated. The anisoelastic bias coefficient may be calculated by finding the averaged increase in the output bias caused by the vibratory motion at each frequency and dividing it by the appropriate square of the applied acceleration. The various values of this bias can then be averaged to give the estimate of the acceleration squared (g^2) bias.

Quadrature effects are deduced by repeating the above tests, but with the gyroscope orientated with the input axis perpendicular to the axis of motion. Noise introduced into the output signals of the gyroscope can be analysed by examination of the statistics calculated from the signals produced by the sensor during its vibratory tests and comparing them with those deduced when it was stationary before and after testing.

To test the endurance, survivability and resilience of a sensor, it is usual to use random motion of the vibration table. The tests are undertaken, as described above, but the frequency and amplitude of the motion are continuously varied randomly across the spectrum defined by the frequency bandwidth of the power spectral density used to specify this form of vibration. This power spectral density defines the maximum acceleration that the sensor mounted on the vibrator will experience at any frequency. This is the so-called random vibration testing.

Depending on the aspect of application of the sensor that is being investigated, the sensor may or may not be operational. If, for example, its susceptibility to damage during transportation is being investigated, the sensor may be vibrated for weeks, or even months, whilst in a quiescent state. In this case the sensor would be characterised before and after the tests, to establish any changes in performance, as described above, for quiescent centrifuge or shock testing. If a particular regime of flight of a tactical missile is being examined when, for example, the dynamic forces are high, the sensor will be operational, but the duration of application of the appropriate vibration spectrum may be short, possibly only ten seconds or less. During this form of testing, the output signals from the gyroscope would be recorded before and after the vibratory tests to establish the significance of any observed changes in the response of the sensor during the vibratory tests.

Care must be exercised when evaluating mechanically dithered ring laser gyroscopes in order to prevent any unrepresentative interaction between the vibratory motion and the sensor leading to spurious bias errors. It is important that the sensor is attached to a rigid base plate and also that these sensors are tested individually.

7.5.9 Combination tests

Having analysed the response of a sensor to a single disturbance such as rotation or acceleration, it may be appropriate to undertake some limited evaluation of the sensor by combining, for example:

(i) rotation and acceleration by mounting a 'rugged' rate table on a large centrifuge so that the gyroscope can be rotated about its input axis whilst an acceleration is applied along that axis;

(ii) vibration and acceleration by mounting a small vibrating table on a large centrifuge enabling an acceleration to be applied along an axis while vibrating the sensor about an orthogonal axis or axes;

(iii) rotation and vibration by mounting a rugged rate table on a vibrating table and either applying a vibratory motion along one axis whilst simultaneously applying a rotational disturbance, or applying those disturbances about orthogonal axes.

Various other combinations are possible including the application of shock accelerations or thermal cycling the sensor during these combination tests.

This form of testing is usually valuable for assessing the response of the gyroscope when experiencing the total environment to which it is likely to be subjected in the host vehicle. It is useful to compare the response of the sensor resulting from the combination of stimuli with that expected from the sensor when experiencing the individual stimuli, such as the linear acceleration or rotation used in the combination test. Any anomalies can then be investigated in the laboratory and the actual basis of the behaviour established before the ultimate combination test series takes place, such as actual flight trials. Care must be taken to ensure that combinations of disturbances applied in the laboratory are realistic and not beyond the capability of the host vehicle, and consequently beyond the specified capability of the sensor.

Flight trials allow a qualitative assessment rather than a definitive analysis, owing to the difficulty of defining all the input stimuli very precisely. It is not normally possible to define the stimuli with the same precision as for the normal single stimulus testing such as rotation or multi-position testing.

7.5.10 Ageing and storage tests

Many project applications require a gyroscope to have a shelf life of many years, fifteen is not untypical, and still provide performance within specification at the end of that period. It is not easy to provide realistic accelerated ageing or storage tests other than through multiple thermal cycling and enhanced vibration testing to simulate transportation. One method of evaluating ageing is by analogy or reference to the behaviour of similar known components in other devices, such as permanent magnets and establishing the change in magnetic flux expected over the life of the sensor. This would then enable the designer to allow a suitable margin for the anticipated degradation in performance.

A further method for evaluating ageing is to characterise a large number of gyroscopes using, say, multi-position tests and then storing them in a typical environment. Periodic withdrawal of a few, at say one or two year intervals, for re-evaluation should provide a guide to the form of ageing or change in performance of the sensor and thus allowing a comparison with the predicted changes.

7.6 Accelerometer testing

The performance of an accelerometer is usually investigated using a series of static and dynamic test procedures similar to those already described for a gyroscope. However, a reduced scale of testing is required to characterise the performance of this device [14]. For example, rate table testing is generally unnecessary and the multi-position tests are undertaken using a precision dividing head. This piece of equipment, which usually has a setting accuracy of about one second of arc, enables the sensitive (input) axis of an accelerometer to be rotated with respect to the gravity vector. Hence, the component of gravity acting along the input axis of the sensor may be varied very precisely. A photograph of such equipment is shown in Figure 7.16.

When a pendulous accelerometer is under test, care should be used when mounting the sensor to ensure that whenever possible the hinge axis is not vertical. This avoids the effects of frictional forces in the hinge influencing the output signals.

Before conducting a series of tests to evaluate the performance of an accelerometer, preliminary tests are usually undertaken to ensure that the accelerometer is functioning as designed by the manufacturer. Typical tests include observation of the output for a short period (10–20 minutes) after switch-on to check the warm-up trends and the determination of the threshold acceleration level which produces an output signal. Small accelerations may be applied along the input axis of the sensor using the precision dividing head described above.

As in the case of gyroscope testing, the IEEE has issued a number of test procedure documents [7, 8, 9] for the testing of acceler-ometers. The following text concentrates on general test techniques for accelerometers, examples of test technique and methods of analysis, enabling the performance of an instrument to be evaluated.

Figure 7.16   Photograph of a precision dividing head
Courtesy of DRA, Farnborough

7.6.1 Multi-position tests

The purpose of these tests is to determine the following parameters of an accelerometer:

(i)   scale factor;
(ii)   scale factor linearity;
(iii) null bias error;
(iv)   axis alignment error;
(v)   switch-on to switch-on repeatability.

The device to be tested is fastened firmly to a precision dividing head and, generally, the output signals from the accelerometer are recorded for four different attitudes of the sensitive axis corresponding to 0 g, 1 g, 0 g, −1 g acting along this axis. At each setting, a series of data points is recorded and the whole test may be repeated a number of times, from which the above performance parameters may be calculated. Typical data showing the residual error between the calculated and measured value of the gravity vector for a fixed temperature are shown in Figure 7.17.

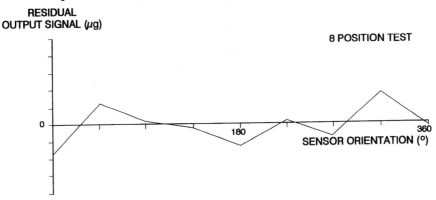

Figure 7.17 Sample data from multi-position testing

In the case of switch-on to switch-on repeatability evaluation, the tests in a given position are repeated at least 12 times, the whole procedure being repeated usually in each of the normal cardinal positions for the tests. Sample data for a single switch-on stability test are shown in Figure 7.18. It is usual to monitor the temperature of the sensor during the warm-up period as shown in the figure. These tests are sometimes known as tombstone tests.

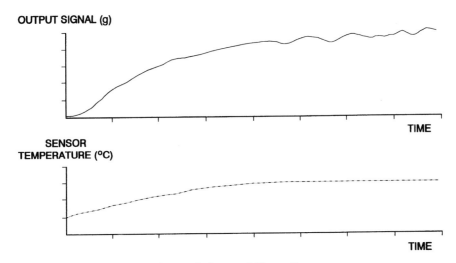

Figure 7.18 Sample data from switch-on stability testing

7.6.2 Long-term stability

For these tests, the output signals from the accelerometer are recorded with the sensor fixed in a particular orientation. The duration of the test may be hours, weeks or even longer. As in the case of gyroscope testing, the sensor is switched off for a period between tests before the procedure is repeated with the accelerometer in a different orientation, or series of different orientations. At each position, the ambient temperature is monitored and recorded for the duration of the test, enabling the accelerometer output signal to be corrected as shown in Figure 7.19.

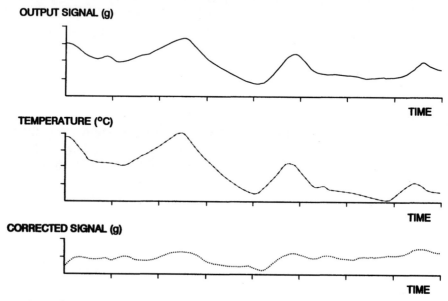

Figure 7.19 Sample data from long-term stability testing

7.6.3 Thermal tests

The purpose of the thermal tests is to establish the variation in the basic parameters used to describe the performance of an accelerometer with temperature; either when the sensor is at a uniform elevated or depressed temperature, known as soaking, or with a temperature gradient across the sensor. The sensor under test on a precision dividing head is usually enclosed in a climatic chamber enabling tests to be carried out at sub-zero temperatures, typically down to −55°C, and elevated to high temperatures, usually up to 75 or 80°C. Figure 7.20 shows a typical experimental arrangement. Various tests can be undertaken to monitor the behaviour of the accelerometer when soaked at elevated and depressed temperatures as the output signals are recorded at the four principal orientations of the accelerometer. This enables variations in the parameters of performance to be estimated and, where possible, correlated with the temperature of the sensor during the test. Sample data are shown in Figure 7.21.

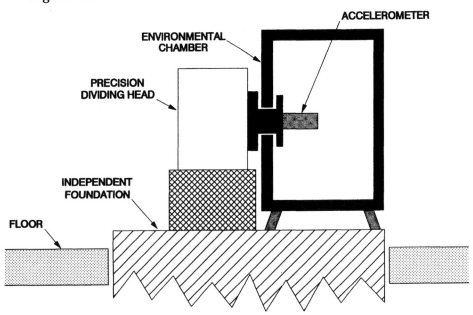

Figure 7.20   Schematic diagram of thermal test equipment

Additionally, the response of the accelerometer can be monitored during different rates of temperature decrease or increase. In this case, the accelerometer is set to a given orientation and the ambient temperature is varied linearly over a given range during a given period of time. During this period, the temperature is recorded and the tests are then repeated for a different orientation of the accelerometer.

7.6.4 Magnetic sensitivity tests

The purpose of these tests is to establish the effect that external magnetic fields have on the performance of an accelerometer. The tests take a similar form to those described for a gyroscope in Section 7.5.5. The sensor is set to the desired orientation with respect to the gravity vector and a pair of Helmholtz coils are arranged to enable a magnetic field to be applied along a chosen axis of the accelerometer.

The response of the accelerometer is recorded before, during and after the application of the magnetic field. This series of tests is usually repeated for different field strengths and different orientations of the accelerometer with respect to the gravity vector and the magnetic field. From these tests, the behaviour of the accelerometer in the presence of magnetic fields, and the change in characteristics it may exhibit, can be assessed from any correlation between changes in performance and the applied magnetic field vector. An example of magnetic sensitivity is given in Figure 7.22.

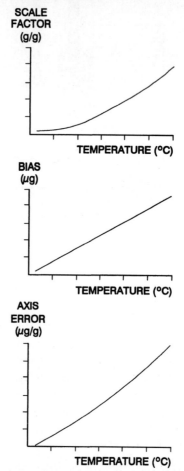

Figure 7.21   Temperature sensitivity of accelerometer parameters

7.6.5 Centrifuge tests

As noted in the case of the centrifuge testing of gyroscopes, this type of testing must be undertaken with care. It is normal for these tests to

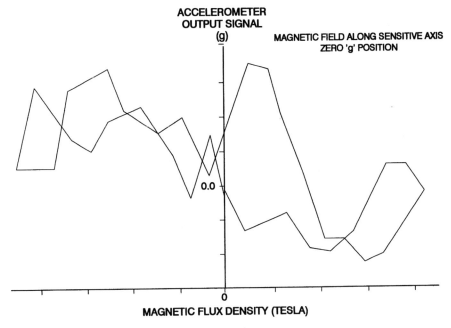

Figure 7.22   Magnetic sensitivity of an accelerometer

be taken towards the end of the test schedule. The purpose of these tests is to record the response of the accelerometer to high values of acceleration compared with gravitational acceleration, and evaluate the linearity of the scale factor at maximum values of the input acceleration [9]. This equipment can also be used to examine the resilience and tolerance of the sensor to applied accelerations in excess of the recommended or maximum acceleration that the sensor is designed to measure along its input axis. This form of testing is commonly referred to as 'over range' testing. Great care must be taken with these latter tests to avoid permanent damage to the sensor under test.

The equipment used and the form of testing is very similar to that already described for the testing of a gyroscope on a centrifuge in Section 7.5.6, where the sensor is either operational or in a quiescent state. Again it is crucial that the sensor under test is secured firmly and rigidly to the test platform. Testing can be undertaken with the sensitive axis parallel to the applied acceleration for the scale factor linearity response tests. Sample data showing the residual error, the difference between the measured and actual acceleration, are shown in Figure 7.23.

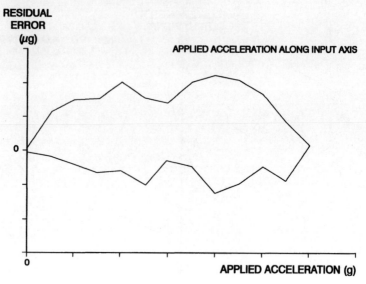

Figure 7.23   Accelerometer data from a centrifuge test

When the sensitive axis is set orthogonal to the applied acceleration, then the cross-talk or orthogonality of the axes can be examined. In these cases, the accelerometer will be operating in an energised state and the acceleration applied to the sensor is increased in small steps up to the maximum value it should measure and then reduced to zero, step by step. At each step in this cycle, the response of the accelerometer is recorded for a given period and the response compared with that predicted by the multi-position tests with the sensor on the dividing head. This cycle is repeated as many times as is necessary to generate confidence in the data [10–12], and where appropriate, the sensor is rotated through 180° to record the response of the accelerometer to accelerations in the opposite direction with respect to the sensitive axis.

The sensor will also be operational for the 'over range' testing where the applied acceleration is increased in small steps above the normal permitted maximum, the response of the accelerometer being recorded at each incremental step. Usually, an acceleration in the region of 10–20% above the maximum value is used in these tests unless the objective is to test the sensor to destruction. After an over range test, it is usual to repeat at least part of the multi-position testing to establish that the accelerometer has not been damaged permanently by the over range testing. If the objective was to damage the sensor, then multi-position tests can establish the new characteristics of the sensor after damage.

The sensor can be mounted on the centrifuge at an angle of 45° to the applied acceleration vector to establish the response of the sensor when a large acceleration is applied simultaneously along two orthogonal axes. As with the conventional arrangement of the axes, the response of the sensor is recorded at each step as the applied acceleration is incremented to a maximum value, and then reduced in steps to zero. This form of testing allows cross-axis sensitivity to be estimated at high values of acceleration through comparison of recorded data with those predicted from the earlier tests.

For some types of resilience testing, the sensor will not be operational. In this case, the sensor is characterised by, for example, multi-position testing and then tested in the centrifuge in a quiescent state. It is then re-characterised to establish the form of any change in the performance characteristics of the accelerometer.

Another form of dynamic test is to mount the accelerometer on a wedge, usually with a 30° or 60° angle, with the normal to the slope pointing away from the axis of rotation. The orientation of the input axis of the sensor is then aligned so that it senses a particular or desired component of the gravity vector when the centrifuge is at rest. When the accelerometer is rotated on the centrifuge, this component of the gravity vector can be nulled by the applied acceleration and hence the acceleration can be taken through the zero position dynamically.

7.6.6 Shock tests

The purpose of these tests is to examine the response of the accelerometer to an applied shock, or to evaluate the resilience of the device to such an applied acceleration of very short duration, typically of the order of milliseconds [13]. As in the case of the centrifuge testing described above, the accelerometer may be operating or quiescent during these tests.

The form of the testing and the equipment used is very similar to that already described for evaluating the response of a gyroscope to applied shocks in Section 7.3.7 and will not be repeated here.

7.6.7 Vibration tests

As in the case of gyroscope testing, this is normally the last series of tests undertaken on the sensor owing to the possibility of permanent and irreversible damage to the accelerometer. The type of equipment used is identical to that discussed in Section 7.5.8.

The purpose of these tests is normally five-fold:

(i)   investigation of any resonant responses of the sensor;

(ii)   evaluation of the vibro-pendulous error of the accelerometer;

(iii) examination of the frequency response of the instrument;

(iv) examination of the resilience of the device in a particular vibratory environment, depending on the reason for the test the accelerometer may be quiescent or operational;

(v)   estimation of the change in noise characteristics of the signals generated by an accelerometer experiencing a vibratory environment.

The testing is undertaken using equipment identical to that already described in Section 7.5.8 for the testing of gyroscopes in a vibratory environment. Additionally, as already described in that section, the motion imparted to the sensor can be random or sinusoidal.

An initial series of tests is usually undertaken with the accelerometer operational in order to establish the frequency and magnitude of any resonances which can be excited in the sensor. This enables the investigation of the sensor to be undertaken at frequencies well away from those that will excite resonances in the instrument. This so-called resonance search is undertaken in an identical manner to that described already for gyroscopes, in Section 7.5.8, using the so-called sine sweep. The warnings given in Section 7.5.8 regarding bolting the interface or mounting block firmly to the test table, and also undertaking a thorough investigation of this mounting block to detect any resonances, should be observed.

Vibro-pendulous rectification is a bias produced by the forced motion of the pendulum in a pendulous accelerometer when it is subjected to linear vibratory motion. This bias appears when the motion is applied along an axis at 45° to the input axis and lying in a plane containing both the input and pendulum axes. The magnitude of this error is evaluated by aligning the accelerometer to be tested so that the applied vibratory motion is along an axis, as described above, enabling the bias to be observed. Careful alignment is necessary, and it is usual to mount the accelerometer on a precision machined interface block that positions the sensor in the requisite orientation.

The accuracy of the mounting can be checked with the sensor operational by rotating the vibrating table through exactly 45° to bring the sensitive, or input, axis to the horizontal and ensuring it registers a signal equating to a zero g output. The position is adjusted to give a precise zero g output. The vibration table should then be rotated back through exactly 45°.

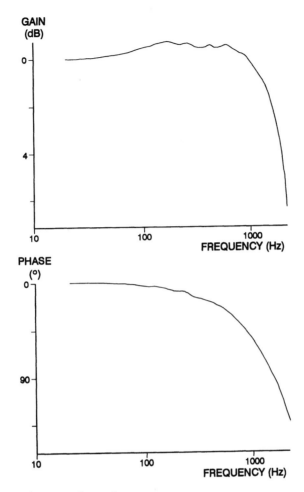

Figure 7.24   Accelerometer dynamic response

A control, or 'feedback', accelerometer is also used to monitor the applied motion and other sensors may be applied to the test block, to monitor the motion about the orthogonal axes.

Initially, the bias of the accelerometer is observed by collecting data over a short period with the sensor stationary. The vibratory motion is then applied to the sensor, at a given frequency and acceleration value. The output signals from the accelerometer under test are recorded over a period of typically one minute, having been filtered if necessary. The change in bias is evaluated and divided by the peak acceleration squared to give the value of the vibro-pendulous error. The test is repeated a number of times to give confidence in the estimate of this error, and repeated at different frequencies and peak acceleration values. Each frequency is chosen to be well removed from any resonant frequencies. Finally, an average value can be estimated from the matrix of values calculated.

The frequency response of an accelerometer is established by mounting the accelerometer firmly on a vibration table and applying sinusoidal vibratory motion in the direction of its sensitive axis. During any one test, the peak acceleration is held constant and the frequency is varied across a given bandwidth, typically 25 Hz to 2 kHz. The output signal from the accelerometer is recorded and compared with a reference signal from a piezoelectric crystal using a transfer function analyser. This enables gain and phase responses to be derived, as shown in Figure 7.24.

The resilience of an accelerometer to a vibratory environment may be evaluated using techniques identical to those already described for a gyroscope. This is discussed in detail in Section 7.5.8. Observations of the recorded data, along with calculated statistical data, enable estimates to be made of changes in the noise characteristics of a sensor resulting from the vibratory motion experienced by that device.

7.6.8 Combination tests

This type of testing is sometimes appropriate for estimating the performance of an accelerometer in a realistic environment where there are combinations of forces or disturbances acting on the sensor. The basic philosophy for this form of testing has already been described in Section 7.5.9 for the evaluation of gyroscopes and will not be repeated. The warnings given in Section 7.5.9 regarding damage to the sensor through over stressing are equally important here.

7.6.9 Ageing and storage tests

The problems, basic philosophy and constraints described for the evaluation of gyroscopes in such tests in Section 7.5.10 also apply to the testing of accelerometers. It is possible that an agency will undertake these studies on behalf of a project and evaluate simultaneously both the accelerometers and the gyroscopes used in a particular vehicle. Generally, the storage and ageing problems associated with pendulous accelerometers centre around deterioration of the permanent magnets, so some prediction of the change in scale factor can be made from knowledge of the magnetic material used.

7.7 Calibration and error compensation

7.7.1 Introduction

The sensors are calibrated by comparing the analogue or digital signals produced by the sensor with the known applied motion. Thus, from the rate transfer tests, the output signals from a gyroscope can be compared with the accurately known rotation rate and the scale factor, defined as so many milli-volts per degree per second of rotation rate, for instance. Similarly, using the gravity vector as an accurate standard, the scale factor of an accelerometer can be defined.

There are several different levels at which error compensation can be applied. However, the fundamental idea is the same for all; to correct the effects of a predictable systematic error, or errors, on the accuracy of a sensor. Additionally, a basic requirement is that an error process can be represented by an equation and hence modelled mathematically, and that a signal corresponding to the disturbing effect, such as temperature or acceleration, is available.

Predictable error components can be estimated from observations of performance and used in the opposite sense to correct, or compensate, for the imperfections in the sensor performance. Often this technique relies on the use of a constant coefficient in the error representation, but for more demanding applications, or complex error behaviour, it is common to use polynomial representations.

The scope of possible error compensations can range from the correction of a single error parameter, using a constant applied to a whole batch of instruments, to multi-parameter compensation of individual instruments using complex time varying polynomials. In the case of the former type of compensation, averaged data are used from a production line evaluation, for example; it is essentially for compensating a systematic trend of error displayed by a particular class or design of sensor. The other extreme requires the characterisation of each sensor through a series of laboratory tests and then correction of the performance of each sensor according to the test results. In this case, the error coefficients for the observed errors on each sensor are evaluated quantitatively and thus the observed measurements are corrected for the known systematic errors.

The former type of error compensation is usually very easy to implement at the system level, particularly as each unit is treated identically. In the latter case, the compensation is more complex and difficult to handle as the degree of compensation can vary from system to system. Of course, each sensor will have its own set of error coefficients which have to be read into the system processor. The coefficients for correction may be stored electronically in a chip located in each sensor. However, this technique does offer considerable scope for a significant improvement in the performance of each sensor and, consequently, for the system in which they are used. The choice of compensation technique is usually a trade-off between various factors, but particularly the balance between the ease of implementation and the benefits gained from improved sensor accuracy. Another consideration, of course, is the extra cost of the detailed characterisation, which is the consequence of the more detailed testing compared with the direct benefit of using a more accurate instrument, or instruments.

The common types of errors which occur in gyroscopes and accelerometers were described in Chapters 4, 5 and 6. Some examples are given below to illustrate how such errors can be modelled and thus compensated in systems using different types of sensor.

7.7.2 Gyroscope error compensation

As described in Chapter 4, the measurement of turn rate ($\tilde{\omega}_x$) provided by a conventional gyroscope may be expressed in terms of the applied rate about its input axis (ω_x) as:

$$\omega_x = (1 + S_x)\omega_x + M_y\omega_y + M_z\omega_z + B_f + B_{gx}a_x + B_{gz}a_z + B_{axz}a_ya_z + n_x \quad (7.3)$$

where a_x and a_z are the accelerations of the gyroscope along its input and spin axes respectively.

$$
\begin{aligned}
B_f &= \text{g-insensitive bias} \\
B_{gx}, B_{gz} &= \text{g-sensitive bias coefficients} \\
B_{axz} &= \text{anisoelastic bias coefficient} \\
n_x &= \text{zero-mean random bias} \\
M_y, M_z &= \text{cross-coupling coefficients} \\
S_x &= \text{scale factor error which may be expressed as a} \\
&\quad \text{polynomial in } \omega_x \text{ to represent scale factor} \\
&\quad \text{non-linearities}
\end{aligned}
$$

As discussed in Chapter 4, each of the error coefficients has repeatable and non-repeatable components. The repeatable components of S_x, M_y, M_z, B_f, B_{gx}, B_{gz} and B_{axz} are measurable and their effects can therefore be compensated. It is not practical to attempt to compensate the in-run random error as such effects can only be controlled by careful sensor design and manufacture.

Similar forms of the above expression may be used to model the bias and scale factor errors arising in two-axis gyroscopes such as the dynamically tuned gyroscope. For optical gyroscopes, such as the ring laser gyroscope, the acceleration sensitive errors are negligible and measured rate may be modelled for many applications as:

$$\tilde{\omega}_x = (1+S_x)\omega_x + M_y\omega_y + M_z\omega_z + B_f + n_x \qquad (7.4)$$

As with mechanical sensors, fixed bias, cross-coupling and fixed scale factor errors can generally be measured to sufficient accuracy to allow some effective compensation to take place. Other errors are less predictable and therefore cannot be controlled using on-line correction methods of the type considered here.

7.7.3 Accelerometer error compensation

As described in Chapter 6, a measurement provided by an accelerometer ($\tilde{a}_x$) may be expressed in terms of an applied acceleration and the sensor error coefficients as follows:

$$\tilde{a}_x = (1+S_x)a_x + M_ya_y + M_za_z + B_f + B_va_xa_y + n_x \qquad (7.5)$$

where a_x represents the acceleration applied in the direction of the sensitive axis and a_y and a_z are the accelerations applied perpendicular to the sensitive axis,

S_x = scale factor error, usually expressed in polynomial form to include non-linear effects
M_y, M_z = cross-axis coupling factors
B_f = measurement bias
B_v = vibro-pendulous error coefficient
n_x = random bias

In general, the fixed bias, cross-coupling and the scale factor error coefficients can be measured and corrections can therefore be applied to offset the repeatable components of these errors. The error contributed by the vibro-pendulous effect in the presence of slowly varying accelerations may also be compensated to a large extent. However, random biases and vibration dependent errors cannot be compensated accurately.

7.7.4 Further comments on error compensation

Compensation techniques are designed to remove the predictable error terms in the inertial sensors by measuring the appropriate error coefficients and using these values to apply corrections to the measurements. These are usually implemented in the system software. In addition to the compensation of sensor errors, it is also common practice to compensate for any system errors such as sensor mounting misalignments which can be measured and are thus predictable in their effects on system performance.

It will be remembered from the discussions in Chapters 4, 5 and 6 that the measured coefficients vary with time, temperature, vibration, applied motion and from switch-on to switch-on, and it is these variations which ultimately determine system performance. Whilst it is not possible to compensate for most of these effects, temperature compensations are often essential to achieve a given performance goal.

Thermal effects on bias and scale factor errors can be very significant and are often difficult to model accurately. This is because in some sensors, especially mechanical gyroscopes, temperature gradients within the sensor can alter the performance of many of its components. Therefore, it is usual for a very accurate inertial system to control the temperature of its sensors very precisely. Consequently, such a system may have a significant warm-up time. However, this does of course alleviate the need for complex and difficult thermal compensation.

The accuracy that may be achieved from the application of compensation techniques is dependent on precisely how the coefficients in the error equation represent the actual sensor errors. This representation can often vary as a function of time, the environment in which the sensor is used and how often it is used. For the more demanding applications, it may be necessary to re-calibrate the sensor regularly, to ensure that the compensation routines are as effective as required by the particular application.

7.8 Testing of inertial navigation systems

Depending on the form and nature of the inertial navigation system, it may be appropriate to test either the complete inertial navigation system, or just the inertial measurement unit in the laboratory. Usually, when the development of a system has reached the point for laboratory testing, the characteristics of the component sensors used will be very well known and the purpose of the testing will be somewhat different from the objectives of the component tests. Often similar tests are undertaken, but with the aim of checking the performance of the system, for example, that it behaves as predicted from the knowledge of the performance of its component inertial sensors. Sometimes, there can be adverse interaction between these components. Examples include cross-talk between vibratory sensors or changes in the lock-in characteristics of mechanically dithered ring laser gyroscopes mounted on the same structure. A manufacturer will wish to check that the units built on a production line will meet the design specification and the customer will also wish to confirm that the inertial system will fulfil the requirements of his or her particular application.

Typically, such tests involve mounting an inertial measurement unit, or a full navigation system, on a multi-axis table or on a suitable rig. The unit may then be rotated through a series of accurately known angles and positioned in different orientations with respect to the local gravity vector, as shown in Figure 7.25. The dominant sensor errors may then be determined from static measurements of acceleration and turn rate taken in each orientation of the unit. An example of this type of approach is described by Joos and Krogmann [15] in relation to a system which uses conventional sensors. The unit under test is placed on a precision three-axis table, and a series of constant rate tests and multi-position tests are then used to allow the major sources of error to be identified.

Alternatively, estimates of the system errors may be obtained by testing the inertial measurement unit as a component of a full strapdown inertial navigation system. In this case, the estimates made by the navigation system of angle turned through and/or linear acceleration in the navigation reference frame may be used to deduce the various system errors as discussed below. An example of the latter approach is described by Brown, Mark and Ebner [16] for the testing of an inertial system containing ring laser gyroscopes. In the scheme described, the unit is made to rotate through a sequence of turns using a two degree-of-freedom table. Immediately prior to each turn, the system performs a self alignment exercise with respect to the navigation frame. The system itself then keeps track of the table rotations throughout each revolution. Components of acceleration in the navigation reference frame are computed immediately on completion of each revolution, by resolving the measured accelerations into the navigation frame using the computed attitude information. Errors in the computed accelerations may then be attributed to a combination of acceleration and angular rate measurement errors. By rotating the unit through a carefully chosen set of turns, it is possible to obtain estimates of the dominant sensor errors. A possible advantage of this type of approach is the avoidance of any need to use a highly accurate and expensive test table, since the positioning of the unit with respect to the local gravity vector is less critical.

A scheme of tests is suggested below which allows a detailed investigation of an inertial measurement unit or a full inertial navigation system.

Static acceleration tests

By mounting the navigation system on a level table with each sensitive axis pointing alternately up and down (six position test), it is possible to extract estimates of the accelerometer biases, scale factor errors and the sensitive axis misalignments with respect to a set of datum mounting faces. These estimates can be computed by summing and differencing various combinations of accelerometer measurements.

Static rate tests

By monitoring the angular rate measurements provided by the system for a pre-defined period of time, for a number of different orientations of the unit, it is possible to extract estimates of the gyroscope fixed biases and g-dependent biases. As in the previous tests, sum and differencing techniques may be used to separate the various error contributions.

Angle tests

Using a precision multi-position test table, the inertial measurement unit may be rotated through very accurately known angles. By comparing these known rotations with estimates of the angles turned through by the system, derived by integrating the rate outputs provided by the gyroscopes, it is possible to derive estimates of the various errors in the gyroscopic measurements. For instance, if the table is rotated clockwise and counter-clockwise through the same angle, estimates of gyroscope to case mounting error may be derived along with gyroscope biases and scale factor errors.

Inertial navigation system multi-position tests

This method makes use of the propagation of sensor imperfections as errors in the components of acceleration derived in the navigation reference frame. Provided that a sufficient number of test rotations takes place, this method may be used to extract estimates of most of the dominant sensor errors associated with a strapdown system containing conventional gyroscopes.

In selecting a suitable set of rotations for the testing of a particular unit, it is often useful to include separate 90° and 180° turns about each axis. For the testing of systems containing conventional gyroscopes, compliance with this requirement allows many different linear combinations of the gyroscope's fixed biases, g-dependent errors and mounting misalignments to appear in the measurement equations, and hence to become observable. Further, it is necessary to ensure that each axis of the unit is aligned with the local vertical in separate test orientations, both the up and down positions, to allow the accelerometer biases, scale factor errors and mounting misalignment to be identified. A sample set of rotations which satisfies these requirements, and which may be implemented using a two degree-of-freedom test table, is shown in Figure 7.25.

In practice, the series of tests outlined above may be carried out in a recursive manner, using the error estimates from one test to update or correct the error model used in subsequent tests. Various signal processing techniques are used to extract estimates of the system errors. For laboratory testing of the type described here, it is often sufficient to use least squares methods to compute the magnitude of the errors. Kalman filtering techniques are often used as an alternative to the least squares method.

On completion of these test procedures, a sequence of separate rate, centrifuge and vibration tests would normally be used in order to assess system performance over a range of dynamic conditions commensurate with those expected during operational use.

7.9 Hardware in the loop tests

One form of evaluation that is currently very popular with missile designers and engineers, and is a consequence of the enormous advances in computer technology, is called hardware in the loop testing. This form of testing is often used by a project during its research and development phase in order to establish an accurate estimate of the performance of a missile, for example. Sometimes an inertial sensor, but usually a complete inertial system, is mounted on a test rig. This is usually a multi-axis test table, as shown in Figure 7.26, that can reproduce accurately the angular motion that the system would experience during its operational life, such as a flight in a missile. Normally, the test rig is limited to angular rotations. The output signals from the device under test are connected, through a suitable interface, to a computer simulating the motion and

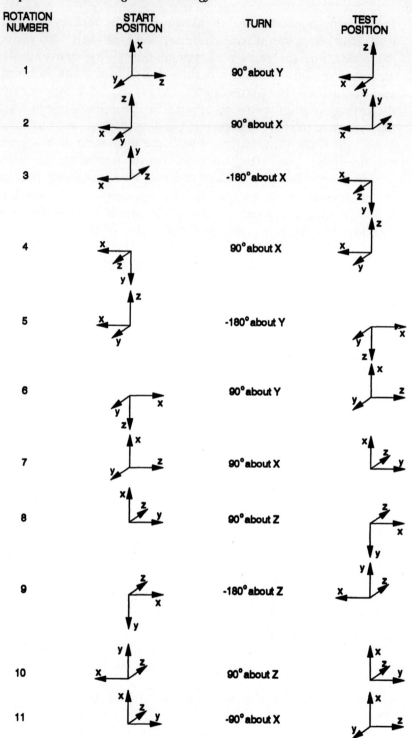

Figure 7.25   Test rotations for a strapdown inertial measurement unit (IMU)

performance of the vehicle. This mathematical model in the computer also generates the signals that control the test rig, hence a simulation operating in real time, using actual hardware can be configured to enable realistic performance assessments to be established for complex systems operating in various flight regimes.

Figure 7.26 *Photograph of a three axis table used for navigation system testing*
Courtesy of CPE

References

1 ANDREWS, T.R.: 'Standard gyro and accelerometer terminology'. Compiled by the Gyro and Accelerometer Panel, Aerospace and Electronic Systems Society, IEEE, 1975

2 'IEEE test procedure for single degree of freedom spring restrained rate gyros'. ANSI IEEE Standard 293, 1969

3 'IEEE test procedure for single degree of freedom spring restrained rate gyros'. IEEE Standard 517, 1974

4 'IEEE test procedure for single degree of freedom spring restrained rate gyros', supplement for strapdown applications. IEEE Standard 529, 1980

5 'IEEE specification format guide and test procedure for two degree of freedom dynamically tuned gyros'. ANSI IEEE Standard 813, 1988

6 'IEEE specification format guide and test procedure for single axis laser gyros'. ANSI IEEE Standard 647, 1981

7 'IEEE specification format guide and test procedure for linear single axis, pendulous, analogue, torque balance accelerometer'. IEEE Standard 337, 1972

8 'IEEE specification format guide and test procedure for linear single axis, pendulous, analogue, torque balance accelerometer'. IEEE Standard 530, 1978

9 'IEEE recommended practice for precision centrifuge testing of linear accelerometers'. IEEE Standard 836, 1991

10 KIRKPATRICK, E.G.: *'Introductory statistics and probability for engineering science and technology'* (Prentice-Hall, 1974)

11 TOPPING, J.: *'Errors of observation and their treatment'* (Chapman and Hall, 1975)

12 BOX, G.E.P., HUNTER, W.G. and HUNTER, J.S.: *'Statistics for experimenters'* (Wiley, 1978)

13 HARRIS, C.M. and CREDE, C.E. (Eds.): *'Shock and vibration handbook'* (McGraw-Hill, 1961), **1-3**

14 SMITHSON, T.G.: 'A review of the mechanical design and development of a high performance accelerometer', *Proc. Inst. Mech. Eng.*, Conference on mechanical technology of inertial sensors, 1987, **C49/87**

15 JOOS, D.K. and KROGMANN, U.K.: 'Estimation of strapdown sensor parameters for inertial system error compensation'. NATO AGARD symposium on *'Precision positioning and inertial guidance sensors*, technology and operational aspects', October 1980

16 BROWN, A., EBNER, R. and MARK, J.: 'A calibration technique for a laser gyro strapdown inertial navigation system'. Proceedings of DGON symposium on *Gyro technology*, Stuttgart, Germany, 1982

Chapter 8
Strapdown system technology

8.1 Introduction

The preceding chapters have described the fundamental principles of strapdown navigation systems and the sensors required to provide the necessary measurements of angular rate and specific force acceleration. In this chapter, aspects of strapdown system technology are discussed.

8.2 The components of a strapdown navigation system

As indicated in the earlier discussion, a strapdown inertial navigation system is basically formed of a set of inertial instruments and a computer. However, for reasons which will shortly become clear, such a system may be sub-divided further into the following component parts:

- instrument cluster;
- instrument electronics;
- attitude computer;
- navigation computer.

These components, which form the basic building blocks of a full strapdown navigation system, are shown schematically in Figure 8.1. The units will be mounted in a case, together with the necessary electrical power supplies and interface electronics, which may then be installed in a vehicle requiring an onboard navigation capability. Although it is often assumed that the unit will be fixed rigidly in the vehicle, it is usually necessary for it to be installed on anti-vibration

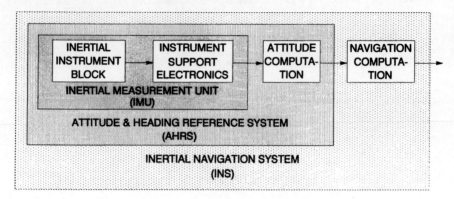

Figure 8.1   Strapdown inertial navigation system building blocks

(AV) mounts to provide isolation from vehicle motion at frequencies to which the unit is particularly sensitive.

We are primarily concerned here with the implementation of a full inertial navigation system, but applications arise in which the full navigation function is not required. For example, in some short range missile applications, inertial measurements, typically of angular rate and specific force, are required purely for flight control purposes. In such cases, the instrument cluster and instrument electronics blocks alone are used to form what is known as an inertial measurement unit or IMU. For other applications requiring attitude and heading information alone, the IMU is combined with a processor in which the attitude equations are solved. The resulting system is known as an attitude and heading reference system or AHRS, the processor being referred to here as the attitude computer. An AHRS is sometimes used in combination with a Doppler radar to form a navigation system. Finally, the addition of a further computer in which the navigation equations are solved provides a full inertial navigation capability.

In the following Sections, the components of the strapdown inertial navigation system defined above are described separately in more detail. This includes some discussion of the requirements for internal power supplies and anti-vibration mounts.

8.3 The instrument cluster

8.3.1 Orthogonal sensor configurations

The instrument cluster usually includes a number of gyroscopes and accelerometers which provide measurements of angular rate and specific force respectively. The unit may contain either three single-axis gyroscopes or two dual-axis gyroscopes, as well as three single-axis accelerometers, all attached to a rigid block which can be mounted in the body of the host vehicle, either directly or on AV mounts. The sensitive axes of the instruments are most commonly mutually orthogonal in a Cartesian reference frame, as illustrated in Figure 8.2.

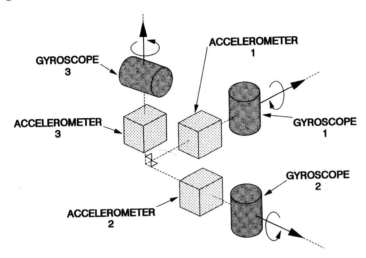

Figure 8.2   Orthogonal instrument cluster arrangement

This arrangement of the instruments allows the components of angular rate and specific force in three mutually orthogonal directions to be measured directly, thus providing the information required to implement the strapdown computing tasks.

As indicated above, systems using dual-axis sensors, such as the dynamically tuned gyroscope, as an alternative to the single-axis rate integrating gyroscope require one less sensor. A dual-axis sensor configuration also provides an additional rate measurement. Through careful choice of the relative orientation of the two dual-axis sensors, the redundant measurement provided by one gyroscope may be used to monitor the performance of the other as part of a built in test facility. In practice of course, many other factors will influence the choice between these two gyroscopes, and their different character-istics are discussed in Chapter 4.

Other instrument arrangements are possible using modern sensing techniques which offer various novel approaches. For instance, a pair of multi-sensors mounted orthogonally or a single ring laser triad with suitable accelerometers may be used to form an instrument cluster. Such sensors are described in Chapters 5 and 6.

8.3.2 Skewed-sensor configurations

Theoretically, it is possible to mount the instruments in orientations other than the orthogonal arrangement illustrated in Figure 8.2, provided that the measurements may be expressed as independent linear combinations of the orthogonal components of angular rate and specific force required by the navigation system. The orthogonal components may then be extracted from the measurements as part of the strapdown processing task. Such instrument configurations are referred to as skewed-sensor configurations and may be used to advantage in certain applications.

Skewed-sensor arrangements are used primarily in applications requiring on-line failure detection and fail-safe operation, as discussed in Section 8.3.4. However, they may also be used in situations where the turn rate about a single axis of a vehicle may exceed the normal operating range of a gyroscope having performance characteristics suitable for the particular application. By mounting the gyroscopes so that their sensitive axes form an angle with the high rate axis of the vehicle it is possible to ensure that the resolved component of the turn rate falls within the maximum range of the sensor. Given knowledge of the skew angle, it is possible to calculate the turn rate about the vehicle axis using the measurements provided by the skewed arrangement of inertial sensors. An example of this technique based on a system using two dua-axis gyroscopes is discussed below.

8.3.3 A skewed-sensor configuration using dual-axis gyroscopes

Two dual-axis gyroscopes may be configured in the symmetrical form illustrated below. A body axis frame $Ox_by_bz_b$ is indicated along with gyroscope axes $Ox_1y_1z_1$ and $Ox_2y_2z_2$. The gyroscope spin axes lie in the Ox_by_b plane in the directions Ox_1 and Ox_2, inclined at angle Φ to Ox_b.

The turn rates about the respective body axes are denoted ω_x, ω_y and ω_z. This particular configuration may be used where the rotation rate (ω_x) about the axis x_b exceeds the normal operating range of the gyroscopes. For the arrangement shown in the figure, the turn rates ω_A, ω_B, ω_C and ω_D, sensed by the gyroscopes may be expressed in terms of the body rates as follows:

Gyro 1 :

$$\omega_A = \frac{\left(\omega_x \sin\Phi + \omega_y \cos\Phi\right)}{\sqrt{2}} - \frac{\omega_z}{\sqrt{2}}$$

$$\omega_B = \frac{\left(\omega_x \sin\Phi + \omega_y \cos\Phi\right)}{\sqrt{2}} + \frac{\omega_z}{\sqrt{2}}$$

$$\omega_C = \frac{\left(\omega_x \sin\Phi - \omega_y \cos\Phi\right)}{\sqrt{2}} + \frac{\omega_z}{\sqrt{2}} \qquad (8.1)$$

Gyro 2 :

$$\omega_D = \frac{\left(\omega_x \sin\Phi - \omega_y \cos\Phi\right)}{\sqrt{2}} - \frac{\omega_z}{\sqrt{2}}$$

where Φ is the absolute value of the angular displacement between the spin axis of each gyroscope and the body axis x_b. Estimates of the body rates may be derived by summing and differencing the measurements of the turn rates provided by the gyroscopes, as shown below, where the '^' notation is used to denote an estimated quantity.

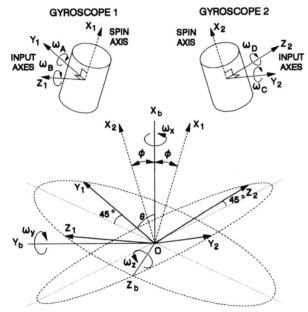

Figure 8.3   Dual-axis gyroscope skewed configuration

$$\hat{\omega}_x = \frac{\left(\omega_A + \omega_B + \omega_C + \omega_D\right)}{2\sqrt{2} \sin\Phi}$$

$$\hat{\omega}_y = \frac{\left(\omega_A + \omega_B - \omega_C - \omega_D\right)}{2\sqrt{2} \cos\Phi} \qquad (8.2)$$

$$\hat{\omega}_z = \frac{\left(-\omega_A + \omega_B + \omega_C - \omega_D\right)}{2\sqrt{2}}$$

This equation is in fact a least squares solution to the measurement eqn. 8.1.

The component of ω_x sensed by each gyroscope is equal to ω_x times the cosine of the direction angle, θ, between the body axis x_b and the gyroscope's input axes. For the instrument configuration considered here:

$$\theta = \cos^{-1}\left\{\frac{\sin \Phi}{\sqrt{2}}\right\} \tag{8.3}$$

If the maximum body rate about the axis x_b is 1200°/second and the maximum rate which can be measured by the gyroscopes is 600°/second, then in the absence of any motion about the other axes, the angle θ must be greater than 60°, i.e. the angular displacement of the spin axis (Φ) as shown in the figure should not exceed 45°. In general, the value of Φ will need to be less than this figure to cope with turn rates about the other axes of the body.

In order to satisfy a particular set of performance objectives using a skewed-sensor configuration of this type, it will be necessary to use higher quality gyroscopes or to compensate the sensors more precisely than would be required for a conventional strapdown arrangement. It can be shown that biases on the measurements of turn rate provided by the sensors and the accuracy of mounting alignment become more critical in inertial systems which use a skewed arrangement of sensors.

From eqn. 8.2, it can be shown that biases in the four rate measurements, denoted $\delta\omega_A$, $\delta\omega_B$, $\delta\omega_C$ and $\delta\omega_D$, and an error in the skew angle Φ will give rise to biases in the estimates of the rates about the body axes, $\delta\omega_x$, $\delta\omega_y$ and $\delta\omega_z$, given by:-

$$\delta\omega_x = \frac{(\delta\omega_A + \delta\omega_B + \delta\omega_C + \delta\omega_D)}{2\sqrt{2}\sin\Phi} - (\omega_A + \omega_B + \omega_C + \omega_D)\frac{\cos\Phi}{2\sqrt{2}\sin^2\Phi}$$

$$\delta\omega_y = \frac{(\delta\omega_A + \delta\omega_B - \delta\omega_C - \delta\omega_D)}{2\sqrt{2}\cos\Phi} + (\omega_A + \omega_B - \omega_C + \omega_D)\frac{\sin\Phi}{2\sqrt{2}\cos^2\Phi}\delta\Phi$$

$$\delta\omega_z = \frac{(-\delta\omega_A + \delta\omega_B + \delta\omega_C - \delta\omega_D)}{2\sqrt{2}}$$

$$\tag{8.4}$$

with the result that the effects of the measurement biases on the estimates of turn rate in the x or y directions can be magnified through the use of the skewed-sensor configuration. The system is also particularly susceptible to misalignment of the sensor mounts. Hence, accurate knowledge of the skew angles is required to obtain precise estimates of the body rates. This is particularly true in the situation illustrated in Figure 8.3 above, where the sensitive axes of the gyroscopes are displaced by large angles with respect to a potentially high-rate axis.

In general, skewed-sensor systems based on conventional angular momentum gyroscopes are expected to be applicable in situations where body rates exceed the sensor maximum angular rate capability only transiently. In many applications where high turn rates are likely to be sustained, it is considered that optical rate sensors, such as the ring laser gyroscope or the fibre gyroscope, now offer the best solution because of the high rotation rate capability and excellent scale factor linearity offered by these types of sensors.

8.3.4 Redundant sensor configurations

For reasons of safety and reliability, many applications require navigation systems with on-line failure detection and fail-safe operation [1–3]. To satisfy this objective using a strapdown system, additional sensors are required to provide a level of redundancy in the measurements. This may be achieved using an orthogonal sensor arrangement by adding additional sensors to detect the angular rates and accelerations in each vehicle axis. Alternatively, and more commonly, a skewed-sensor configuration is often proposed. For example, a skewed-sensor system employing four dual-axis gyroscopes and eight accelerometers may be used to provide quadruplex redundancy for an aircraft flight control and avionics sensor unit. The gyroscopic input axes are equally distributed on a cone, the axis of which is aligned with the pitch axis of the aircraft for the purposes of this example. The accelerometers may be oriented in a similar manner.

Table 8.1

Sources of information		
Pitch rate	Roll rate	Yaw rate
$K_1(\omega_{1x} + \omega_{1y})$	$K_1(\omega_{1x} - \omega_{1y})$	$K_1(\omega_{3x} - \omega_{3y})$
$K_1(\omega_{2x} + \omega_{2y})$	$(\omega_{2x} - \omega_{2y})$ $-K_1(\omega_{3x} - \omega_{3y})$	$(\omega_{2x} - \omega_{2y})$ $-K_1(\omega_{1x} - \omega_{1y})$
$K_1(\omega_{3x} + \omega_{3y})$	$K_2(\omega_{2x} - \omega_{2y})$ $-K_2(\omega_{4x} - \omega_{4y})$	$(\omega_{4x} - \omega_{4y})$ $+K_1(\omega_{1x} - \omega_{1y})$
$K_1(\omega_{4x} + \omega_{4y})$	$(\omega_{4x} - \omega_{4y})$ $+K_1(\omega_{3x} - \omega_{3y})$	$K_2(\omega_{2x} - \omega_{2y})$ $+K_2(\omega_{4x} - \omega_{4y})$

The four separate sources of rate information provided by such a system are indicated in Table 8.1 where ω_{ix} and ω_{iy} are the rates measured about the x and y input axes of the ith gyroscope, and K_1

and K_2 are geometrical constants. For the instrument configuration considered here, $K_1 = 1/\sqrt{2}$ and $K_2 = 1/2$. Similar equations may be written for the acceleration in aircraft body axes. This arrangement of sensors, which is discussed in detail in Reference 1, offers high reliability with a fail-operational, fail-operational, fail-safe level of fault tolerance. Fail-operational means that a fault must be detected, localised and the system must be dynamically reconfigured. Fail-safe refers to the capability to detect a fault which must not affect system safety. To achieve the same level of redundancy with an orthogonal sensor arrangement would require up to eight two-axis gyroscopes and twelve accelerometers.

The reader interested in redundant strapdown sensor configurations is referred to the many excellent papers on the subject which include References 2 and 3 given at the end of this chapter.

8.4 Instrument electronics

The instrument electronics unit contains the dedicated electronics needed to operate the inertial sensors. Typically, this includes instrument power supplies, read-out electronics to provide signals in the form needed by a navigation processor and possibly a computer. The precise requirements vary in accordance with the types of instruments used and the level of performance needed.

For the vast majority of applications, the electronic signals provided by the inertial sensors are required in a digital format for input directly to a computer. Whilst many sensors naturally provide output in digital form, this is not always the case. Where analogue output is provided, this will need to be converted, and the analogue to digital conversion process forms part of the instrument electronics.

The output signals from the inertial sensors are often provided in incremental form, i.e. as measurements of incremental angle and incremental velocity corresponding to the integral of the measured angular rate and linear acceleration, respectively, over a short period of time, τ. The incremental angle output ($\Delta\theta$) which may be provided by a gyroscope may be expressed mathematically as:

$$\Delta\theta = \int_{t}^{t=\tau} \omega \, dt \tag{8.5}$$

where ω is the measured turn rate. Similarly, the incremental velocity output (Δv) from an accelerometer may be written as:

$$\Delta v = \int\limits_{t}^{t+\tau} f \, dt \qquad (8.6)$$

where *f* is the measured acceleration.

This form of sensor output is very convenient since it eases the tasks of updating attitude and velocity. Many contemporary sensors, ring laser gyroscopes for example, naturally provide output signals in this form, whilst for others it is a result of the digitising process carried out within the inertial measurement unit. The way in which the incremental measurements are used is discussed in Chapter 10 together with other aspects of the strapdown processing tasks.

Many conventional sensors such as spinning mass gyroscopes and pendulous accelerometers typically operate in a null seeking or rebalance loop mode in order to achieve a linear and accurate response characteristic. In such cases, instrument rebalance electronics will form part of the instrument electronics block along with gyroscope spin motor and pick-off power supplies.

The use of a computer within the inertial measurement unit (IMU) [4] enables some form of on-line compensation of the instrument outputs to be performed based on instrument characterisation data obtained during laboratory or production testing, as described in Chapter 7. Since such computing tasks are very specific to the type of instrument used, they may well be implemented here rather than as part of the subsequent attitude and navigation processing.

Because instrument characteristics can be temperature dependent, there may also be a need to compensate the instrument outputs for temperature variation in order to achieve satisfactory performance. It follows therefore that instrument temperature monitoring is often required; the dilemma is where to monitor the temperatures. In many sensors, the variation in performance with temperature is the result of the temperature sensitivity of magnetic material used in the very core of the sensor. Hence, it may not be adequate to sense temperature outside the instrument cluster or merely close to the case of the instrument.

Finally, it may well be advisable for most applications to carry out some form of on-line testing of the inertial sensors and associated electronics. This may involve checks to confirm that the outputs of the sensors remain within certain known limits appropriate to the application, and that they continue to vary in the expected manner whilst operational. For instance, a sensor output remaining at a fixed level for an extended period of time may well suggest that a failure has occurred and a warning should be given. Such tasks may also be implemented within the IMU processor which would form part of the built-in test equipment (BITE) within the unit.

It follows from the above remarks that the instrument electronics block could typically comprise the following:

- instrument power supplies;
- rebalance loop electronics;
- temperature monitoring electronics;
- instrument compensation processing;
- analogue to digital conversion electronics;
- output interface conditioning;
- built-in test facility.

These components are illustrated schematically in Figure 8.4 which shows an inertial measurement unit containing two, dual-axis, gyroscopes and three accelerometers.

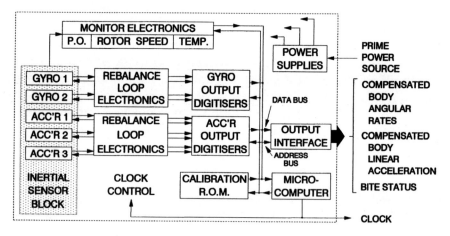

Figure 8.4   Inertial measurement unit functions

8.5 The attitude computer

The attitude computer essentially takes measurements of body rate about three orthogonal axes provided by the inertial measurement unit and uses this information to derive estimates of body attitude by a process of integration. The attitude is usually represented within the computer as a set of direction cosines or quaternion parameters as discussed in Chapter 3, either of which is appropriate for on-line attitude computation. The Euler angle representation described in Chapter 3 is not generally recommended for implementation in strapdown systems. As a result of the preponderance of trigonometric terms in the equations coupled with the presence of a singularities for pitch angles of ±90°, the Euler equations do not lend themselves to real time solution in an onboard navigation processor. However, it should be borne in mind that there may well be a requirement to extract the Euler angles from the direction cosines or quaternion parameters for control purposes in some applications.

The equations to be solved in the attitude computer are summarised below assuming quaternion parameters are to be used to define the attitude of the vehicle body with respect to the navigation reference frame. The quaternion may be expressed as a four element vector $[a \ b \ c \ d]^T$, the elements of which are calculated by solving the following set of differential equations:

$$\dot{a} = -0.5\left(b\omega_x + a\omega_y + d\omega_z\right)$$
$$\dot{b} = 0.5\left(a\omega_x - d\omega_y + c\omega_z\right)$$
$$\dot{c} = 0.5\left(d\omega_x + a\omega_y - b\omega_z\right) \tag{8.7}$$
$$\dot{d} = -0.5\left(c\omega_x - b\omega_y - a\omega_z\right)$$

where ω_x, ω_y and ω_z are estimates of the components of vehicle turn rate with respect to the navigation reference frame. These quantities are computed by differencing the measurements of body rate output by the IMU and estimates of the turn rate of the navigation frame calculated in the navigation computer.

The quaternion parameters may be used to construct the direction cosine matrix which relates the body reference frame to the navigation reference frame ($\mathbf{C}_b^n$) using:

$$\mathbf{C}_b^n = \begin{pmatrix} \left(a^2 + b^2 - c^2 - d^2\right) & 2(bc - ad) & 2(bd + ac) \\ 2(bc + ad) & \left(a^2 - b^2 + c^2 - d^2\right) & 2(cd - ab) \\ 2(bd - ac) & 2(cd + ab) & \left(a^2 - b^2 - c^2 + d^2\right) \end{pmatrix} \tag{8.8}$$

It is customary in most strapdown attitude computation schemes to carry out self consistency checks. In the case of the quaternion, the self consistency check involves confirming that the sum of the squares of the individual quaternion elements remains equal to unity. i.e.

$$a^2 + b^2 + c^2 + d^2 = 1 \tag{8.9}$$

The attitude computation algorithm used for a given application must be able to keep track of vehicle orientation whilst it is turning at its maximum rate and in the presence of the total motion of the vehicle, including vibration. Algorithms which may be used to implement the attitude computation function in the presence of such motion are described in Chapter 10.

8.6 The navigation computer

The solution of the navigation equations is carried out in the navigation computer. To implement the navigation function, it is first necessary to transform, or resolve, the specific force measurements provided by the accelerometers, denoted here by the vector $\mathbf{f}^n$, into the navigation reference frame. This can be accomplished using the attitude information provided by the attitude computer. Using the direction cosine representation of attitude, for instance, the required transformation is achieved using:

$$\mathbf{f}^n = \mathbf{C}_b^n \, \mathbf{f}^b \tag{8.10}$$

where $\mathbf{f}^n$ is the specific force expressed in navigation axes and $\mathbf{C}_b^n$ is the direction cosine matrix described earlier. Both the specific force and the direction cosine matrix are time varying quantities. Therefore, care must be taken to ensure that all significant movements of the vehicle, including turn rates and vibratory motion, can be accommodated in the computer implementation of this equation.

The resolved specific force components form the inputs to the navigation equations which are used to calculate vehicle velocity and position. The navigation equations are described in Chapter 3 but are repeated here for completeness. For a system which is required to navigate in the vicinity of the Earth to provide estimates of north and east velocity, latitude, longitude and height above the Earth, the equations to be solved may be written as follows:

$$\dot{v}_N = f_N - v_E\left(2\Omega + \dot{\lambda}\right)\sin L + v_D \dot{L} \tag{8.11}$$

$$\dot{v}_E = f_E - v_N\left(2\Omega + \dot{\lambda}\right)\sin L + v_D\left(2\Omega + \dot{\lambda}\right)\cos L \tag{8.12}$$

$$\dot{v}_D = f_D - v_E\left(2\Omega + \dot{\lambda}\right)\cos L + v_N \dot{L} + g \tag{8.13}$$

$$\dot{L} = \frac{v_N}{\left(R_0 + h\right)} \tag{8.14}$$

$$\dot{\lambda} = \frac{v_E \sec L}{\left(R_0 + h\right)} \tag{8.15}$$

$$\dot{h} = -v_D \tag{8.16}$$

where $\quad v_N \; v_E \; v_D =$ the north, east and vertical components
of vehicle velocity with respect to the Earth

$f_N \; f_E \; f_D \;$ = the components of specific force resolved in the local geographic reference frame

$L \qquad$ = vehicle latitude

$\lambda \qquad$ = vehicle longitude

$h \qquad$ = vehicle height above ground

$R_0 \qquad$ = mean radius of the Earth

$\Omega \qquad$ = Earth's rate of turn

$g \qquad$ = acceleration due to gravity

Refinements to these equations needed to take account of the shape of the Earth and variation in gravitational attraction over the surface of the Earth are given at the end of Chapter 3.

The turn rate of the vehicle with respect to the local geographic navigation frame $\omega_{nb}^b = [\omega_x, \omega_y, \omega_z]^T$ which is required to implement the attitude computation process described above is given by:

$$\omega_{nb}^b = \omega_{ib}^b - C_n^b \omega_{in}^n \tag{8.17}$$

where $\quad \omega_{ib}^b =$ turn rate of the body with respect to inertial frame as measured by the strapdown gyroscopes in the inertial measurement unit

$\omega_{in}^n =$ turn rate of the navigation frame with respect to the inertial frame, which is computed as follows:

$$\omega_{in}^n = \begin{pmatrix} \Omega \cos L + \dfrac{v_E}{(R_0 + h)} \\[3mm] -\dfrac{v_N}{(R_0 + h)} \\[3mm] -\Omega \sin L - \dfrac{v_E \tan L}{(R_0 + h)} \end{pmatrix} \tag{8.18}$$

Algorithms which may be used to implement the navigation function are described in Chapter 10.

8.7 Power conditioning

The raw power supplies available in the host vehicle, whether it is an aircraft, a missile, a ship or a land vehicle, will not usually be sufficiently stable or provide the particular voltage levels required by the inertial navigation system. Therefore, it will be necessary to include power conditioning within the unit to generate the supply voltages required which are smoothed sufficiently and controlled to the desired amplitude to ensure satisfactory operation of the navigation system.

8.8 Anti-vibration mounts

A strapdown inertial navigation system will usually be installed on anti-vibration (AV) mounts to provide isolation from vehicle motion at frequencies to which the unit is particularly sensitive. In many applications, the unit may need to be isolated from certain frequencies in the vibration spectrum of the vehicle which may excite resonances within the inertial sensors or give rise to computational errors. The design of suitable AV mounts is frequently a complex task requiring careful matching of the mount design to the characteristics of the inertial sensors within the unit. The effects of vibration are discussed in more detail in Chapter 11 in relation to both instrument errors and overall system performance.

8.9 Concluding remarks

A strapdown inertial navigation system providing navigation in three dimensions will have the following components in one form or another:

- instrument cluster   —to sense translational and rotational movements;
- instrument electronics—to provide the control of the sensors and to produce measurement information;
- attitude computer   —to compute the attitude of the vehicle for resolution of the specific force measurements;
- navigation computer   —to resolve the specific force data and to solve the navigation equations to generate estimates of position and velocity;
- gravitational model   —to allow compensation for the effects of gravitational attraction on the translational measurements;
- power conditioning   —to provide smoothed and controlled voltage levels needed for satisfactory system operation;
- input/output interface—to communicate with the host vehicle.

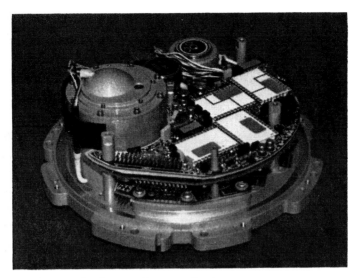

Figure 8.5   *Photograph of a strapdown navigation system*
Courtesy of British Aerospace plc

These units are mounted in a case which is installed in a vehicle. The case is usually attached to the body of the vehicle via AV mounts.

Reduced configurations are possible to produce an attitude and heading reference unit, or a unit for navigation in a single plane. A photograph of an existing strapdown navigation system is shown in Figure 8.5.

In a strapdown system, the inertial sensors provide measurements of angular rates and specific force in axes that are usually aligned with the principal body axes of the vehicle. Skewed-sensor arrangements may be used in some designs to allow the instruments to cope with very high rates about a single axis or in systems employing multiple sensors to provide redundancy for fault tolerance purposes. The inertial measurements have to be transformed to the appropriate axis set for navigation. A variety of reference frames is used depending on the particular application; typically, a local geographic frame is used to provide estimates of latitude, longitude and height, for navigation in the vicinity of the Earth.

A number of methods is available for the transformation procedure, direction cosine matrices or quaternion parameters being most commonly used since both are free from singularities at ±90° pitch angles. Quaternions are generally preferred.

The algorithms required for specific force transformation, the correction for the gravitational attraction and the solution of the navigation equations are implemented in the navigation computer. This processor produces the estimates of vehicle velocity and position in whichever axis set the vehicle is using for its navigation.

References

1 KROGMANN, U.: 'Optimal integration of inertial sensor functions for flight control and avionics', AIAA–DASC, San Jose, October 1988
2 KROGMANN, U.: 'Design considerations for highly reliable hard and software fault tolerant inertial reference systems'. Proceedings of DGON symposium on *Gyro technology*, Stuttgart, Germany, 1990
3 HARRISON, J.V. and GAI, E.G.: 'Evaluating sensor orientations for navigation performance and failure detection', *IEEE Trans.*, 1977, **AES-13** (6)
4 EDWARDs, C.S. and CHAPLIN, R.J.: 'Strapdown dynamically tuned gyroscopes and the use of microprocessors to simplify their application'. Proceedings of DGON symposium on *Gyro technology*, Stuttgart, Germany, 1979

Chapter 9

Inertial navigation system alignment

9.1 Introduction

Alignment is the process whereby the orientation of the axes of an inertial navigation system is determined with respect to the reference axis system. The basic concept of aligning an inertial navigation system is quite simple and straight forward. However, there are many complications that make alignment both time consuming and complex. Accurate alignment is crucial, however, if precision navigation is to be achieved over long periods of time without any form of aiding.

In addition to the determination of initial attitude, it is necessary to initialise the velocity and position defined by the navigation system as part of the alignment process. However, since it is the angular alignment which frequently poses the major difficulty, this chapter is devoted largely to this aspect of the alignment process.

In many applications, it is essential to achieve an accurate alignment of an inertial navigation system within a very short period of time. This is particularly true in many military applications, in which a very rapid response time is often a prime requirement in order to achieve a very short, if not zero, reaction time.

There are two fundamental types of alignment process: self alignment, using gyro-compassing techniques, and the alignment of a slave system with respect to a master reference. There are various systematic and random errors that limit the accuracy to which an inertial navigation system can be aligned, whichever method is used. These include the effects of inertial sensor errors, data latency caused by transmission delays, signal quantisation, vibration effects and other undesirable or unquantifiable motion.

Various techniques have been developed to overcome the effects of the random and systematic errors and enable slave systems in missiles, for example, to be aligned whilst under the wing of an aircraft in flight, or in the magazine of a ship underway on the ocean. Differing techniques, such as angular rate matching or velocity matching, can be used to align the slave system, the actual circumstances determining the technique which produces the more accurate alignment. In general, a manoeuvre of the aircraft or ship speeds up the alignment process and increases the accuracy achieved.

The basic principles of alignment on both fixed and moving platforms are described in Section 9.2; the particular problems encountered when aligning on the ground, in the air and at sea are discussed in Sections 9.3, 9.4 and 9.5 respectively.

9.2 Basic principles

The inertial system to be aligned contains an instrument cluster in which the gyroscopes and accelerometers are arranged to provide three axes of angular rate information and three axes of specific force data in three directions, which are usually mutually perpendicular. In a conventional sensor arrangement, the sensitive axes of the gyroscopes are physically aligned with the accelerometer axes. Essentially, the alignment process involves the determination of the orientation of the orthogonal axis set defined by the accelerometer input axes with respect to the designated reference frame.

Ideally, we would like the navigation system to be capable of aligning itself automatically following switch-on, without recourse to any external measurement information. In the situation where the aligning system is mounted in a rigid stationary vehicle, a self alignment may be carried out based solely on the measurements of specific force and angular rate provided by the inertial system, as described below.

9.2.1 Alignment on a fixed platform

Consider the situation where one needs to align an inertial navigation system to the local geographic co-ordinate frame defined by the directions of true north and the local vertical. For the purposes of this analysis, it is assumed that the navigation system is stationary with respect to the Earth. In this situation, the accelerometers measure three orthogonal components of the specific force needed to overcome gravity whilst the gyroscopes measure the components of the Earth's turn rate in the same directions.

Consider first the alignment of a stabilised platform system in which the instrument cluster can be rotated physically into alignment with the local geographic reference frame. In this situation, it is usual to refer to the accelerometers whose sensitive axes are to be aligned with the north, east and vertical axes of the reference frame as the north, east and vertical accelerometers respectively. Similarly, north, east and vertical gyroscopes may be defined.

In a stable platform mechanisation, alignment is achieved by adjusting the orientation of the platform until the measured components of specific force and Earth's rate become equal to the expected values. The horizontal components of gravity acting in the north and east directions are nominally zero. The instrument cluster is therefore rotated until the outputs of the north and east accelerometers reach a null, thus levelling the platform. Since the east component of Earth's rate is also known to be zero, the platform is then rotated about the vertical until the east gyroscope output is nulled, thus achieving an alignment in azimuth. This type of process is referred to as gyrocompassing and is described extensively in the literature [1]. An equivalent alignment process, sometimes referred to as analytic gyro-compassing, can be used to align a strapdown inertial navigation system.

In a strapdown system, attitude information may be stored either as a direction cosine matrix or as a set of quaternion parameters, as described in Chapter 3. The objective of the angular alignment process is to determine the direction cosine matrix or the quaternion parameters which define the relationship between the inertial sensor axes and the local geographic frame. The measurements provided by the inertial sensors in body axes may be resolved into the local geographic frame using the current best estimate of the body attitude with respect to this frame. The resolved sensor measurements are then compared with the expected turn rates and accelerations to enable the direction cosines or quaternion parameters to be calculated correctly. The principles of the method are illustrated below with the aid of single plane examples to show how the attitude of the strapdown inertial sensors with respect to the local geographic reference frame may be extracted from the inertial measurements.

Since the true components of gravity in the north and east directions are nominally zero, any departure from zero in the accelerometer measurements resolved in these directions may be interpreted as an error in the stored attitude data, and in particular as an error in the knowledge of the direction of the local vertical. A single plane illustration is given in Figure 9.1.

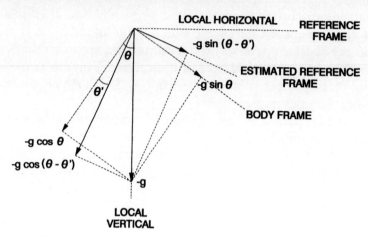

Figure 9.1   Alignment to the gravity vector in a single plane

The accelerometers provide measurements of the true acceleration in body axes, $-g \sin \theta$ and $-g \cos \theta$ respectively. These measurements are resolved through an angle θ' which is an estimate of the true body angle θ, or the angle that the body makes with the estimated reference frame shown in the Figure 9.1. It can be seen from the Figure that the resolved component in the estimated horizontal plane, denoted g_x, is given by:

$$g_x = -g \sin (\theta - \theta') \tag{9.1}$$

θ' may be adjusted until g_x becomes zero, at which time $\theta' = \theta$, i.e. the estimated body angle becomes equal to the true body angle and the estimated reference frame becomes coincident with the true reference frame.

Given accurate measurements of the specific force acceleration, this process allows the orientation of the axis set defined by the accelerometers with respect to the local vertical to be defined accurately, and is analogous to the process of levelling the stable element in a platform inertial navigation system.

Having defined the local horizontal plane, and so effectively achieved a level in the alignment process, it is then necessary to determine the heading or azimuthal orientation of the inertial instrument frame in the horizontal plane, i.e. to determine direction with respect to true north. This is achieved from knowledge of the true components of Earth's rate in the local geographic frame. Assuming that the gyroscopes are of sufficient precision to detect Earth's rate accurately, the stored attitude information is now adjusted until the resolved component of the measured rate in the east direction reduces to zero. A diagram illustrating the alignment in azimuth is shown in Figure 9.2.

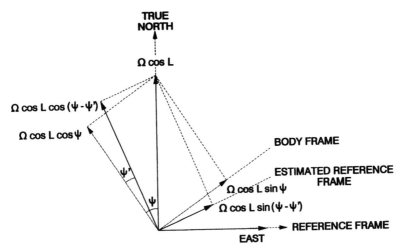

Figure 9.2   Alignment in azimuth

In this case, ψ is the true orientation of the x-axis of the instrument frame with respect to true north whilst ψ' is the estimate of that quantity. The components of Earth's rate (Ω) detected by the x and y axis gyroscopes shown in the figure are $\Omega \cos L \sin \psi$ and $\Omega \cos L \cos \psi$ respectively, where L is the latitude of the aligning system. The east component of Earth's rate as determined by the navigation system, denoted ω_E, may be expressed as follows:

$$\omega_E = \Omega \cos L \sin(\psi - \psi') \tag{9.2}$$

ψ' is adjusted until ω_E becomes zero, in which case $\psi' = \psi$.

9.2.2 Alignment on a moving platform

In order to align a strapdown inertial navigation system in a moving vehicle, a technique which is similar in principle to that described above may be used. However, when aligning in a moving vehicle, the accelerations and turn rates to which the system is subjected are no longer well defined in the way that they are when the system is stationary. It therefore becomes necessary to provide some independent measure of these quantities against which the measurements generated by the aligning system may be compared.

Consider the situation depicted in **Figure 9.3** in which the axes defined by the strapdown sensors are shown rotated through an angle θ in a single plane with respect to the navigation reference frame.

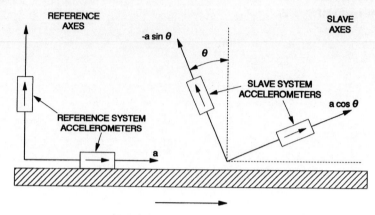

APPLIED ACCELERATION (a)

Figure 9.3   Measurement matching alignment in a single plane

If the acceleration of the vehicle in the reference x direction is a, then the accelerations sensed by the strapdown system accelerometers will be as follows:

$$a_x = a \ \cos \theta$$
$$a_y = a \ \sin \theta \qquad\qquad (9.3)$$

In the absence of any instrument measurement inaccuracies, alignment of the strapdown system may be achieved by resolving the accelerometer measurements through an angle θ' and adjusting its magnitude using a feedback process so as to null the difference between the resolved components of the slave system measurements and the accelerations measured by the reference system.

Mathematically, θ' is adjusted to allow the following relationships to be satisfied:

$$a_x \cos \theta' + a_y \sin\theta' = a$$
$$a_x \sin \theta' - a_y \cos \theta' = 0 \qquad\qquad (9.4)$$

Substituting for a_x and a_y from eqn. 9.3 yields:

$$a \cos(\theta - \theta') = a$$
$$a \sin(\theta - \theta') = 0 \qquad\qquad (9.5)$$

It can be seen these relationships will be satisfied when $\theta' = \theta$.

Therefore, it is possible to determine the orientation of the strapdown sensors by comparing the accelerometer measurements resolved into the reference frame with independent measurements of these same quantities. An estimate of θ can also be derived in a similar manner by comparing angular rate measurements. Whichever method is adopted, it will be noted that alignment about a given axis is dependent on the measurement of an acceleration or turn rate taking place along or about an axis which is orthogonal to the axis in which the misalignment exists.

As an alternative to the type of procedure described above, alignment may be achieved by comparing estimates of velocity or position generated by the strapdown system with similar estimates provided by an external source over a period of time. Velocity and position errors will propagate with time as a result of the angular alignment errors, therefore, any difference in the velocity and position estimates generated between the aligning system and the external source over this time will be partially the result of an alignment error. Such methods are discussed in more detail below in the context of in-flight and shipboard alignment.

With aircraft and shipboard systems, the independent measurement information may be provided by a separate inertial navigation system onboard the same vehicle. By comparing the two sets of inertial measurements it is possible to deduce the relative orientation of the two frames on a continuous basis. The precise measurements available will be dependent on the reference system mechanisation onboard the ship or aircraft. As a rule, a stable platform navigation system will only output estimates of position, velocity, attitude and heading. A strapdown reference system offers greater flexibility, potentially providing linear acceleration and angular rate information in addition to the usual navigation outputs listed above. Alternatively, position fixes may be derived onboard the vehicle from signals transmitted by a radio beacon or a satellite system.

9.3 Alignment on the ground

9.3.1 Introduction

Attention is now turned to the alignment of an inertial navigation system in a ground-based vehicle. Clearly, the scope for carrying out manoeuvres or applying motion to aid the process of alignment is very limited in such applications. Attention is focused here on a requirement which often arises in practice, that of determining the orientation of a set of sensor axes with respect to the local geographic frame. For convenience, the local geographic axis set is often chosen to be the reference frame.

In the past, a site survey would be carried out to establish a north line. Heading information would then be transferred to the aligning navigation system using theodolites and a prism attached to the aligning system. Although high accuracy can be obtained using this approach, it is both time consuming and labour intensive. The methods discussed below are usually more convenient to implement and avoid such problems.

9.3.2 Ground alignment methods

In principle, the techniques outlined in Section 9.2 for the self alignment of a strapdown inertial system on a stationary platform can be used. We now look in more detail at the computation required to implement that alignment process. As described above, the objective of the angular alignment process is to determine the direction cosine matrix, C_n^b, or its quaternion equivalent, which relates the body and geographic reference frames. The body mounted sensors will measure components of the specific force needed to overcome gravity and components of Earth's rate, denoted by the vector quantities g^b and ω_{ie}^b respectively. These vectors are related to the gravity and Earth's rate vectors specified in the local geographic frame, g^n and ω_{ie}^n respectively, in accordance with the following equations:

$$g^b = C_n^b g^n \tag{9.6}$$

$$\omega_{ie}^b = C_n^b \omega_{ie}^n \tag{9.7}$$

where $g^n = [0 \ 0 \ -g]^T$ and $\omega_{ie}^n = [\Omega \cos L \ 0 \ -\Omega \sin L]^T$ in which Ω and L denote Earth's rate and latitude respectively. Given knowledge of these quantities, estimates of the elements of the direction cosine matrix may be computed directly from the measurements of $g^b = [g_x \ g_y \ g_z]^T$ and $\omega_{ie}^b = [\omega_x \ \omega_y \ \omega_z]^T$ as follows:

$$
\begin{aligned}
c_{31} &= -\frac{g_x}{g} & c_{11} &= \frac{\omega_x}{\Omega \cos L} - \frac{g_x \tan L}{g} \\
c_{32} &= -\frac{g_y}{g} & c_{12} &= \frac{\omega_y}{\Omega \cos L} - \frac{g_y \tan L}{g} \\
c_{33} &= -\frac{g_z}{g} & c_{13} &= \frac{\omega_z}{\Omega \cos L} - \frac{g_z \tan L}{g}
\end{aligned}
\tag{9.8}
$$

where c_{11}, c_{12}, ... c_{33} are elements of the direction cosine matrix $\mathbf{C}_n^b$. The remaining direction cosine elements (c_{21}, c_{22} and c_{23}) may be determined by making use of the orthogonality properties of the direction cosine matrix which yield:

$$c_{21} = -c_{12}\,c_{33} + c_{13}\,c_{32}$$

$$c_{22} = c_{11}\,c_{33} - c_{31}\,c_{13}$$

$$c_{23} = -c_{11}\,c_{32} + c_{31}\,c_{12} \qquad (9.9)$$

It can be seen from the above equations that the direction cosine matrix is uniquely defined provided that L is not equal to $\pm 90°$, i.e. provided that the aligning system is not located at either the north or south poles of the Earth. This clearly would lead to singularities in the equations for some of the direction cosine elements which therefore become indeterminate. However, over much of the Earth's surface, a single set of inertial measurements can provide all of the information needed to compute the direction cosine matrix, and so achieve a strapdown system alignment.

The accuracy with which such an alignment can be accomplished is largely determined by the precision of the available measurements and the resolution of the instrument outputs. As a result of instrument biases, the above procedure will yield an estimate of the direction cosine matrix $\tilde{\mathbf{C}}_b^n$ which will be in error. As described in Chapter 11, $\tilde{\mathbf{C}}_b^n$ may be expressed as the product of the true matrix $\mathbf{C}_b^n$ and a matrix $\mathbf{B}$ which represents the misalignment between the actual and computed geographic frames:

$$\tilde{\mathbf{C}}_b^n = \mathbf{B}\mathbf{C}_b^n \qquad (9.10)$$

For small angular misalignments, $\mathbf{B}$ can be written in skew symmetric form as:

$$\mathbf{B} = \mathbf{I} - \mathbf{\Psi} \qquad (9.11)$$

where $\mathbf{I}$ is a 3×3 identity matrix and

$$\mathbf{\Psi} = \begin{pmatrix} 1 & -\delta\gamma & \delta\beta \\ \delta\gamma & 1 & -\delta\alpha \\ -\delta\beta & \delta\alpha & 1 \end{pmatrix} \qquad (9.12)$$

$\delta\alpha$, $\delta\beta$ and $\delta\gamma$ are the misalignments about the north, east and vertical axes of the geographic frame respectively, and are equivalent to the physical misalignments of the instrument cluster in a stable platform navigation system. The 'tilt' errors ($\delta\alpha$ and $\delta\beta$) which result are predominantly determined by the accelerometer biases, whilst the azimuth or heading error ($\delta\gamma$) is a function of gyroscopic bias as described next.

The direction cosine matrix, $\tilde{\mathbf{C}}_b^n$, is adjusted through the alignment process until the residual north and east components of accelerometer bias are offset by components of g in each of these directions, effectively nulling the estimates of acceleration in these directions. The resulting attitude errors correspond to the 'tilt' errors which arise when aligning a stable platform system. In azimuth, the platform rotates about the vertical to a position where a component of the Earth's horizontal rate ($\Omega \cos L$) appears about the east axis to null the east gyroscopic bias. An equivalent process takes place in a strapdown system, again through appropriate adjustment of the direction cosine matrix.

The resulting attitude and heading errors may be expressed as follows for the particular situation in which the body frame is nominally aligned with the geographic frame, i.e. where $\mathbf{C}_b^n = \mathbf{I}$, it can be shown that:

$$\delta\alpha = \frac{B_y}{g}$$

$$\delta\beta = -\frac{B_x}{g} \tag{9.13}$$

$$\delta\gamma = \frac{D_y}{\Omega \cos L}$$

More generally, where the system is not aligned with the geographic frame, the sensor biases arising in each of the above equations will be made up of a combination of the biases in all three gyroscopes or all three accelerometers.

It can be shown using these equations that a 1 mg accelerometer bias will give rise to a level error of 1 mrad (~3.4 arc minutes) whilst a gyroscopic drift of 0.01°/hour will result in an azimuthal alignment error of 1 mrad at a latitude of 45°. It is clear that good quality gyroscopes are needed to achieve an accurate alignment in azimuth. It is noted that for some inertial system applications, it is the alignment requirements which can dictate the specification of the inertial sensors rather than the way in which the sensor errors propagate during navigation.

The alignment method as described here, using a single set of instrument measurements, would allow only a coarse alignment to take place. To achieve a more accurate estimate of the direction cosine matrix, sequential measurements would be used to carry out a self alignment over a period of time. Some Kalman filtering of the measurement data would normally be applied under these circumstances.

In addition to the alignment error mechanisms described above, errors in azimuth also arise as a result of gyroscopic random noise (n) and accelerometer bias instability (b). Noise on the output of the gyroscopes (random walk in angle), which is of particular concern in systems using mechanically dithered ring laser gyroscopes, gives rise to an RMS azimuth alignment error which is inversely proportional to the square root of the alignment time (t_a), viz. $\delta\gamma = n/\Omega \cos L \sqrt{t_a}$. Therefore, given a random walk error of 0.005°/hr, an alignment accuracy of 1 mrad can be achieved at a latitude of 45° in a period of 15 minutes. The effect of this noise can be reduced by extending the alignment period, i.e. extending the time over which the noise is filtered. Small changes in the north component of accelerometer bias with time also introduce an azimuth alignment error which may be expressed as $\delta\gamma = b/g\Omega \cos L$. A bias drift of 1 µg/second will result in an alignment error of 20 mrad at a latitude of 45°. The minimisation of bias shifts with temperature as well as switch-on transients is vital for applications where this effect becomes significant.

Vehicle perturbations

A process very similar to that described above may be adopted to align an inertial navigation system mounted in a vehicle which is not perfectly stationary, but subjected to disturbances. For instance, it may be required to align a navigation system in an aircraft on a runway preparing for take-off which is being buffeted by the wind and perturbed by engine vibration. In such a situation, the mean attitude of the aligning system with respect to the local geographic frame is fixed, and the specific force and turn rates to which the aligning system is subjected are nominally fixed. Some form of base motion isolation is needed to allow the alignment errors to be deduced from the measurements of turn rate and specific force provided by the sensors [1].

A self alignment may be carried out in the presence of the small perturbations using a Kalman filter incorporating a model of the base motion disturbance. Failure to take account of any filter measurement differences caused by the disturbances will result in an incorrect alignment, since the measurements of the disturbance will be interpreted incorrectly as resulting from alignment errors. The application of Kalman filtering techniques for the alignment of strapdown inertial navigation systems is discussed more fully in Sections 9.4 and 9.5 in relation to the alignment of such systems in flight and at sea.

9.3.3 Northfinding techniques

In view of the limitations of both of the aforementioned techniques, various designs for special purpose equipment, which would allow the directions of the local vertical and true north to be defined within a land-based vehicle, have been produced. Such devices, often referred to as northfinders, are designed with a view to establishing the direction of true north within a short period of time using relatively inexpensive inertial sensors.

One possible mechanisation uses measurements of two orthogonal components of Earth's rate to establish a bearing angle of a pre-defined case reference axis with respect to north. The sensing element is a two-degree of freedom gyroscope such as a dynamically tuned gyroscope (DTG) with its spin axis vertical. The DTG assembly is suspended by a wire to provide automatic levelling of the two input axes which are at right angles to one another. Hence, the input axes are maintained in the horizontal plane. The input axes are held in a torque rebalance loop to provide measurements of the rate of turn about each axis. The pendulous assembly is enclosed within a container which is filled with a fluid to provide damping.

In this configuration, the gyroscope measures two horizontal components of the Earth's rotation rate as indicated in Figure 9.4.

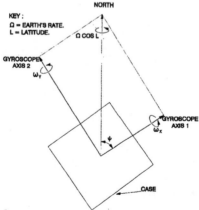

Figure 9.4   A northfinder

The angular rates (ω_x and ω_y) measured about the two input axes of the gyroscope may be expressed as follows:

$$\omega_x = \Omega \cos L \cos \psi$$
$$\omega_y = \Omega \cos L \sin \psi$$

(9.14)

where Ω = Earth's rate, L = latitude, ψ = heading of gyroscope axis with respect to true north.

By taking the ratio of the two independent gyroscopic measurements, the latitude dependent terms cancel, allowing the gyroscope heading angle, ψ, to be computed:

$$\frac{\omega_y}{\omega_x} = \frac{\Omega \cos L \sin \psi}{\Omega \cos L \cos \psi} = \tan \psi$$

$$\psi = \arctan\left(\omega_y / \omega_x\right) \tag{9.15}$$

Heading can be calculated in this way provided $\omega_x \neq 0$. In the event that ω_x is close to zero, the following equation may be used:

$$\psi = 90 - \arctan\left(\omega_x / \omega_y\right) \tag{9.16}$$

It can be seen that the northfinder does not require knowledge of latitude, or prior orientation in any particular direction, to enable a measure of heading to be obtained.

In order to achieve useful accuracy from a device of this type, gyroscope measurement accuracy of 0.005°/hour or better may be required. However, the need for a highly accurate gyroscope may be avoided by rotating the entire sensor assembly through 180° about the vertical, without switching it off, and then taking a second pair of measurements in this new orientation. The measurements obtained in each position are then differenced, allowing any biases on the measurements to be largely eliminated. The heading angle is then computed from the ratio of the measurement differences.

The rotation of the sensor may be accomplished using a small d.c. motor to drive the assembly from one mechanical stop to another which are nominally 180° apart. The stops are positioned so that the gyroscope input axes are aligned with the case reference axis, or at right angles to it, when the measurements are taken. Over the short period of time required to rotate the sensor (typically five seconds) and to take these measurements, all but the gyroscope in-run random measurement errors can be removed. This technique also helps to reduce any errors arising through the sensitive axes of the gyroscope not being perfectly horizontal.

There are a number of variations of this method, one of which involves positioning the gyroscope with one of its input axes vertical and the spin axis in the horizontal plane. Two measurements of the horizontal component of Earth's rate are taken with the gyroscope in two separate orientations 90° apart. An estimate of heading can then be obtained from the ratio of these two measurements in the manner described above. This scheme allows the northfinder to be used as a directional gyroscope after the heading angle has been determined.

Other variations incorporate accelerometers to allow the inclination to the vertical to be determined as well as heading.

9.4 In-flight alignment

9.4.1 Introduction

The requirement frequently arises to align an inertial navigation system in an air launched missile prior to missile release from an aircraft platform. A convenient reference for this purpose may be provided by the aircraft's own inertial navigation system. Such an alignment of the missile system may therefore be achieved by the transfer of data from the aircraft's navigation system to the missile by a process known as transfer alignment. This may be achieved quite simply by the direct copying of data from the aircraft to the missile navigation system, or more precisely by using some form of inertial measurement matching process of the type outlined in Section 9.2.2. Alternatively, the missile inertial navigation system may be aligned in flight using position fixes provided by satellite signals or airborne radar systems. All such methods are discussed below, but with particular emphasis on the use of transfer alignment.

It is noted that it is sometimes neither desirable nor possible to have the inertial system in a guided missile run-up and aligned waiting for the launch command. In this situation, it is necessary to align the missile's inertial navigation system very rapidly, immediately prior to launch.

9.4.2 Sources of error

As a result of physical misalignments between different mounting locations on an aircraft, the accuracy with which inertial data can be transferred from one location to another onboard the aircraft will be restricted. Such errors may be categorised in terms of static and dynamic components as follows:

static errors will exist as a result of manufacturing tolerances and imprecise installation of equipment leading to mounting misalignments between different items of equipment on the aircraft.

dynamic errors will exist because the airframe will not be perfectly rigid and will bend in response to the aerodynamic loading on the wings and launch rails to which a missile is attached. Such effects become particularly significant in the presence of aircraft manoeuvres. Significant error contributions can also be expected to arise as a result of vibration.

Methods of alleviating such problems are discussed in the following section.

9.4.3 In-flight alignment methods

Attention is focused here on the alignment of an inertial navigation system contained in an air-launched missile which may be attached to a fuselage or wing pylon beneath an aircraft.

9.4.3.1 One-shot transfer alignment

One of the simplest alignment techniques which may be adopted in this situation is to copy position, velocity and attitude data from the aircraft's own navigation system directly to the missile system. This is sometimes referred to as a one-shot alignment process and is depicted in Figure 9.5.

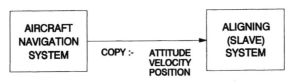

Figure 9.5   One-shot transfer alignment

Clearly, any angular displacement between the aircraft and missile systems which exists at the instant that the data are transferred will appear as an alignment error in the missile system. Therefore, the success of such a scheme is reliant on the two systems being physically harmonised to high accuracy, or on accurate knowledge of their relative orientation being available when the alignment takes place. In the latter situation, the data from the aircraft's navigation system may be resolved accurately in missile axes before being passed to the missile navigation system.

In general, the precise harmonisation of one system with respect to the other will not be known, for the reasons outlined in the previous section. Further, the aircraft navigation system will be positioned some distance from the aligning system in the missile and there will be relative motion between them should the aircraft turn or manoeuvre; the so-called lever-arm motion. In this situation the velocity information passed to the missile will be in error. As a result, the accuracy of alignment which can be achieved using a one-shot alignment procedure will be extremely limited and more precise methods are usually sought.

9.4.3.2 Airborne inertial measurement matching

An alternative method of transfer alignment, which has received much attention in recent years [2–5], is that of inertial measurement matching. This technique relies on the comparison of measurements of applied motion obtained from the two systems to compute the relative orientation of their reference axes, as introduced in the discussion of basic principles in Section 9.2 and depicted in Figure 9.6. An initial coarse alignment may be achieved by the one-shot process, discussed above, before initiating the measurement matching process which is described below.

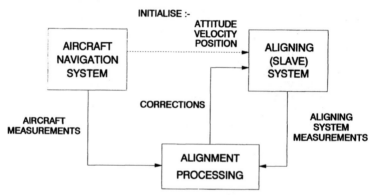

Figure 9.6   Inertial measurement matching alignment scheme

In theory, a transfer alignment between two inertial navigation systems on an aircraft can be achieved most rapidly by comparing measurements generated by the aircraft system and the missile system of the fundamental navigation quantities of specific force acceleration and angular rate, resolved into a common co-ordinate frame. In the absence of measurement errors, and assuming that the two systems are mounted side by side on a perfectly rigid platform, the measurement differences arise purely as a result of alignment errors. Under such conditions, it is possible to identify accurately the misalignments between the two systems.

In practice this approach is often impractical for a number of reasons. The reference system may use stable platform technology, in which linear acceleration and turn rate data are not standard outputs.

This is particularly true in the case of many military aircraft, although the situation is likely to change in the near future with the wider use of strapdown technology for combat aircraft inertial navigation systems.

There are also technical reasons which may preclude the use of linear acceleration and angular rate matching procedures as a viable option for airborne transfer alignment. This is particularly true where the physical separation between the reference and aligning systems is large, and where significant flexure motion is present. The turn rates and linear accelerations sensed by the reference and aligning systems will differ as a result of the flexure motion which is present. These differences will then be interpreted incorrectly as errors in the stored attitude data, and so degrade the accuracy of alignment which can be achieved. Acceleration matching and angular rate matching are particularly sensitive to the effects of flexure. Although it is possible theoretically to model the flexural motion, and thus separate the components of the measurement differences caused by flexure from those attributable to alignment errors, adequate models of such motion are rarely available in practice.

Even when attempting to carry out an alignment on a perfectly rigid airframe, the translational motion sensed at the reference and the aligning system locations will differ, as the aircraft rotates, as a result of lever-arm motion. The measurement differences which arise as a result of lever-arm motion as the aircraft manoeuvres will also be interpreted incorrectly as alignment inaccuracies and therefore inhibit the alignment process. These additional measurement differences are functions of aircraft turn rate, angular acceleration and the physical separation between the two systems. Although it is theoretically possible to correct one set of measurements before comparison with the other, such corrections are dependent on the availability of sufficiently precise estimates of these quantities. It is reasonable to assume that distance would be known to sufficient accuracy and the actual turn rates may be provided directly by a strapdown system, but angular acceleration measurements are not usually available and, without the use of angular accelerometers, are not easy to estimate.

For the reasons outlined above, acceleration and rate matching are not generally recommended for alignment of inertial systems onboard aircraft, even when both the reference and aligning (slave) systems are configured in a strapdown form. An alternative approach is the use of velocity matching described in Section 9.4.3.3. Velocity errors propagate in an inertial navigation system as a result of alignment inaccuracy, as well as through inertial instrument imperfections. By comparing the velocity estimates provided by the reference and aligning systems, it may therefore be possible to obtain estimates of the alignment errors and, under some circumstances, estimates of the sensor biases. Hence, it is possible to achieve a measure of sensor calibration as part of the same process.

Because of the smoothing effect of the integration process which takes place between the raw measurements from the instruments and the velocity estimates within an inertial navigation system, the effects of flexure and sensor noise on the process of alignment are much less severe than experienced with acceleration matching. Further, it has the advantage of allowing lever-arm corrections to be implemented more easily, such corrections at the velocity level being purely functions of turn rate and separation distance.

9.4.3.3 Velocity matching alignment

As suggested in the preceding section, an in-flight alignment may be achieved by comparing estimates of velocity generated by the aligning system with estimates of the same quantities provided by the aircraft's own navigation system. The nature of the alignment problem, which involves the identification of a number of interrelated and time varying error sources using measurements which are corrupted with noise, is well suited to statistical modelling techniques. These techniques include Kalman filtering, the principles of which are discussed in Appendix A.

This section outlines the system and measurement equations required to construct a Kalman filter which may be used to process the velocity information and so obtain estimates of the alignment errors. For the purposes of this Kalman filter illustration, a number of simplifying assumptions have been made in the formulation given here and these are described below.

The system equations

It is necessary to determine accurately the attitude and velocity of the aligning system with respect to a designated reference frame. Typically, this may be a body fixed axis set within the aircraft or the local geographic navigation frame. The aligning system and reference frames are denoted here by the superscripts and subscripts b and n respectively. Following the notation used in Chapter 3, the propagation of the direction cosine matrix (C_b^n) which relates the sensor axes of the aligning system to the reference frame is governed by the following differential equation:

$$\dot{C}_b^n = C_b^n \Omega_{nb}^b \qquad (9.17)$$

where Ω_{nb}^b is a skew symmetric matrix formed from the turn rates of the aligning system with respect to the reference frame. This turn rate is obtained by differencing the angular rates sensed by the aligning system (ω_{ib}^b) and the turn rate of the reference frame (ω_{in}^n). An estimate of the direction cosine matrix, denoted $\hat{C}_b^n$, is calculated using measurements of the turn rate to which the aligning system is subjected $(\hat{\omega}_{ib}^b)$ and an estimate of the reference frame rate $(\hat{\omega}_{in}^n)$ to determine $\hat{\Omega}_{nb}^b$, updating from some initial estimate using:

$$\dot{\hat{C}}_b^n = \hat{C}_b^n \hat{\Omega}_{nb}^b \qquad (9.18)$$

As described in Section 9.3.2, for small angle misalignments, the true and estimated direction cosine matrices may be related by the equation:

$$\hat{C}_b^n = [I - \Psi]C_n^b \qquad (9.19)$$

where I is the identity matrix and Ψ is a skew symmetric matrix which may be written as:

$$\Psi = \begin{pmatrix} 0 & -\delta\gamma & \delta\beta \\ \delta\gamma & 0 & -\delta\alpha \\ -\delta\beta & \delta\alpha & 0 \end{pmatrix}$$

in which the off diagonal elements $\delta\alpha$, $\delta\beta$ and $\delta\gamma$ represent the attitude errors in the aligning system.

It can be shown that the attitude errors propagate according to:

$$\dot{\psi} = -\omega_{in}^n \times \psi - C_b^n \delta\omega_{ib}^b + \delta\omega_{in}^n \qquad (9.20)$$

where $\quad \psi = [\delta\alpha \ \delta\beta \ \delta\gamma]^T$ the alignment error vector

$\delta\omega_{ib}^b = (\hat{\omega}_{ib}^b - \omega_{ib}^b)$ the gyroscopic measurement errors in the aligning system

$\delta\omega_{in}^n = (\hat{\omega}_{in}^n - \omega_{in}^n)$ the errors in the reference frame rate estimates

$\times$ denotes the cross product of two vector quantities

For the purposes of this example of Kalman filter formulation, the gyroscopic errors are modelled in the filter as additive Gaussian white noise and the reference rate errors are assumed to be zero. The derivation of this equation is given in Chapter 11 where the propagation of errors in strapdown inertial navigation systems is discussed in greater detail.

The velocity equations may be expressed approximately as:

$$\dot{v}^n = C_b^n f^b - g \qquad (9.21)$$

where $\mathbf{v}^n$ is the velocity of the aircraft, $\mathbf{f}^b$ is the specific force sensed by the accelerometers in the aligning system in body axes and $\mathbf{g}$ is the local gravity vector. The propagation of the errors in the estimates of velocity computed by the aligning system ($\delta\mathbf{v}$) may be expressed as:

$$\dot{\delta\mathbf{v}} = \mathbf{f}^n \times \mathbf{\psi} + \mathbf{C}_b^n \, \delta\mathbf{f}^b \tag{9.22}$$

where $\mathbf{f}^n$ is the specific force measured by the aligning system resolved in reference axes and $\delta\mathbf{f}^b$ represents the errors in the accelerometer measurements. This is modelled in the Kalman filter as additive Gaussian white noise.

Eqns. 9.20 and 9.21 may combined and expressed in state space form as:

$$\dot{\delta\mathbf{x}} = \mathbf{F}\,\delta\mathbf{x} + \mathbf{D}\mathbf{w} \tag{9.23}$$

where $\delta\mathbf{x}$ is the error state vector, $\mathbf{F}$ is the system error matrix, $\mathbf{D}$ is the noise input matrix and $\mathbf{w}$ is the system noise which represents the instrument noise together with any unmodelled biases. The error state vector may be expressed in component form as:

$$\delta\mathbf{x} = [\delta\alpha \quad \delta\beta \quad \delta\gamma \quad \delta v_N \quad \delta v_E]^T \tag{9.24}$$

where $\delta\alpha$, $\delta\beta$, $\delta\gamma$ are the components of the vector $\mathbf{\psi}$, the attitude errors, and δv_N, δv_E are the north and east velocity errors.

The error equation may be expressed in full as follows:

$$\begin{pmatrix} \dot{\delta\alpha} \\ \dot{\delta\beta} \\ \dot{\delta\gamma} \\ \dot{\delta v_N} \\ \dot{\delta v_E} \end{pmatrix} = \begin{pmatrix} 0 & \omega_D & -\omega_E & 0 & 0 \\ -\omega_D & 0 & \omega_N & 0 & 0 \\ \omega_E & -\omega_N & 0 & 0 & 0 \\ 0 & -\mathbf{f}_D & \mathbf{f}_E & 0 & 0 \\ \mathbf{f}_D & 0 & -\mathbf{f}_N & 0 & 0 \end{pmatrix} \begin{pmatrix} \delta\alpha \\ \delta\beta \\ \delta\gamma \\ \delta v_N \\ \delta v_E \end{pmatrix}$$

$$+ \begin{pmatrix} -c_{11} & -c_{12} & -c_{13} & 0 & 0 & 0 \\ -c_{21} & -c_{22} & -c_{23} & 0 & 0 & 0 \\ -c_{31} & -c_{32} & -c_{33} & 0 & 0 & 0 \\ 0 & 0 & 0 & c_{11} & c_{12} & c_{13} \\ 0 & 0 & 0 & c_{21} & c_{22} & c_{23} \end{pmatrix} \begin{pmatrix} w_{gx} \\ w_{gy} \\ w_{gz} \\ w_{ax} \\ w_{ay} \\ w_{az} \end{pmatrix} \tag{9.25}$$

where $\omega_N = \Omega \cos L + v_E/(R_0 + h)$
$\omega_E = -v_N/R_0$
$\omega_D = -\Omega \sin L - v_E \tan L/(R_0 + h)$

Ω	= Earth's rate
L	= latitude
R_0	= radius of the Earth
h	= aircraft altitude
f_N f_E f_D	= north, east and vertical components of vehicle acceleration
c_{11} c_{12}....	= direction cosine elements of the matrix $\mathbf{C}_b^n$
w_{gx} w_{gy} w_{gz}	= gyroscope noise components
w_{ax} w_{ay} w_{az}	= accelerometer noise components

It can be seen from the system error eqn. 9.22 that an acceleration of the aircraft in the north or east direction is required to cause the azimuthal misalignment ($\delta\gamma$) to propagate as a velocity error.

The error model may be augmented by modelling the gyroscope and accelerometer errors explicitly. For example, additional states may be included to represent the fixed biases in the sensor measurements.

To enable the Kalman filter to be mechanised in discrete form, the system error model is converted to a difference equation by integrating between successive measurement instants to give:

$$\delta\mathbf{x}_{k+1} = \mathbf{\Phi}_k \delta\mathbf{x}_k + \mathbf{w}_k \tag{9.26}$$

where $\mathbf{\Phi}_k = \exp[\mathbf{F}_k(t_{k+1} - t_k)]$, the system transition matrix between time t_k and t_{k+1} and $\mathbf{w}_k$ is a zero-mean white noise sequence.

The measurement equations

The measurements of north and east velocity provided by the aircraft's navigation system constitute the Kalman filter measurements ($\tilde{\mathbf{z}}$):

$$\tilde{\mathbf{z}} = \begin{pmatrix} \tilde{v}_N \\ \tilde{v}_E \end{pmatrix} \tag{9.27}$$

Estimates of these measurements ($\hat{\mathbf{z}}$) are obtained from the aligning system:

$$\hat{\mathbf{z}} = \begin{pmatrix} \hat{v}_N \\ \hat{v}_E \end{pmatrix} \tag{9.28}$$

Where the reference and aligning systems are installed some distance apart on the aircraft, it will be necessary to compensate for the rotation induced velocity components, $\mathbf{v}_r$, the lever-arm motion. Such corrections are calculated using measurements of the aircraft's turn rate ($\boldsymbol{\omega}_a$) and knowledge of the physical separation between the two systems ($\mathbf{r}$) using $\mathbf{v}_r = \boldsymbol{\omega}_a \times \mathbf{r}$ resolved in the reference frame.

Measurements of ω_a may be provided either by the aircraft's navigation system or by the aligning system with sufficient accuracy.

The velocity measurements are compared at each measurement update to generate the filter measurement differences or innovations, denoted as δz, where:

$$\delta z = \begin{pmatrix} \tilde{v}_N - \hat{v}_N \\ \tilde{v}_E - \hat{v}_E \end{pmatrix} = \begin{pmatrix} -\delta v_N \\ -\delta v_E \end{pmatrix} \tag{9.29}$$

The measurement differences at time t_k (δz_k) may be expressed in terms of the error states (δx_k) as follows:

$$\delta z_k = H_k \, \delta x_k + v_k \tag{9.30}$$

where H_k is the Kalman filter measurement matrix which takes the following form:

$$H_k = \begin{pmatrix} 0 & 0 & 0 & -1 & 0 \\ 0 & 0 & 0 & 0 & -1 \end{pmatrix} \tag{9.31}$$

and v_k is the measurement noise vector. This represents the noise on the reference measurements and model mismatch introduced through aircraft flexure and lever-arm motion.

The Kalman filter

In eqns. 9.23 and 9.30, we have the necessary system and measurement equations with which to construct a Kalman filter. The form of the filter equations is given in Appendix A at the end of the book.

The filter provides estimates of the attitude errors and the north and east velocity errors. These estimates are used to correct the aligning system estimates of attitude and velocity after each measurement update. Where instrument bias states are included in the error model, the bias estimates so generated may be used to correct the sensor outputs as part of the alignment process. A block diagram representation of the alignment scheme is given in Figure 9.7.

Although it is often recommended that the aircraft should perform a well defined manoeuvre to aid the alignment process. An example is the weave trajectory illustrated in Figure 9.8. Analysis of the problem has shown that an alignment can often be achieved in the presence of relatively small perturbations, as would be experienced normally during flight.

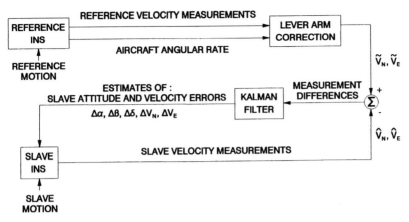

Figure 9.7   Velocity matching alignment scheme

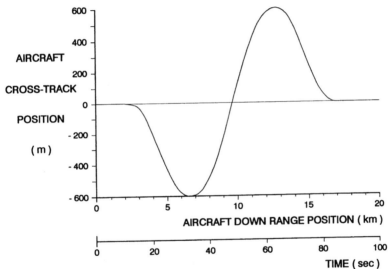

Figure 9.8   Aircraft alignment/calibration manoeuvre

Example results

Some simulation results which illustrate the alignment that may be achieved using velocity matching are given in Figure 9.9. The results show the reduction in the alignment error of an airborne navigation system, over a period of 100 seconds, as the aircraft executes a weave manoeuvre, and have been obtained using a filter formulation similar to that described, but with the addition of instrument bias states. These results were obtained using a typical aircraft grade inertial system, capable of navigating to an accuracy of one nautical mile per hour, to provide the reference measurements. The aligning system was of sub-inertial quality incorporating gyroscopes and accelerometers with 1σ biases of 10°/hour and 2 milli-g respectively.

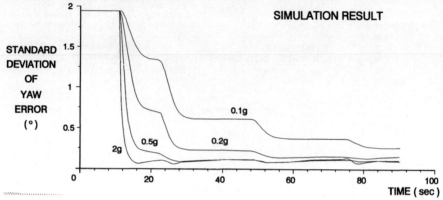

Figure 9.9   Alignment by velocity matching in the presence of an aircraft weave manoeuvre

The Figure shows the reduction in the standard deviation of the yaw error as a function of time. The roll and pitch errors, which are not shown here, converge very rapidly as the system effectively aligns itself to the local gravity vector. The accuracy of alignment in the horizontal plane is limited by any residual bias in the accelerometer measurements. In the case shown here, the accelerometer bias is 2 milli-g which results in tilt errors of approximately 0.1°. The yaw alignment error does not begin to converge until the aircraft commences its manoeuvre, since it only propagates as a velocity error and therefore only becomes observable when the aircraft manoeuvres. The effects of the manoeuvres are clearly shown in the Figure. It can be seen that the yaw alignment error falls each time the aircraft starts to change direction.

   In the presence of more severe manoeuvres, mean errors also arise which are correlated with the motion of the aircraft. These errors must be summed with the standard deviations shown in the Figure to give the full alignment error. The bias terms are principally the result of geometric effects induced as the aircraft banks to turn. Alignment information can only be deduced about axes which are perpendicular to the direction of the applied acceleration, with the result that some redistribution of the alignment errors tends to take place as the aircraft manoeuvres.

9.4.3.4 Position update alignment

An aircraft may be equipped with various sensors or systems capable of providing position fix information which may be used to align an onboard inertial navigation system during flight. Suitable data may be provided by satellite updates [6] or generated through the use of a ground tracking radar or a terrain referenced navigation system of the type discussed in Chapter 12.

As described earlier, position errors will propagate in an inertial navigation system as a result of alignment inaccuracies. By comparing the external position fixes with the estimates of position generated by the aligning navigation system, estimates of the position errors are obtained. Based on a model of the errors in the aligning system it is possible to deduce the alignment errors from these differences in position. A block diagram of such a scheme is given in Figure 9.10.

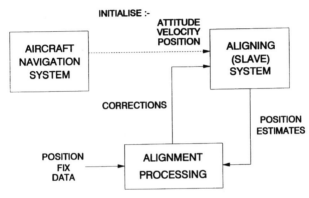

Figure 9.10   Position update alignment scheme

This method of alignment is precisely equivalent to the inertial aiding process described in Chapter 12. In the context of integrated navigation systems, or aided inertial navigation systems, the external measurements are assumed to be available throughout all or much of the period for which the navigation system is required to navigate. In the context of pre-flight alignment, it refers to the use of the external measurement data purely to carry out an alignment prior to a period of navigation. Since the principles of the method are as described in Chapter 12, no further discussion of this topic appears in this chapter.

9.5 Alignment at sea

9.5.1 Introduction

A modern warship contains a wide variety of sensors and weapon systems. In order that the ship can deploy the forces at its disposal and use them in an effective manner, all such equipment must operate in harmony. For example, information about an attacking missile or aircraft derived from a sensor at one location must be in a form that can be used to direct or control a weapon system at a different remote location.

9.5.2 Sources of error

It is common practice to set up a series of datum levels and training marks at strategic locations around the ship to which all equipment is referenced or harmonised when it is installed on the ship. In this way, it is hoped to ensure that all equipment will operate in a common frame of reference. It has long been suspected that, although the accuracy to which equipment is harmonised during the construction of the ship is very high, the accuracy of this harmonisation degrades when the ship goes to sea. This view has been reinforced by observations of ships at sea and the results of ship trials which have attempted to measure the amount by which ships bend or flex in different sea conditions. Such errors may be categorised as follows:

long-term deformations occurring through the action of ageing and the effects of solar heating. A gradual movement of the structure takes place as the ship ages and as the load state changes. It has also been observed that significant bending of the ship structure can occur under the action of solar heating. Angular variations of the order of 1° are believed to take place over the period of a day as the sun moves around the vessel;

ship flexure can occur in heavy seas as the ship moves in response to the motion of the waves, the magnitude of the angular displacement between any two locations becoming larger as the separation increases. Attempts to measure the amount by which ships flex when at sea have revealed significant angular displacements at typical ship motion frequencies of 0.1 Hz to 0.3 Hz, the dominant flexure motion being the twisting of the hull about the roll axis of the vessel. The magnitude of ship flexure is a function of sea state and the direction in which the waves are approaching the vessel. Further transient distortion may occur as the ship manoeuvres, or through the action of the stabilisers;

other abrupt changes which are expected to arise from underwater shock, induced for instance by a depth charge, and as a result of slamming in heavy seas, where the bows leave the water and impact on re-entry.

In addition, battle damage will introduce potentially very large distortions of a ship's structure, probably rendering some weapon systems ineffective unless a static re-harmonisation takes place.

9.5.3 Shipboard alignment methods

To overcome the problems outlined in the previous section, it is necessary to devise means by which the harmonisation of the various shipboard systems can be maintained under all operational conditions. Although an accurate reference is provided on naval ships by the ship's attitude and heading reference system (AHRS) or even more precisely by a ship's inertial navigation system (SINS), the accuracy with which that reference may be transferred about the ship is limited by bending and flexure of the ship. For this reason, other means are sought for the alignment of equipment onboard ships.

9.5.3.1 Shipboard transfer alignment methods

Assuming that a master reference can be maintained accurately, slave systems may be aligned to that reference. There are various methods which can be adopted to achieve this end. The simplest technique is to transfer data (attitude, velocity and position) directly from the master system to the slave using the one-shot alignment scheme described for airborne alignment. However, as with airborne alignment, any physical misalignments resulting from ship flexure will contribute directly to the errors in the aligning system if this approach is adopted.

One possible method of overcoming this limitation on a ship is to use an optical harmonisation scheme to determine the relative orientation of the master reference of the launch platform and a missile system directly. An autocollimator, fixed in one co-ordinate reference frame, may be used to determine the rotation of a reflector which is attached to the second reference frame. Although such techniques have been used in some applications, they are not generally feasible because of the difficulty of maintaining line of sight contact between the two locations which could be some considerable distance apart. For example, a missile silo in a ship may be installed 50 metres, or more, away from the ship's inertial reference system.

Alternatively, alignment may be achieved on a ship by comparing inertial measurements generated by the aligning system with similar measurements provided by an inertial reference unit [7, 8]. The velocity matching scheme described in Section 9.4 for in-flight alignment is of limited use for shipboard applications since it is dependent on a manoeuvre of the vehicle, particularly if an alignment is to take place within a short period of time. In many circumstances this may be totally impractical. Studies of shipboard alignment methods have suggested that the use of velocity and pitch rate matching offers a possible solution [8]. Such a scheme is discussed in more detail in the following section.

9.5.3.2 Shipboard inertial measurement matching

In this section, the scope for achieving an alignment at sea using velocity and angular rate matching is discussed. The application of velocity matching alone is of limited use for shipboard alignment because ships are clearly unable to manoeuvre in the way that aircraft can to aid the alignment process. However, velocity matching may be used to achieve a level alignment, since errors in the knowledge of the local vertical will cause the measurements of specific force needed to overcome gravity to be resolved incorrectly and to propagate as apparent components of north and east velocity.

On a ship, an alignment in azimuth may be achieved within a relatively short period of time by comparing angular rate measurements, provided the ship exhibits some motion in pitch or roll. The measurements may be processed using a Kalman filter based on an error model of the aligning system, as described in the context of in-flight alignment in Section 9.4. The form of the measurement equation is described below.

The measurements of turn rate provided by the reference and aligning systems are assumed to be generated in local co-ordinate frames denoted a and b respectively. The rates sensed by a triad of strapdown gyroscopes mounted at each location with their sensitive axes aligned with these reference frames may be expressed as $\boldsymbol{\omega}_{ia}^a$ and $\boldsymbol{\omega}_{ib}^b$ in line with the nomenclature used in Chapter 3. The measurements provided by the gyroscopes in the reference and aligning systems are resolved into a common reference frame, the a-frame for instance, before comparison takes place.

Hence, the reference measurements may be expressed as:

$$\tilde{\mathbf{z}} = \boldsymbol{\omega}_{ia}^a \qquad (9.32)$$

assuming errors in the measurements to be negligible. The estimates of these measurements generated by the aligning system are denoted by the $^\wedge$ notation.

$$\hat{\mathbf{z}} = \hat{\mathbf{C}}_b^a \hat{\boldsymbol{\omega}}_{ib}^b \qquad (9.33)$$

The gyroscope outputs $(\boldsymbol{\omega}_{ib}^b)$ may be written as the sum of the true rate $(\boldsymbol{\omega}_{ib}^b)$ and the error in the measurement $(\delta\boldsymbol{\omega}_{ib}^b)$ whilst the estimated direction cosine matrix may be expressed as the product of a skew symmetric error matrix, $[\mathbf{I} - \boldsymbol{\Psi}]$, and the true matrix $\mathbf{C}_b^a$ to give:

$$\hat{\mathbf{z}} = \left[\mathbf{I} - \boldsymbol{\Psi}\right]\mathbf{C}_b^a\left[\boldsymbol{\omega}_{ib}^b + \delta\boldsymbol{\omega}_{ib}^b\right]$$

Expanding the right hand side of this equation, writing $\boldsymbol{\Psi} = [\boldsymbol{\psi} \times]$ and ignoring error product terms gives:

$$\hat{\mathbf{z}} = \mathbf{C}_b^a\boldsymbol{\omega}_{ib}^b - \boldsymbol{\psi}\times\mathbf{C}_b^a\boldsymbol{\omega}_{ib}^b + \mathbf{C}_b^a\delta\boldsymbol{\omega}_{ib}^b \qquad (9.34)$$

$$= \boldsymbol{\omega}_{ib}^a + \boldsymbol{\omega}_{ib}^a\times\boldsymbol{\psi} + \mathbf{C}_b^a\delta\boldsymbol{\omega}_{ib}^b$$

The turn rate of the aligning system may be expressed as the sum of the turn rate sensed by the reference system and any ship flexure which may be present $(\boldsymbol{\omega}_f)$. Hence, eqn. 9.34 may be rewritten as follows:

$$\hat{\mathbf{z}} = \boldsymbol{\omega}_{ia}^a + \boldsymbol{\omega}_f + \boldsymbol{\omega}_{ib}^a\times\boldsymbol{\psi} + \mathbf{C}_b^a\delta\boldsymbol{\omega}_{ib}^b \qquad (9.35)$$

The measurement differences may then be written as:

$$\delta\mathbf{z} = \tilde{\mathbf{z}} - \hat{\mathbf{z}}$$

$$= -\boldsymbol{\omega}_{ib}^a\times\boldsymbol{\psi} - \mathbf{C}_b^a\delta\boldsymbol{\omega}_{ib}^b - \boldsymbol{\omega}_f \qquad (9.36)$$

The measurement differences $(\delta\mathbf{z}_k)$ at time t_k may be expressed in terms of the error states $(\delta\mathbf{x}_k)$ as follows:

$$\delta\mathbf{z}_k = \mathbf{H}_k\,\delta\mathbf{x}_k + \mathbf{v}_k \qquad (9.37)$$

where $\mathbf{H}_k$ is the Kalman filter measurement matrix which takes the following form:

$$\mathbf{H}_k = \begin{pmatrix} 0 & \omega_Z & -\omega_Y & 0 & 0 \\ -\omega_Z & 0 & \omega_X & 0 & 0 \\ \omega_Y & -\omega_X & 0 & 0 & 0 \end{pmatrix} \qquad (9.38)$$

where ω_x, ω_y and ω_z are the components of the vector $\boldsymbol{\omega}_{ib}^a$ and $\mathbf{v}_k$ is the measurement noise vector. This represents the noise on the measurements and model mismatch introduced through ship flexure.

A Kalman filter may now be constructed using the measurement eqn. 9.37 and a system equation of the form described earlier, Section 9.4.3.3 eqn. 9.23. A block diagram of the resulting alignment scheme is given in Figure 9.11.

Example result

The simulation result shown in Figure 9.12 illustrates the accuracy of alignment which may be achieved using a combination of velocity and angular rate matching. The results show the convergence of the

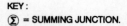

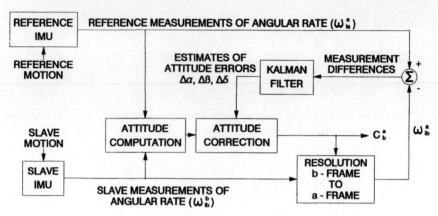

Figure 9.11   Angular rate matching alignment scheme

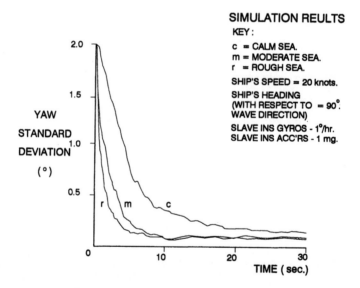

Figure 9.12   Illustration of measurement matching at sea

azimuthal alignment error in calm, moderate and rough sea conditions where the waves are approaching the ship from the side.

These results were obtained assuming no knowledge of the ship's flexure characteristics. However, the measurements of velocity included compensation for relative motion of the reference and aligning systems caused by the rotation of the ship. The aligning system contained medium grade inertial sensors with accelerometer biases of one milli-g and gyroscope biases of 1°/hour; a higher quality reference system was used. The Kalman filter used here was found to be robust in that it is able to cope with initial alignment errors of 10° or more.

The effects of ship flexure

Although it is possible in theory to model the ship's flexure explicitly in the Kalman filter and so derive estimates of the flexure rates, a sufficiently precise model is unlikely to be available in practice. Besides, this will result in a highly tuned filter which will be very sensitive to parametric variations. For these reasons, a sub-optimal Kalman filter may be used in which the flexure is represented as a noise process, as described above. The way in which ship flexure limits the accuracy of the alignment that can be achieved when using a filter of this type is demonstrated by the simplified analysis which follows.

Consider the two axis sets shown in Figure 9.13 which correspond to the orientations of the reference and aligning systems at two locations remote from each other on a ship. The reference frame is taken to be aligned perfectly with the roll, pitch and yaw axes of the ship whilst the aligning system, denoted here as the slave system, is misaligned in yaw by an angle $\delta\psi$.

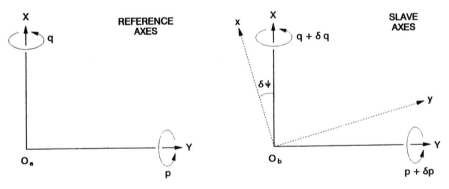

Figure 9.13   Illustration of the effects of ship flexure on axis alignment
   O_aXY denotes axes at reference system origin
   O_bXY denotes parallel reference axes at slave system origin
   O_ixy denotes slave system axes to be brought into alignment with O_bXY

The angular rates p and q sensed about the reference axes are the roll and pitch rates of the vessel respectively. The slave system senses rates $p + \delta p$ and $q + \delta q$ resolved in slave system axes, where δp and δq represent the relative angular rates between the two systems, the rates at which the ship is bending or flexing.

Consider first the mechanism by which alignment occurs in the absence of flexure. Using pitch rate matching, the rate measured by the reference system, q, is compared with the slave system rate, $q \cos \delta\psi - p \sin \delta\psi$, to yield a measurement difference δz, where:

$$\delta z = q(1 - \cos \delta\psi) + p \sin \delta\psi \qquad (9.39)$$

It can be seen from the above equation that δz becomes zero when the misalignment is zero. Hence, by adjusting $\delta\psi$ in order to null this measurement difference, it is possible to align the slave system perfectly in the absence of ship's flexure.

In the presence of ship's flexure, additional turn rates δp and δq are present at the slave system and the rate sensed about the nominal pitch axis of the slave system becomes $(q + \delta q) \cos \psi - (p + \delta p) \sin \psi$. The measurement difference is now:

$$\delta z = q(1 - \cos \delta\psi) + p \sin \delta\psi - \delta q \cos \delta\psi + \delta p \sin \delta\psi \qquad (9.40)$$

which may be expressed to first order in $\delta\psi$ as:

$$\delta z = (p + \delta p) \, \delta\psi - \delta q \qquad (9.41)$$

In this case, the measurement difference settles to zero when:

$$\delta\psi = \delta q / (p + \delta p) \qquad (9.42)$$

It is clear from this result that the magnitude of the residual yaw misalignment will reduce as the roll rate of the ship becomes larger, or as the flexure about the measurement axis, pitch in this case, becomes smaller. By a similar argument, it can be shown that the accuracy of the estimate of yaw error obtained using roll rate matching will be limited by the relative magnitude of the roll flexure and the pitch rate of the vessel. Since flexure about the roll axis is believed to be larger than the pitch rate flexure in general, and ships tend to roll more rapidly than they pitch, pitch rate matching is the preferred option.

Figure 9.14 shows the azimuthal alignment accuracy achieved as the ratio of pitch rate flexure to roll rate is varied. In line with theoretical expectations discussed here, the accuracy of alignment is shown to improve as this ratio becomes smaller.

9.5.3.3 Shipboard alignment using position fixes

Accurate harmonisation between different items of equipment on a ship may be achieved using inertial navigation systems installed alongside each item, or system, to maintain a common reference frame at each location. Such a scheme is shown in Figure 9.15.

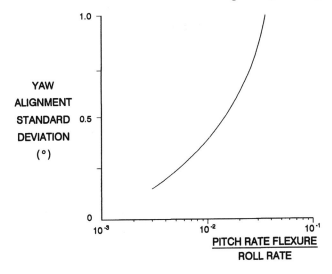

Figure 9.14   Azimuth alignment accuracy as a function of the ratio pitch rate flexure to roll rate

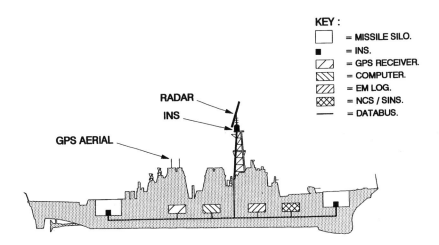

Figure 9.15   Shipboard harmonisation scheme

The reference may be maintained at each location by using accurate position fixes, provided by satellite signal updates for example. It is envisaged that each system could be equipped with a GPS satellite receiver and antenna to facilitate its alignment to the local geographic frame. Alternatively, with appropriate filtering and lever-arm corrections, a single GPS antenna and receiver could provide data for all of the inertial systems on the ship. It is noted that the GPS receiver gives the location of the phase centre of the antenna which is likely to be located at the top of a mast.

The alignment of each inertial system may be accomplished independently of ship motion, although the speed of convergence is greatly increased in the presence of a ship's manoeuvres. This technique would enable the accurate alignment of each system to be achieved, largely irrespective of any relative motion between the different locations resulting from bending of the ship's structure. Clearly, this approach is dependent on the continuing availability of satellite signals. In the event of loss of such transmissions, the period of time for which alignment can subsequently be maintained is dependent on the quality and characteristics of the sensors in each inertial unit.

References

1 BRITTING, K.R.: 'Inertial navigation system analysis' (John Wiley and Sons, 1971)
2 DEYST, J.J. and SUTHERLAND, A.A.: 'Strapdown inertial system alignment using statistical filters: a simplified formulation', *AIAA Journal*, 11 (4)
3 HARRIS, R.A. and WAKEFIELD, C.D.: 'Co-ordinate alignment for elastic bodies' (NAECON, 1977)
4 SCHULTZ, R.L. and KEYS, C.L.: 'Airborne IRP alignment using acceleration and angular rate matching'. Proceedings of joint conference on *Automatic control*, June 1973
5 BAR-ITZHACK, I.Y. and PORAT, B.: 'Azimuth observability enhancement during inertial navigation system in-flight alignment', *J. Guid. Control*, 1980, 3 (4)
6 TAFEL, R.W. and KRASNJANSKI, D.: 'Rapid alignment of aircraft strapdown inertial navigation systems using Navstar GPS'. Proceedings of NATO AGARD conference on *Precision positioning and guidance systems*, 1980
7 BROWNE, B.H. and LACKOWSKI, D.H.: 'Estimation of dynamic alignment error in shipboard fire control systems'. Proceedings of IEE conference on *Decision and control*, 1976
8 TITTERTON, D.H. and WESTON, J.L.: 'Dynamic shipboard alignment techniques'. Proceedings of DGON Symposium on *Gyro technology*, Stuttgart, Germany, 1987

Chapter 10
Strapdown navigation system computation

10.1 Introduction

The analytical equations which must be solved in order to extract attitude, velocity and position information from the inertial measurements provided by the gyroscopes and accelerometers in a strapdown system have been described in Chapter 3. This chapter is concerned with the real time implementation of these equations in a computer. The main computing tasks, those of attitude determination, specific force resolution and solution of the navigation equation, are indicated in the block diagram in Figure 10.1.

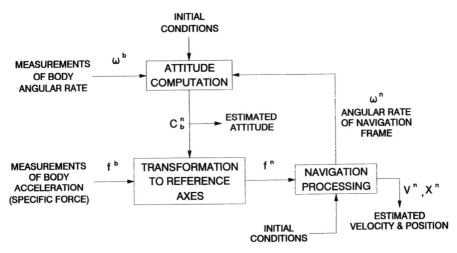

Figure 10.1   Strapdown inertial navigation computing tasks

The most demanding of the processing tasks, in terms of computer loading, are the attitude computation and the specific force vector resolution. In the presence of high frequency motion, the implementation of these tasks in real time creates a substantial computing burden for the strapdown system computer. The navigation processing task, which is common to all types of inertial navigation system, both strapdown and stable platform mechanisations, is less demanding computationally. In addition to these tasks, the determination of attitude in terms of Euler angle rotations is required in some applications.

Early attempts to produce a strapdown inertial navigation system were limited, in part, by the computer technologies available at the time. Apart from the physical size of early computers which delayed the development of strapdown systems, particularly for airborne applications, the lack of computing speed, or through-put, was a major obstacle to the achievement of fast and accurate attitude computation. As a result, the performance which could be achieved in early strapdown systems was limited, particular under high frequency motion conditions.

Such difficulties prompted much effort to be directed towards the development of efficient computing algorithms and, in particular, the splitting of the strapdown computing processes into low and high speed segments. The low speed calculations are designed to take account of low frequency, large amplitude, body motions arising from vehicle manoeuvres, whilst the high speed section involves a relatively simple algorithm which is designed to keep track of the high frequency, low amplitude, motions of the vehicle. Contemporary algorithms often adopt this approach for both the attitude computation and the specific force resolution.

10.2 Attitude computation

As indicated previously, it is the computation of attitude which is particularly critical in a strapdown system. It should therefore come as no surprise to find that it is this topic which has been the subject of much study [1–5]. In many applications, the dynamic range of the angular motions to be taken account of can be very large, varying from a few degrees per hour to 2000 degrees per second or more. In addition, the system may be subjected to high frequency dynamic motion in some applications. For example, a strapdown system in a guided missile may be subjected to such motion as a result of body bending and rocket motor induced vibration. The ability of the strapdown algorithm to keep track of body attitude accurately in a severe vibratory environment may well be the critical factor in determining its performance, if accurate navigation is to be achieved.

The conventional approach to attitude determination is to compute the direction cosine matrix, relating the vehicle body reference frame to the reference co-ordinate system, or its quaternion equivalent, using a numerical integration scheme. Theoretically it is possible to calculate body attitude sufficiently accurately, even in the presence of high frequency angular motion, provided that the computational frequency is sufficiently high. However, in practice this may impose an intolerable burden on the processor.

An alternative formulation advocated by Bortz [1] involves the representation of changes in attitude as a rotation vector. As described below, this approach allows the attitude computation to be split conveniently into low and high speed sections, denoted here as the *k*-cycle and *j*-cycle rates. The lower speed part of the calculation is designed to take account of the relatively low frequency, large amplitude, body motion arising as a result of vehicle manoeuvres. The high speed section involves a simple algorithm, which is designed to track the high frequency, low amplitude, motions of the vehicle.

Using this approach, coning motion (see Section 10.2.3) at frequencies near to the lower computational rate may be accounted for without the need to increase the speed at which the bulk of the computation is implemented.

10.2.1 Direction cosine algorithms

In order to update the direction cosine matrix, C, defined in Chapter 3, it is necessary to solve a matrix differential equation of the form:

$$\dot{C} = C\Omega \tag{10.1}$$

where Ω is a skew symmetric matrix formed from the elements of the turn rate vector ω. For clarity, the subscripts and superscripts used earlier have been omitted in the following development.

Over a single computer cycle, from time t_k to t_{k+1}, the solution of the equation may be written as follows:

$$C_{k+1} = C_k \exp \int_{t_k}^{t_{k+1}} \Omega \, dt \tag{10.2}$$

Provided that the orientation of the turn rate vector, ω remains fixed in space over the update interval, we may write:

$$\int_{t_k}^{t_{k+1}} \Omega \, dt = [\sigma \times] \tag{10.3}$$

Hence, eqn. 10.2 becomes:

$$C_{k+1} = C_k \exp [\sigma \times]$$

$$= C_k A_k \tag{10.4}$$

where C_k represents the direction cosine matrix which relates body to reference axes at the kth computer cycle, and A_k the direction cosine matrix which transforms a vector from body co-ordinates at the $k+1$th computer cycle to body co-ordinates at the kth computer cycle. The variable σ is an angle vector with direction and magnitude such that a rotation of the body frame about σ, through an angle equal to the magnitude of σ, will rotate the body frame from its orientation at computer cycle k to its position at computer cycle $k+1$. The components of σ are denoted by σ_x, σ_y and σ_z and its magnitude is given by:

$$\sigma = \sqrt{\left(\sigma_x^2 + \sigma_y^2 + \sigma_z^2\right)} \tag{10.5}$$

$$\text{and} \quad \sigma \times = \begin{pmatrix} 0 & -\sigma_z & \sigma_y \\ \sigma_z & 0 & -\sigma_x \\ -\sigma_y & \sigma_x & 0 \end{pmatrix} \tag{10.6}$$

Expanding the exponential term in eqn. 10.4 gives:

$$A_k = I + [\sigma \times] + \frac{[\sigma \times]^2}{2!} + \frac{[\sigma \times]^3}{3!} + \frac{[\sigma \times]^4}{4!} + \dots \tag{10.7}$$

and using eqn. 10.6 it can be shown

$$[\sigma \times]^2 = \begin{pmatrix} -\left(\sigma_y^2 + \sigma_z^2\right) & \sigma_x \sigma_y & \sigma_x \sigma_z \\ \sigma_x \sigma_y & -\left(\sigma_x^2 + \sigma_z^2\right) & \sigma_y \sigma_z \\ \sigma_x \sigma_z & \sigma_y \sigma_z & -\left(\sigma_x^2 + \sigma_y^2\right) \end{pmatrix}$$

$$[\sigma \times]^3 = -\left(\sigma_x^2 + \sigma_y^2 + \sigma_z^2\right)[\sigma \times]$$

$$[\sigma \times]^4 = -\left(\sigma_x^2 + \sigma_y^2 + \sigma_z^2\right)[\sigma \times]^2$$

$$\begin{matrix} \cdot \\ \cdot \end{matrix} \tag{10.8}$$

Thus, we may write:

$$
\mathbf{A}_k = \mathbf{I} + [\boldsymbol{\sigma} \times] + \frac{[\boldsymbol{\sigma} \times]^2}{2!} + \sigma^2 \frac{[\boldsymbol{\sigma} \times]}{3!} + \frac{[\boldsymbol{\sigma} \times]^2}{4!} + \dots
$$

$$
= \mathbf{I} + \left(1 - \frac{\sigma^2}{3!} + \frac{\sigma^4}{5!} - \dots\right)[\boldsymbol{\sigma} \times] + \left(\frac{1}{2!} - \frac{\sigma^2}{4!} + \frac{\sigma^4}{6!} - \dots\right)[\boldsymbol{\sigma} \times]^2 \quad (10.9)
$$

which may be written as follows:

$$
\mathbf{A}_k = \mathbf{I} + \frac{\sin \sigma}{\sigma}[\boldsymbol{\sigma} \times] + \frac{(1 - \cos \sigma)}{\sigma^2}[\boldsymbol{\sigma} \times]^2 \quad (10.10)
$$

Provided that $\boldsymbol{\sigma}$ is the angle vector as defined above, eqn. 10.10 provides an exact representation of the attitude matrix which relates body attitude at times t_{k+1} and t_k. If it were possible to implement this equation perfectly, it would yield an orthogonal matrix which need only be evaluated when transformation of the measured specific force vector is required. In practice of course, it is necessary to truncate the mathematical functions in eqn. 10.10, in order to produce an algorithm which can be implemented in real time. Following eqn. 10.9, $\mathbf{A}_k$ may be calculated using:

$$
\mathbf{A}_k = \mathbf{I} + a_1[\boldsymbol{\sigma} \times] + a_2[\boldsymbol{\sigma} \times]^2 \quad (10.11)
$$

where

$$
a_1 = 1 - \frac{\sigma^2}{3!} + \frac{\sigma^4}{5!} - \dots
$$

and

$$
a_2 = \frac{1}{2!} - \frac{\sigma^2}{4!} + \frac{\sigma^4}{6!} - \dots
$$

The direction cosine matrix may therefore be updated for body motion, as sensed by the strapdown gyroscopes, using a recursive algorithm based on eqns. 10.4 and 10.9. The order of the algorithm will be determined by the number of terms included in eqn. 10.9. For example, both of the infinite series would be truncated at the σ^2 level in a fourth order algorithm. The computation rate should be selected to ensure that the magnitude of $\boldsymbol{\sigma}$ remains small at the maximum turn rate, thus avoiding the need to include a large number of terms in the expression for $\mathbf{A}_k$.

The definition of attitude computation errors

The computed attitude matrix, written here as $\hat{\mathbf{A}}$, may be expressed in terms of the true attitude matrix, $\mathbf{A}$, and an error matrix $\mathbf{E}$ as follows:

$$\hat{\mathbf{A}} = \mathbf{A}[\mathbf{I} + \mathbf{E}] \tag{10.12}$$

re-arranging, we have:

$$\mathbf{E} = \mathbf{A}^T \hat{\mathbf{A}} - \mathbf{I} \tag{10.13}$$

Substituting for **A** and $\hat{\mathbf{A}}$ following eqns. 10.10 and 10.11, we may write:

$$
\begin{aligned}
\mathbf{E} &= \left(\mathbf{I} - \frac{\sin\sigma}{\sigma}[\boldsymbol{\sigma}\times] + \frac{(1-\cos\sigma)}{\sigma^2}[\boldsymbol{\sigma}\times]^2 \right) \left(\mathbf{I} + a_1[\boldsymbol{\sigma}\times] + a_2[\boldsymbol{\sigma}\times]^2 \right) - \mathbf{I} \\
&= \left(\sigma a_1 \cos\sigma - \sin\sigma + \sigma^2 a_2 \sin\sigma \right) \frac{[\boldsymbol{\sigma}\times]}{\sigma} \\
&\quad + \left(1 - \cos\sigma - \sigma a_1 \sin\sigma + \sigma^2 a_2 \cos\sigma \right) \frac{[\boldsymbol{\sigma}\times]^2}{\sigma^2} \tag{10.14}
\end{aligned}
$$

The first term in eqn. 10.14 represents a skew symmetric matrix, following the form of $[\boldsymbol{\sigma}\times]$, and is denoted below by the symbol **U**. The second term is symmetric, following $[\boldsymbol{\sigma}\times]^2$, and may be represented by a symmetric matrix **S**. Hence, we may write:

$$\mathbf{E} = \mathbf{U} + \mathbf{S} \tag{10.15}$$

as shown in Reference 5.

A is an orthogonal matrix, if the equation $\mathbf{A}^T\mathbf{A} = \mathbf{I}$ is satisfied. For the computed matrix, we may write:

$$\hat{\mathbf{A}}^T\hat{\mathbf{A}} = [\mathbf{I} + \mathbf{E}]^T[\mathbf{I} + \mathbf{E}] \tag{10.16}$$

Ignoring second and higher order terms, gives:

$$\hat{\mathbf{A}}^T\hat{\mathbf{A}} = \mathbf{I} + \mathbf{E} + \mathbf{E}^T \tag{10.17}$$

Substituting for **E** and $\mathbf{E}^T$ in terms of their symmetric and skew symmetric components, and noting that $\mathbf{S}^T = \mathbf{S}$ and $\mathbf{U}^T = -\mathbf{U}$, gives:

$$\hat{\mathbf{A}}^T\hat{\mathbf{A}} = \mathbf{I} - 2\mathbf{S} \tag{10.18}$$

We can also write:

$$\mathbf{A}^T\hat{\mathbf{A}} = \mathbf{I} + \mathbf{S} + \mathbf{U} \tag{10.19}$$

From the last two equations, the following conclusions can be drawn:

S represents the deviation of the matrix $\hat{\mathbf{A}}$ from the orthogonal form. The diagonal elements of **S** are termed the scale errors, whilst the off diagonal terms represent the skew errors [5].

If **S** is zero, $\hat{\mathbf{A}}$ becomes an orthogonal matrix representing a co-ordinate rotation, which is different from that defined by **A**. **U** provides a measure of the difference between the two rotations.

A single parameter, D_{dc}, the root sum square of the upper or lower off diagonal elements of **U**, divided by the computer update interval δt, may be used as a measure of the drift in the computed attitude matrix. The parameter D_{dc} may be used to assess the accuracy of the various orders of attitude algorithm considered.

The matrix **U** is defined as:

$$\mathbf{U} = \left(\sigma a_1 \cos\sigma - \sin\sigma + \sigma^2 a_2 \sin\sigma\right)\frac{[\boldsymbol{\sigma} \times]}{\sigma} \tag{10.20}$$

In the case of a single x axis rotation, where $\boldsymbol{\sigma} = [\sigma \ \ 0 \ \ 0]^T$, we have:

$$D_{dc} = \frac{1}{\delta t}\left(\sigma a_1 \cos\sigma - \sin\sigma + \sigma^2 a_2 \sin\sigma\right) \tag{10.21}$$

where $a_1 = 1$, $a_2 = 0$ is a 1st order algorithm

$\qquad\quad a_1 = 1$, $a_2 = 0.5$ is a 2nd order algorithm

$\qquad\quad a_1 = 1 - \dfrac{\sigma^2}{6}$, $a_2 = 0.5$ is a 3rd order algorithm

Example

Consider cases in which the maximum size of the angular increment (σ_{max}) is 0.1 radian and 0.05 radian. If the maximum angular rate of the body is 10 rad/s, these figures correspond to update intervals of 0.01 second and 0.005 second.

The drift errors in the computed attitude are shown in Table 10.1 for these two cases using different orders of algorithm.

Table 10.1

Order of algorithm	Attitude drift error (°/hour)	
	$\sigma_{max} = 0.1$ radian	$\sigma_{max} = 0.05$ radian
1	6870	1720
2	3430	860
3	7	0.4
4	1.7	0.1

The substantial improvement in accuracy which may be achieved by reducing the update interval is clearly seen. The large reduction in the drift rate obtained through the inclusion of third order terms occurs as a result of the cancellation of both the third and fourth order terms in the expression for drift error with this level of truncation. In many applications, high turn rates will not normally be sustained for long periods, the mean rates expected being substantially lower than the figure of 10 rad/s considered in the above analysis. In such cases, the mean drift errors resulting from imprecise attitude computation will be considerably smaller than the figures quoted in the table. It may therefore be feasible to use first or second order update algorithms for some applications.

It is assumed here that the major part of the attitude computation described above, along with some other aspects of the navigation processing to be described later, will be implemented at the lower k-cycle data rate. However, some parts of the computation may need to be performed at higher rates, whilst others can be carried out less frequently, as discussed later.

10.2.2 Rotation angle computation

A further limitation on the accuracy of the direction cosine matrix updates is the accuracy with which the rotation angle, σ, can be determined. Consider first the case where the direction of the angular rate vector, ω, remains fixed in space over an update interval. In this case, σ is determined quite simply as the integral of ω over the computer cycle, k:

$$\sigma = \int_{t_k}^{t_{k+1}} \omega \, dt \tag{10.22}$$

i.e. σ is the sum of the incremental measurements provided directly by some gyroscopes over the time interval t_k to t_{k+1}. The relationship between σ and ω can also be expressed as $d\sigma/dt = \omega$ in this situation.

In general, however, it is not possible to determine σ precisely by simply summing the measurements of incremental angle. If the direction of ω does not remain fixed in space but is rotating, as occurs in the presence of coning motion for example, then following Bortz [1], we may write:

$$\dot{\sigma} = \omega + \epsilon \tag{10.23}$$

in which ω represents the inertially measurable angular motion and $\dot{\epsilon}$ is a component of $\dot{\sigma}$ owing to the non-inertially measurable motion, known as the non-commutativity rate vector.

An expression for $\dot{\sigma}$ under general motion conditions, i.e. where the motion is not restricted to a single axis, may be derived by differentiating eqn. 10.10, writing $dA/dt = A[\omega \times]$ and manipulating vectors formed from the resulting expressions as described in Reference 1 to give:

$$\dot{\sigma} = \omega + \frac{1}{2}\sigma \times \omega + \frac{1}{\sigma^2}\left\{1 - \frac{\sigma \sin \sigma}{2(1 - \cos \sigma)}\right\}\sigma \times (\sigma \times \omega) \qquad (10.24)$$

As a result of non-commutativity effects, the final orientation after a series of rotations is dependent on both the individual rotations and the order in which they have occurred. The above equation indicates how the history of previous rotations (σ) and the current angular rate (ω) affect the build up of the non-commutativity term (ϵ).

A practical implementation would require the right hand side of eqn. 10.24 to be truncated. For instance, writing the sine and cosine terms as series expansions and ignoring terms higher than fourth order in σ, the above equation may be written:

$$\dot{\sigma} = \omega + \frac{1}{2}\sigma \times \omega + \frac{1}{12}\sigma \times \sigma \times \omega \qquad (10.25)$$

In Reference 3, the following algorithm is proposed in which σ is approximated as the integral of ω:

$$\delta\alpha(t) = \int_{t_k}^{t} \alpha \times \omega \, dt$$

$$\text{where} \quad \alpha(t) = \int_{t_k}^{t} \omega \, dt$$

$$\sigma = \alpha(t = t_{k+1}) + \delta\alpha(t = t_{k+1}) \qquad (10.26)$$

In general, where significant levels of angular vibration are present, it will be necessary to solve eqn. 10.26 at a higher rate, denoted the *j*- cycle update rate, than that at which the direction cosine matrix is updated. It will be necessary to select a *j*-cycle update rate which is sufficiently fast to ensure that the value of σ obtained using eqn. 10.26 agrees well with the true value, given by eqn. 10.24, in the presence of the maximum body rates and vibratory motion.

10.2.3 Rotation vector compensation

In this section, an expression is derived for the drift error in the computed attitude ($\delta\alpha$) arising in the presence of coning motion. Coning refers to the motion which arises when a single axis of a body describes a cone, or some approximation to a cone, in space. Such motion results from the simultaneous application of angular oscillations about two orthogonal axes of the system where the oscillations are out of phase.

It is assumed here that the body is oscillating at a frequency f about the x and y axes. The amplitudes of the x and y motions are θ_x and θ_y respectively. Additionally, a phase difference of ϕ is assumed to exist between the two channels. Thus, we may write:

$$\omega = 2\pi f[\theta_x \cos 2\pi ft \quad \theta_y \cos(2\pi ft + \phi) \quad 0]^T$$

and

$$\alpha = [\theta_x\{\sin 2\pi ft - \sin 2\pi ft_k\} \quad \theta_y\{\sin(2\pi ft + \phi) - \sin(2\pi ft_k + \phi)\} \quad 0]^T$$

Substituting for α and ω in the $\delta\alpha$ equation gives:

$$\delta\alpha = \pi f \int_{t_k}^{t_{k+1}} \begin{pmatrix} \theta_x\{\sin 2\pi ft - \sin 2\pi ft_k\} \\ \theta_y\{\sin(2\pi ft + \phi) - \sin(2\pi ft_k + \phi)\} \\ 0 \end{pmatrix} \times \begin{pmatrix} \theta_x \cos 2\pi ft \\ \theta_y \cos(2\pi ft + \phi) \\ 0 \end{pmatrix} dt$$

which yields a z component :

$$\delta\alpha_z = \pi f \theta_x \theta_y \sin\phi \int_{t_k}^{t_{k+1}} \{1 - \cos 2\pi f(t - t_k)\} dt$$

integrating between the appropriate limits, we have :

$$\delta\alpha_z = \pi f \theta_x \theta_y \sin\phi \left\{ t_{k+1} - t_k - \frac{\sin 2\pi f(t_{k+1} - t_k)}{2\pi f} \right\}$$

writing $t_{k+1} - t_k = \delta t$ gives :

$$\delta\alpha_z = \pi f \theta_x \theta_y \delta t \sin\phi \left\{ 1 - \frac{\sin 2\pi f \delta t}{2\pi f \delta t} \right\} \tag{10.27}$$

Thus, although the rate ω is cyclic about the x and y axes, a z component of $\delta\alpha$ arises, which is a constant proportional to the sine of the phase angle and the amplitude of the motion. It can be seen, from the above equation, that $\delta\alpha_z$ is maximised when $\phi = \pi/2$. Under such conditions, the motion of the body is referred to as coning, owing to the motion of the z axis which describes a cone is space.

Over the time interval δt, the drift error in the computed attitude, which arises if the above correction term is not applied, may be expressed as:

$$\delta\dot{\alpha}_z = \pi f \theta_x \theta_y \sin\phi \left\{ 1 - \frac{\sin 2\pi f \delta t}{2\pi f \delta t} \right\} \tag{10.28}$$

If $\delta\dot{\alpha}_z$ is small compared with the overall system performance requirement, the need to implement the correction terms described is avoided. Further, $\delta\dot{\alpha}_z$ may be minimised by reducing the size of the time step δt.

Example

Consider a situation where the body exhibits coning motion at a frequency, f, of 50 Hz. The angular amplitudes of the motion in x and y is taken to be 0.1°. If the attitude update frequency is 100 Hz, i.e. δt = 0.01 second, the resulting drift is 100°/hour. By increasing the computational frequency to 500 Hz, the drift figure falls to ~6°/hour. A more general development of the above equation allows a root mean square (RMS) value for $\delta\alpha_z$ to be calculated in the presence of a given vibration spectrum [3].

10.2.4 Body and navigation frame rotations

Returning now to the continuous form of the attitude eqn. 10.1, the vector $\boldsymbol{\omega}$ represents the turn rate of the body with respect to the navigation reference frame. When navigating with respect to the local geographic frame, this equation takes the following form, as discussed in Chapter 3:

$$\dot{\mathbf{C}}_b^n = \mathbf{C}_b^n \boldsymbol{\Omega}_{ib}^b - \boldsymbol{\Omega}_{in}^n \mathbf{C}_b^n \tag{10.29}$$

The first term, $\mathbf{C}_b^n \boldsymbol{\Omega}_{ib}^b$, is a function of the body rates, as sensed by the strapdown gyroscopes, and the second term, $-\boldsymbol{\Omega}_{ib}^b \mathbf{C}_b^n$, is a function of the lower navigation frame rates. The updating of the direction cosine matrix to take account of the body motion, i.e. the solution of the equation $d\,\mathbf{C}_b^n / dt = \mathbf{C}_b^n \boldsymbol{\Omega}_{ib}^b$ may be accomplished using eqns. 10.4 and 10.11, as described above.

A similar algorithm may be used to take account of navigation frame rotations. In order to update the direction cosine matrix for navigation frame rotations, an equation similar to eqn. 10.4 may be used, in which $\mathbf{A}$ is replaced with a navigation frame rotation direction cosine matrix, $\mathbf{B}$, as follows:

$$\mathbf{C}_{l+1} = \mathbf{B}_l \mathbf{C}_l \tag{10.30}$$

where $\mathbf{B}_l$ represents the direction cosine matrix relating body to navigation axes at time t_{l+1} to navigation axes at time t_l. $\mathbf{B}_l$ may be expressed in terms of a rotation vector $\boldsymbol{\theta}$ as follows:

$$\mathbf{B}_l = \mathbf{I} + \frac{\sin\theta}{\theta}[\boldsymbol{\theta}\times] + \frac{(1-\cos\theta)}{\theta^2}[\boldsymbol{\theta}\times]^2 \tag{10.31}$$

where $\boldsymbol{\theta}$ is a rotation vector with magnitude and direction such that a rotation of the navigation frame about $\boldsymbol{\theta}$, through an angle equal to the magnitude of $\boldsymbol{\theta}$, will rotate the navigation frame from its orientation at time t_l to its position at time t_{l+1}. The angle $\boldsymbol{\theta}$ may be written as:

$$\boldsymbol{\theta} = \int_{t_l}^{t_{l+1}} \boldsymbol{\omega}_{in}^n \, dt \tag{10.32}$$

In view of the fact that the navigation frame turn rates will generally be much slower than the body rates, direction cosine updates for rotations of the navigation frame may be implemented at a much slower rate, denoted the *l*-cycle update rate. Additionally, the mathematical functions on which the algorithm is based can be truncated at a lower level.

Example

Considering first order terms only in eqn. 10.31, and applying navigation frame updates at one second intervals to take account of rotation of the Earth, the net drift error can be maintained at a negligibly low level. In some short range missile applications, in which angular rate measurement accuracies of several hundred degrees per hour may be acceptable, the need to take account of navigation frame rotations, e.g. Earth's rate at 15°/hour, becomes superfluous.

10.2.5 Quaternion algorithms

Using the quaternion attitude representation, it is necessary to solve the equation:

$$\dot{q} = \frac{1}{2} q \cdot p \tag{10.33}$$

where $p = [0, \boldsymbol{\omega}^T]$ as defined in Section 3.4.3.2. This equation may be expressed in matrix form as:

$$\dot{q} = \frac{1}{2} W q \tag{10.34}$$

$$\text{where} \quad W = \begin{pmatrix} 0 & -\omega_x & -\omega_y & -\omega_z \\ \omega_x & 0 & \omega_z & -\omega_y \\ \omega_y & -\omega_z & 0 & \omega_x \\ \omega_z & \omega_y & -\omega_x & 0 \end{pmatrix} \tag{10.35}$$

and ω_x, ω_y, and ω_z are the components of $\boldsymbol{\omega}$.

For the situation in which the orientation of the rate vector, ω, remains fixed over a computer update interval, the solution to the above equation may be written as:

$$\mathbf{q}_{k+1} = \left[\exp \frac{1}{2} \int\limits_{t_k}^{t_{k+1}} \mathbf{W}\, dt \right] \mathbf{q}_k \tag{10.36}$$

$$\int\limits_{t_k}^{t_{k+1}} \mathbf{W}\, dt = \Sigma = \begin{pmatrix} 0 & -\sigma_x & -\sigma_y & -\sigma_z \\ \sigma_x & 0 & \sigma_z & -\sigma_y \\ \sigma_y & -\sigma_z & 0 & \sigma_x \\ \sigma_z & \sigma_y & -\sigma_x & 0 \end{pmatrix} \tag{10.37}$$

$$\mathbf{q}_{k+1} = \left[\exp(\Sigma / 2) \right] \mathbf{q}_k \tag{10.38}$$

By expanding the exponential term and following a development similar to that used to obtain the direction cosine solution above, it can be shown that the exponential term may be written in quaternion form as:

$$\mathbf{q}_{k+1} = \mathbf{q}_k \cdot \mathbf{r}_k \tag{10.39}$$

where

$$\mathbf{r}_k = \begin{pmatrix} a_c \\ a_s \sigma_x \\ a_s \sigma_y \\ a_s \sigma_z \end{pmatrix} \tag{10.40}$$

$$a_c = \cos(\sigma / 2) = 1 - \frac{(0.5\sigma)^2}{2!} + \frac{(0.5\sigma)^4}{4!} - \ldots \tag{10.41}$$

$$a_s = \frac{\sin(\sigma / 2)}{\sigma} = 0.5 \left(1 - \frac{(0.5\sigma)^2}{2!} + \frac{(0.5\sigma)^4}{4!} - \ldots \right) \tag{10.42}$$

and $(0.5\sigma)^2 = 0.25(\sigma_x^2 + \sigma_y^2 + \sigma_z^2)$

By comparison with the quaternion definition given in Chapter 3, eqn. 3.32a , it can be seen that $\mathbf{r}_k$ is a quaternion representing a rotation of magnitude σ, about a vector σ. This is the quaternion which transforms from body axes at time t_{k+1} to body axes at time t_k, whilst $\mathbf{q}_k$ represents the quaternion relating body to navigation axes at time t_k. Therefore, the quaternion $\mathbf{q}$ may be updated for body motion, as sensed by the strapdown gyroscopes, using eqns. 10.39 to 10.42 recursively. The parameter σ is determined as described in Section 10.2.1.

As for the direction cosine algorithm, the update interval is selected to ensure that s remains small at the maximum body rate, thus avoiding the need to retain a large number of terms in the expressions for a_c and a_s. The order of the quaternion updating algorithm will be determined by the truncation point selected in eqns. 10.41 and 10.42.

Definition of attitude errors

In order to quantify the performance of the quaternion update algorithm, a drift parameter D_q is defined following a development parallel to that used in Section 10.2.1 to determine the drift error in the direction cosine update algorithm. The error in the computed quaternion δr may be expressed in terms of the true and computed quaternions, denoted r and $\hat{r}$ respectively, as follows:

$$\delta r = r^* . \hat{r} \tag{10.43}$$

For a single x axis rotation, $r = [\cos(0.5\sigma) \ \sin(0.5\sigma) \ 0 \ 0]$, $\hat{r} = [a_c \ \sigma a_s \ 0 \ 0]$ and

$$\delta r = \begin{pmatrix} a_c \cos(0.5\sigma) + \sigma a_s \sin(0.5\sigma) \\ \sigma a_s \cos(0.5\sigma) - a_c \sin(0.5\sigma) \\ 0 \\ 0 \end{pmatrix} \tag{10.44}$$

This may be expressed as a direction cosine error matrix in accordance with eqn. 3.59. Following the procedure used to define the direction cosine drift in Section 10.2.1, an expression for the quaternion drift may be defined in terms of the off diagonal elements of the error matrix:

$$D_q = \frac{2}{\delta t} \left\{ a_c \cos(0.5\sigma) + \sigma a_s \sin(0.5\sigma) \right\} \left\{ \sigma a_s \cos(0.5\sigma) - a_c \sin(0.5\sigma) \right\}$$

$$= \frac{1}{\delta t} \left\{ 2\sigma a_c a_s \cos\sigma - a_c^2 \sin\sigma + \sigma^2 a_s^2 \sin\sigma \right\} \tag{10.45}$$

where a_c and a_s are selected according to the order of the algorithm needed as follows:

$a_c = 1, a_s = 0$ is a 1st order algorithm

$a_c = 1 - \dfrac{(0.5\sigma)^2}{2}, a_s = 0.5$ is a 2nd order algorithm

$a_c = 1 - \dfrac{(0.5\sigma)^2}{2}, a_s = 0.5\left\{1 - \dfrac{(0.5\sigma)^2}{6}\right\}$ is a 3rd order algorithm

Example

The drift in the attitude computed using different orders of the quaternion algorithm is summarised in Table 10.2, using the same conditions considered in Section 10.2.1 to evaluate the performance of the direction cosine algorithm.

Table 10.2

Order of algorithm	Attitude drift error (°/hour)	
	$\sigma_{max} = 0.1$ radians	$\sigma_{max} = 0.05$ radians
1	1720	430
2	860	215
3	0.4	0.06
4	0.2	0.06

Comparing these results with those tabulated in Section 10.2.1, it can be seen that the quaternion drift figures are smaller than those obtained using direction cosines. This arises principally because the quaternion equations involve the expansion of terms in $\sin(\sigma/2)$ and $\cos(\sigma/2)$, whereas the direction cosine equations contain similar terms in σ. This also accounts for the correspondence between the quaternion drift figures which arise when σ is 0.1 radians and the direction cosine drifts when σ is 0.05 radians. It is therefore apparent that the quaternion representation yields the more accurate attitude solution, for a given level of truncation, in the presence of a single axis rotation.

In order to update the quaternion for navigation frame rotations, an equation similar to eqn. 10.3 may be used, in which $\mathbf{r}$ is replaced with a navigation frame rotation quaternion, $\mathbf{p}$, as follows:

$$\mathbf{q}_{l+1} = \mathbf{p}_l \cdot \mathbf{q}_l \tag{10.46}$$

where $\mathbf{q}_l$ represents the quaternion relating body to navigation axes at computer cycle-l, and $\mathbf{p}_l$ is the quaternion which transforms from body axes at time t_{l+1} to body axes at time t_l. The quaternion $\mathbf{p}_l$ may be expressed in terms of the rotation vector θ, as follows:

$$\mathbf{p}_l = \begin{pmatrix} b_c \\ b_s \sigma_x \\ b_s \theta_y \\ b_s \theta_z \end{pmatrix} \tag{10.47}$$

where
$$b_c = \cos(\theta/2) = 1 - \frac{(0.5\theta)^2}{2!} + \frac{(0.5\theta)^4}{4!} - \dots \qquad (10.48)$$

$$b_s = \frac{\sin(\theta/2)}{\theta} = 0.5\left(1 - \frac{(0.5\theta)^2}{3!} + \frac{(0.5\theta)^4}{5!} - \dots\right) \qquad (10.49)$$

As in the case of the direction cosines, it is assumed that these equations may be implemented at the reduced rate, the *k*-cycle, compared with the body motion updates. Additionally, a low order algorithm will usually be sufficient to provide accurate navigation frame updates.

In the situation where the quaternion attitude representation is selected, the following equation may be used to compute the direction cosine matrix $\mathbf{C}_b^n$ for use in the acceleration vector transformation algorithm, which is discussed in Section 10.3:

$$\mathbf{C}_b^n = \begin{pmatrix} 1 - 2(c^2 + d^2) & 2(bc - ad) & 2(bd + ac) \\ 2(bc + ad) & 1 - 2(b^2 + d^2) & 2(cd - ab) \\ 2(bd - ac) & 2(cd + ab) & 1 - 2(b^2 + c^2) \end{pmatrix} \qquad (10.50)$$

10.2.6 Orthogonalisation and normalisation algorithms

It is common practice, in the implementation of strapdown attitude algorithms, to apply self consistency checks in an attempt to enhance the accuracy of the computed direction cosines or quaternion parameters. The rows of the direction cosine matrix represent the projection of unit vectors which lie along each axis of the orthogonal reference co-ordinate frame in the body frame. It follows, therefore, that the rows of the direction cosine matrix should always be orthogonal to one another and that the sum of the squares of the elements in each row should equal unity. In the case of the quaternion representation, the self consistency test is to check that the sum of the squares of the four parameters is unity.

Self consistency checks may be applied as part of the attitude algorithm to ensure that the above conditions are satisfied. If required, it is usually sufficient to carry out these checks at a relatively low rate. This part of the computation may be carried out at the *k*-cycle frequency.

10.2.6.1 Direction cosine checking

The condition for orthogonality of the ith and jth rows of the direction cosine matrix, denoted $\mathbf{C}_i$ and $\mathbf{C}_j$, is that their dot product should equal zero, i.e. $\mathbf{C}_i\mathbf{C}_j^T = 0$. In practice, this is not necessarily the case, and we define:

$$\Delta_{ij} = \mathbf{C}_i\mathbf{C}_j^T \tag{10.51}$$

where Δ_{ij} is an angle error defined about an axis perpendicular to $\mathbf{C}_i$ and $\mathbf{C}_j$, the orthogonality error between the two rows.

Since either row, $\mathbf{C}_i$ or $\mathbf{C}_j$, is equally likely to be in error, the correction is apportioned equally between them using:

$$\hat{\mathbf{C}}_i = \mathbf{C}_i - \frac{1}{2}\Delta_{ij}\mathbf{C}_j \tag{10.52}$$

$$\hat{\mathbf{C}}_j = \mathbf{C}_j - \frac{1}{2}\Delta_{ij}\mathbf{C}_i \tag{10.53}$$

where the $\wedge$ notation denotes the corrected quantity.

Normalisation errors may be identified by comparing the sum of the squares of the elements in a row with unity, i.e:

$$\Delta_{ij} = 1 - \mathbf{C}_i\mathbf{C}_j^T \tag{10.54}$$

and corrections applied using:

$$\hat{\mathbf{C}}_i = \mathbf{C}_i - \frac{1}{2}\Delta_{ii}\mathbf{C}_i \tag{10.55}$$

An alternative approach is to operate on the columns of the direction cosine matrix, in a similar manner to that described above for the rows, as described in Reference 3.

10.2.6.2 Quaternion normalisation

The quaternion can be normalised by comparing the sum of the squares of its elements with unity. The normalisation error is given by:

$$\Delta\mathbf{q} = 1 - \mathbf{q} \cdot \mathbf{q}* \tag{10.56}$$

The quaternion may be normalised by dividing each element by $\sqrt{(\mathbf{q} \cdot \mathbf{q}*)}$. Thus, we may write:

$$\begin{aligned}
\mathbf{q} &= \frac{\mathbf{q}}{\sqrt{(\mathbf{q} \cdot \mathbf{q}*)}} \\
&= \{1 - \Delta q\}^{-0.5}\mathbf{q} \\
&\approx \left\{1 + \frac{1}{2}\Delta q\right\}\mathbf{q}
\end{aligned} \tag{10.57}$$

It should be noted that the orthogonalisation and normalisation cannot correct for errors that have occurred in the previous computation cycle. For example, an error arising in a single element of the quaternion will be spread amongst all of the elements, as a result of a normalisation correction. In the opinion of the authors, such techniques should be used with caution. Emphasis should be placed on the design of the basic attitude update algorithm rather than placing any reliance on the normalisation process described here, which may serve only to compound fundamental errors in the update algorithm.

10.2.7 The choice of attitude representation

The relative benefits of using direction cosines or quaternion parameters to represent attitude has received considerable attention in the published literature [3–5]. The results have largely been found to be inconclusive, although the quaternion method does offer some advantage because, inherently, it gives rise to an orthogonal attitude matrix. In addition, the analysis, given in the preceding sections, has shown the accuracy of attitude computation obtained using quaternions to be superior to the accuracy which may be achieved using the direction cosine representation. These factors account, in part at least, for the popularity of the quaternion method in recent years.

In practice, it is possible to design direction cosine and quaternion algorithms with very similar performance capabilities for which the computing burdens are comparable. In the past, computer loading and memory requirements were of major concern when selecting a suitable algorithm. However, continuing developments in computer technology have reduced these concerns considerably. In any case, the differences in the computing requirements for the two algorithm types are relatively small, especially when account is taken of the overall strapdown computing task, including the resolution of the specific force vector.

10.3 Acceleration vector transformation algorithm

This section is concerned with the computational algorithm required to resolve the measured specific force acceleration components into the navigation reference frame. Care must be taken to allow for changes in body rotation occurring over a computer update interval and a two speed algorithm may be required for applications where the system is called upon to operate in a highly vibratory environment [6]. For many applications, a relatively low speed algorithm may be sufficient to resolve accelerations associated with vehicle manoeuvres, although a high speed correction term may be included to take account of vibration.

10.3.1 Acceleration vector transformation using direction cosines

The measured specific force vector, $\mathbf{f}^b$, is expressed in the navigation co-ordinate frame using:

$$\mathbf{f}^n = \mathbf{C}_b^n \mathbf{f}^b \tag{10.58}$$

where $\mathbf{C}_b^n$ is the direction cosine matrix which transforms from body to reference axes, as defined earlier.

An algorithm to implement this function, which accepts incremental velocity measurements, may be developed by first integrating eqn. 10.58 to give:

$$\mathbf{u}^n = \int_{t_k}^{t_{k+1}} \mathbf{f}^n \, dt \tag{10.59}$$

where $\mathbf{u}^n$ represents the change in velocity, expressed in the navigation frame, over the computer cycle t_k to t_{k+1}. The velocity vector, $\mathbf{v}^n$, may be determined by summing the values of $\mathbf{u}^n$ computed at each cycle, and correcting for Coriolis accelerations and the effects of gravity as described later, in Section 10.4. The matrix $\mathbf{C}_b^n$ varies continuously with time over the update interval and may be written in terms of the matrix $\mathbf{C}_k$, the value of $\mathbf{C}_b^n$ at time t_k, and $\mathbf{A}$, a matrix representing the transformation from body axes at time t to body axes at the start of the update interval, t_k, i.e.

$$\mathbf{C}_b^n = \mathbf{C}_k \mathbf{A} \tag{10.60}$$

Substituting for $\mathbf{C}_b^n$ in eqn. 10.58 gives:

$$\mathbf{u}_n = \mathbf{C}_k \int_{t_k}^{t_{k+1}} \mathbf{A} \mathbf{f}^b \, dt \tag{10.61}$$

Following eqn. 10.11, $\mathbf{A}$ may be expressed in the form:

$$\mathbf{A} = \mathbf{I} + [\alpha \times] + 0.5[\alpha \times]^2 - \ldots$$

where $\alpha = \int_{t_k}^{t} \omega^b \, dt$

Substituting for **A** in (10.61) gives:

$$u^n = C_k \int_{t_k}^{t_{k+1}} \left[f^b + \alpha \times f^b + 0.5[\alpha \times]^2 f^b - \ldots \right] dt \tag{10.62}$$

If second and higher order terms are ignored, we may write :

$$u^n = C_k \left[\int_{t_k}^{t_{k+1}} f^b \, dt + \int_{t_k}^{t_{k+1}} \alpha \times f^b \, dt \right] \tag{10.63}$$

If we now write :

$$v = \int_{t_k}^{t} f^b \, dt$$

and integrate the cross product term in eqn. 10.63 by parts, it can be shown that:

$$u^n = C_k \left[v_{k+1} + \frac{1}{2} \alpha_{k+1} \times v_{k+1} + \frac{1}{2} \int_{t_k}^{t_{k+1}} \left(\alpha \times f^b - \omega^b \times v \right) dt \right] \tag{10.64}$$

where $\alpha_{k+1} = \alpha$ evaluated over the time interval t_k to t_{k+1}, by summing the incremental angle measurements over this period. Similarly, $v_{k+1} = v$ evaluated over the same time interval, by summing the incremental velocity measurements provided by the inertial measurement unit.

In eqn. 10.64, we have an expression for u^n containing three terms:

(i)   the sum of the incremental velocity measurements produced by the inertial measurement unit;
(ii)   a cross product of the incremental angle, accumulated from t_k to t_{k+1}, and the incremental velocity change over the same period. In the literature, this term is referred to as the rotation correction;
(iii) a dynamic integral term.

In the situation where f^b and ω^b remain constant over the update interval, we have $\alpha = \omega^b t$ and $v = f^b t$. Substituting for α and v in eqn. 10.64, it can be shown easily that the integral term becomes identically zero under such conditions. However, if f^b and ω^b vary significantly over the update interval, it may be necessary to evaluate the integral in order to provide compensation for the dynamic motion. In this case, the rate at which the integral is evaluated will need to be well above the frequency of the dynamic motion, i.e. at the j-cycle update rate referred to in the previous section. An example to illustrate this effect is given in the following section.

10.3.2 Rotation correction

In order to illustrate the need for the rotation correction term in a missile application using strapdown technology, consider a situation in which such a vehicle is manoeuvring in a lateral plane such that, $\mathbf{f}^b = [0 \ f \ 0]^T$ and $\boldsymbol{\omega}^b = [0 \ 0 \ \omega]^T$, i.e. simultaneously accelerating and rotating in the yaw plane.

If the acceleration transformation update interval is δt, then, following eqn. 10.61, the true velocity change over the time interval from $t = 0$ to $t = \delta t$ is given by:

$$\mathbf{u}^n = \int_0^{\delta t} \begin{pmatrix} \cos \omega t & \sin \omega t & 0 \\ -\sin \omega t & \cos \omega t & 0 \\ 0 & 0 & 1 \end{pmatrix} \begin{pmatrix} 0 \\ f \\ 0 \end{pmatrix} dt$$

$$= \int_0^{\delta t} \begin{pmatrix} f \sin \omega t \\ f \cos \omega t \\ 0 \end{pmatrix} dt$$

$$= \left(\frac{f}{\omega}(1 - \cos \omega \, \delta t) \quad \frac{f}{\omega} \sin \omega \, \delta t \quad 0 \right)^T$$

If no allowance is made for the vehicle's body rotation during the update interval, the computed velocity change simply becomes:

$$\mathbf{u}^{n'} = [0 \ f\delta t \ 0]^T$$

Differencing the expressions for $\mathbf{u}^{n'}$ and $\mathbf{u}^n$ gives:

$$\delta\mathbf{u}^n = \mathbf{u}^{n'} - \mathbf{u}^n$$

$$= \left(\frac{f}{\omega}(1 - \cos \varphi \, \delta t) \quad f\,\delta t\left(1 - \frac{\sin \varphi \, \delta t}{\omega \, \delta t}\right) \quad 0 \right)^T$$

$$= \left(\frac{f\omega \delta t^2}{2} - \cdots \quad \frac{f\omega^2 \delta t^3}{6} - \cdots \quad 0 \right)^T \qquad (10.65)$$

An equivalent acceleration error may be written as:

$$\frac{\delta\mathbf{u}^n}{\delta t} = \left(\frac{f\omega \delta t}{2} - \cdots \quad \frac{f\omega^2 \delta t^2}{6} - \cdots \quad 0 \right)^T 2 \qquad (10.66)$$

Example

Take the case of a missile which is accelerating laterally at 20 g, whilst travelling at a speed of 800 metres/second. The associated turn rate is

~0.25 radians/second. Substituting for f and ω in the above expression, and assuming an update interval of 0.01 seconds, gives an x component of acceleration bias of 25 mg. In the presence of a sustained missile manoeuvre, a bias of this magnitude would give rise to significant velocity and position errors, particularly in the case of medium to long range missile systems. The figure derived above may be put into context by comparing it with the magnitudes of acceleration measurement error which may be acceptable for such applications, typically ~5–10 mg, sometimes less.

It can be shown that the simple rotation correction term, given in eqn. 10.64, is able to compensate for the second order velocity error which appears in the above example. Higher order corrections may be applied if considered necessary, the relevant expressions being derived by including extra terms in the expansion of the transformation matrix **A**, by substituting in eqn. 10.61.

10.3.3 Dynamic correction

In calculating the change in velocity over a computer cycle, expressed in the navigation co-ordinate frame, the following correction term may need to be applied:

$$\delta \mathbf{u}^n = \frac{1}{2} \int\limits_{t_k}^{t_{k+1}} \left(\boldsymbol{\alpha} \times \mathbf{f}^b - \boldsymbol{\omega}^b \times \mathbf{v} \right) dt \tag{10.67}$$

A good test of the performance of the vector transformation algorithm is its ability to cope with sculling motion. Sculling is characterised by the simultaneous application of in phase components of angular and linear oscillatory motion with respect to two orthogonal axes. Such motion can be particularly detrimental to system performance if the computational frequency is too low, or if sculling corrections are not applied.

An expression for $\delta \mathbf{u}^n$ is now derived under conditions where the body performs an angular oscillation about the x axis, whilst simultaneously oscillating linearly in the y direction. Writing:

$$\boldsymbol{\alpha} = \int\limits_{t_k}^{t} \boldsymbol{\omega}^b \, dt$$

where $\boldsymbol{\omega}^b = \begin{bmatrix} 2\pi f \theta_x \cos 2\pi f t & 0 & 0 \end{bmatrix}^T$

and $\mathbf{v} = \int\limits_{t_k}^{t} \mathbf{f}^b \, dt$

where $\mathbf{f} = \begin{bmatrix} 0 & A_y \sin(2\pi f t + \phi) & 0 \end{bmatrix}^T$

in which A_y is the amplitude of the cyclic acceleration. Substituting in the equation for δu^n yields a z-component:

$$\delta u_z^n = \frac{1}{2} \theta_x A_y \cos\phi \int_{t_k}^{t_{k+1}} \left\{ 1 - \cos 2\pi f (t - t_k) \right\} dt$$

$$= \frac{1}{2} \theta_x A_y \cos\phi \left[t_{k+1} - t_k - \frac{\sin 2\pi f (t_{k+1} - t_k)}{2\pi f} \right] \qquad (10.68)$$

Writing $t_{k+1} - t_k = \delta t$ gives:

$$\delta u_z^n = \frac{1}{2} \theta_x A_y \, \delta t \cos\phi \left\{ 1 - \frac{\sin 2\pi f \delta t}{2\pi f \delta t} \right\} \qquad (10.69)$$

It can be seen that δu_z^n is maximised when $\phi = 0$. Over the time interval δt, an effective acceleration error arises which may be expressed as:

$$\delta \dot{u}_z^n = \frac{1}{2} \theta_x A_y \cos\phi \left\{ 1 - \frac{\sin 2\pi f \delta t}{2\pi f \delta t} \right\} \qquad (10.70)$$

If $\delta \dot{u}_z^n$ is small compared with the overall system performance requirement, the need to implement the correction term is avoided.

Example

Consider a situation in which the body exhibits sculling motion at a frequency, f, of 50 Hz. The angular amplitude, θ_x, of the motion is taken to be 0.1°, whilst the cyclic acceleration, with respect to an orthogonal axis, is taken to vary between $\pm A_y$ where $A_y = 10$ g.

If the attitude update frequency is 100 Hz, i.e. $\delta t = 0.01$ second, the resulting acceleration error is 8.7 mg. By increasing the computational frequency to 500 Hz, the magnitude of the error is reduced to ~0.5 mg, the required precision being determined by the actual application.

10.3.4 Acceleration vector transformation using quaternions

In cases where the attitude is computed in quaternion form, the acceleration vector transformation may be effected using the quaternion parameters directly, using the following equation in place of eqn. 10.64:

$$\mathbf{u}^n = \mathbf{q} \cdot \left[\mathbf{v}_{k+1} + \frac{1}{2}\alpha_{k+1} \times \mathbf{v}_{k+1} + \frac{1}{2}\int\limits_{t_k}^{t_{k+1}} \left(\alpha \times \mathbf{f}^b - \omega^b \times \mathbf{v} \right) dt \right] \cdot \mathbf{q}* \quad (10.71)$$

Alternatively, the direction cosine matrix, $\mathbf{C}_b^n$, may be calculated from the quaternion parameters, using eqn. 10.50, and the acceleration transformation performed as described in Section 10.3.1.

10.4 Navigation algorithm

The computational processes required to determine vehicle velocity and position are not unique to strapdown systems and are described in many of the standard inertial navigation texts. More recently, Bar-Itzhack [7] has considered the benefits of using different computational rates for different parts of the navigation computation. For example, terms involving Earth's rate do not need to be evaluated as often as terms which are functions of the body rate. Such considerations are also taken account of in the analysis which follows.

The velocity and position equations, given in Chapter 3, may be expressed in integral form as follows:

$$\mathbf{v}^n = \int\limits_0^t \mathbf{f}^n \, dt - \int\limits_0^t \left[2\omega_{ie}^n + \omega_{en}^n \right] \times \mathbf{v}^n \, dt + \int\limits_0^t \mathbf{g}^n \, dt \quad (10.72)$$

$$\mathbf{x}^n = \int\limits_0^t \mathbf{v}^n \, dt \quad (10.73)$$

It is necessary to evaluate the integral terms within the navigation processor in order to determine vehicle velocity and position. The vectors in the above equations may be expressed in component form, as follows:

$$\mathbf{v}^n = \begin{bmatrix} v_N & v_E & v_D \end{bmatrix}^T$$

$$\mathbf{x}^n = \begin{bmatrix} x_N & x_E & -h \end{bmatrix}^T$$

$$\omega_{ie}^n = \begin{bmatrix} \Omega \cos L & 0 & -\Omega \sin L \end{bmatrix}^T$$

$$\omega_{en}^n = \begin{bmatrix} \dfrac{v_E}{R_0 + h} & \dfrac{-v_N}{R_0 + h} & \dfrac{-v_E \tan L}{R_0 + h} \end{bmatrix}^T$$

$$\mathbf{g}^n = \begin{bmatrix} 0 & 0 & g \end{bmatrix}^T$$

The above expressions apply for navigation in the vicinity of the Earth, in the local vertical geographic frame. The first integral term in eqn.

10.72 represents the sum of the velocity changes over each update cycle, $\mathbf{u}^n$:

$$\mathbf{u}^n = \int_{t_k}^{t_{k+1}} \mathbf{f}^n \, dt$$

$\mathbf{u}^n$ may be determined using eqn. 10.64, developed in the previous section. Since this term is a function of vehicle body attitude, $\mathbf{f}^n = \mathbf{C}_b^n \mathbf{f}^b$, it must be calculated at a rate which is high enough to take account of vehicle dynamic motion. Including the gravity contribution, the velocity vector may be updated over the time interval t_k to t_{k+1} using:

$$\mathbf{v}_{k+1}^n = \mathbf{v}_k^n + \mathbf{u}^n + \mathbf{g}\,\delta t$$

The second integral term in eqn. 10.72 includes the Coriolis correction. In general, the contribution to $\mathbf{v}^n$ is small compared with the other terms in the equation. Since the rate of change of the magnitude of the Coriolis term will be relatively low, it is sufficient to apply this correction at the relatively low l update rate, referred to in Section 10.3, using a fairly simple algorithm, viz:

$$\mathbf{v}_{l+1}^n = \left[\mathbf{I} - 2\mathbf{\Omega}_{ie}\,\delta t_l - \mathbf{\Omega}_{en}\,\delta t_l\right]\mathbf{v}_l^n \tag{10.74}$$

where $\mathbf{\Omega}_{ie} = \left[\boldsymbol{\omega}_{ie} \times\right]$

$\mathbf{\Omega}_{en} = \left[\boldsymbol{\omega}_{en} \times\right]$

$\mathbf{v}_l^n$ = velocity vector at time t_l

$\delta t_l = t_{l=1} - t_l$

Finally, position may be derived by integrating the velocity vector, as shown in eqn. 10.51.

The choice of integration scheme will, of course, be dependent on the application. For relatively short range, low accuracy, applications, a low order scheme such as rectangular or trapezoidal integration is likely to be adequate. Equations for updating position over the time interval t_k to t_{k+1} are shown below:

Rectangular integration:

$$\mathbf{x}_{k+1}^n = \mathbf{x}_k^n + \mathbf{v}_k^n\,\delta t \tag{10.75}$$

Trapezoidal integration:

$$\mathbf{x}_{k+1}^n = \mathbf{x}_k^n + \left(\frac{\mathbf{v}_k^n + \mathbf{v}_{k+1}^n}{2}\right)\delta t \tag{10.76}$$

For aircraft and shipborne inertial system applications, in which the performance requirements are more demanding, a higher order integration scheme such as Simpson's rule or fourth order Runge Kutta integration [8] may be needed.

Simpson's rule:

$$x_{k+1}^n = x_k^n + \left(\frac{v_{k-1}^n + 4v_k^n + v_{k+1}^n}{3} \right) \delta t \tag{10.77}$$

10.5 Summary

Strapdown navigation computation involves the determination of a vehicle's attitude, velocity and position from measurements of angular rate and specific force, obtained from a set of inertial instruments rigidly mounted in the vehicle. Measurements made by the inertial sensors are used in various equations to provide the desired navigation information. In the foregoing discussion, three distinct areas of strapdown computation have been described, namely attitude computation, transformation of the specific force vector and navigation computation. Techniques which may be used to implement each of these functions, in real time, have been outlined, together with some analysis of their application.

In this chapter, the equations to be implemented in an inertial navigation system processor have been described. Algorithms may be developed, based on the equations given, which will accept and process inertial measurements in incremental form directly. In order to achieve the desired real time solution of the various algorithms it is customary for the strapdown computation to be split into low, medium and high frequency sections as follows:

(i) low frequency computation {l-cycle}—certain parts of the strapdown navigation equations involve terms which vary very slowly with time, Earth's rate terms for instance. It is therefore sufficient to implement these sections of the algorithms at relatively low rates. In particular, the updating of the computed attitude for navigation frame rotations, and the application of Coriolis corrections to the navigation calculations, may be carried out at the low rate. In addition, algorithms for attitude orthogonalisation and normalisation, if required, may be implemented at the low rate;

(ii) medium frequency computation {k-cycle}—it is intended that the bulk of the computation should be implemented at the medium rate which will need to be selected to cope with large amplitude dynamic motion, arising as a result of vehicle manoeuvres. This includes the updating of the quaternions or direction cosines, the transformation of the acceleration vector and the solution of the major part of the navigation equations;

(iii) High frequency computation {*j*-cycle}—In order to cope with vibratory motion, coning and sculling, for example, it may be necessary to carry out some relatively simple calculations at high speed. This allows compensation for variations in the applied angular rate and linear acceleration occurring between the updates which take place at a lower frequency.

The various computational functions to be implemented are summarised in Figure 10.2. It is assumed that the inertial measurement unit will be capable of providing outputs at intervals, consistent with the highest computational frequency which is required.

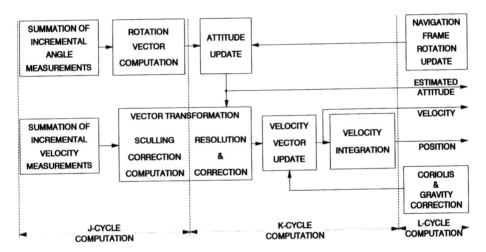

Figure 10.2   Summary of strapdown computational tasks

References

1   BORTZ, J.E.: 'A new mathematical formulation for strapdown inertial navigation', *IEEE Trans. Aerosp. Electron. Syst.*, 1971, **7** (1)
2   MILLER, R.B.: 'A new strapdown attitude algorithm', *J. Guid. Control Dyn.*, 1983, **6**
3   SAVAGE, P.: 'Strapdown system algorithms', *in* 'Advances in strapdown inertial systems', *AGARD Lecture Series No.133*, May 1984
4   'A study of the critical computational problems associated with strapdown inertial navigation systems', NASA CR-968
5   MORTENSEN, R.E.: 'Strapdown guidance error analysis', *IEEE Trans. Aerosp. Electron. Syst.*, 1974, **10** (4)
6   LEVINSON, E.: 'Laser gyro strapdown inertial system applications', *in* 'Strapdown inertial systems', *AGARD Lecture Series No.195*, June 1978
7   BAR-ITZHACK, I.Y.: 'Navigation computation in terrestrial strapdown inertial navigation systems', *IEEE Trans. Aerosp. Electron. Syst.*, 1977, **13** (6)
8   JEFFREY, A.: *'Mathematics for Engineers and Scientists'* (Nelson, 1979)

Chapter 11

Generalised system performance analysis

11.1 Introduction

In a practical implementation, the accuracy to which a strapdown inertial navigation system is able to operate is limited as a result of errors in the data which are passed to it prior to the commencement of navigation, as well as imperfections in the various components which combine to make up the system. The sources of error may be categorised as follows:

- initial alignment errors
- inertial sensor errors
- computational errors

Many of the contributions to the errors in these different categories have been described in Chapters 4, 5, 6, 9 and 10.

Any lack of precision in a measurement used in a dead reckoning system such as an inertial navigation system is passed from one estimate to the next with the overall uncertainty in the precision of the calculated quantity varying or drifting with time. In general, inertial navigation system performance is characterised by a growth in the navigation error from the position co-ordinate values which are initially assigned to it. It is common practice to refer to an inertial navigation system in terms of its mean drift performance; a one nautical mile per hour system is a typical performance class of a system. This perforance would be typical of an inertial navigation system used in a commercial aircraft.

During the early stages of design and system specification, it is necessary to estimate navigation system performance under the conditions in which that system will be called upon to operate. A combination of analysis and simulation techniques is commonly used to predict system performance. In this Chapter, equations are derived relating system performance to sources of error. These equations are used to illustrate the propagation of the various types of error with time. Later in the Chapter the limitations of the analysis techniques are highlighted and an outline of simulation methods which may be used to assess system performance is presented.

11.2 Propagation of errors in a 2-D strapdown navigation system

11.2.1 Navigation in a non-rotating reference frame

The manner in which errors propagate in a strapdown inertial navigation system is discussed first in the context of the simple, two-dimensional, navigator considered at the start of Chapter 3, and illustrated in Figure 3.1. An error block diagram of this system is given in Figure 11.1.

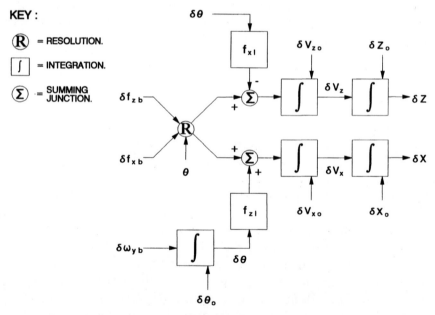

Figure 11.1   Error block diagram of a two-dimensional inertial navigation system

The Figure shows errors in the initial values of position, velocity and attitude as well as biases on the measurements of specific force and angular rate provided by the inertial sensors. For the purposes of simplifying this initial analysis, imperfections in the representation of the gravitational field have been ignored.

These errors propagate throughout the system giving rise to position errors which increase with time. The propagation of the errors can be represented in mathematical form as a set of differential equations which are derived from the system equations given in Chapter 3 by taking partial derivatives. The differential equations for the two-dimensional navigation system, correct to first order in the various system errors, are shown below.

Navigation error equations

$$\delta\dot{\theta} = \delta\omega_{yb}$$

$$\delta\dot{f}_{xi} = \left(-f_{xb}\sin\theta + f_{zb}\cos\theta\right)\delta\theta + \delta f_{xb}\cos\theta + \delta f_{zb}\sin\theta$$

$$= f_{zi}\,\delta\theta + \delta f_{xb}\cos\theta + \delta f_{zb}\sin\theta$$

$$\delta\dot{f}_{zi} = -\left(f_{xb}\cos\theta + f_{zb}\sin\theta\right)\delta\theta - \delta f_{xb}\sin\theta + \delta f_{zb}\cos\theta$$

$$= -f_{xi}\,\delta\theta - \delta f_{xb}\sin\theta + \delta f_{zb}\cos\theta$$

$$\delta\dot{v}_{xi} = \delta f_{xi}$$

$$\delta\dot{v}_{zi} = \delta f_{zi}$$

$$\delta\dot{x}_{i} = \delta v_{xi}$$

$$\delta\dot{z}_{i} = \delta v_{zi} \tag{11.1}$$

Consider now the position errors resulting from the various error sources. An initial error in the estimation of position simply contributes a constant offset to the estimated position which remains fixed whilst the system is navigating and initial velocity errors are integrated to induce position errors which increase linearly with time. The effects of attitude errors and instrument biases on system performance are a little more complex, since individual errors will, in general, affect both channels of the navigation system.

For example, a bias on the output of the x-axis accelerometer, δf_{xb}, introduces an acceleration error which forms error components $\delta f_{xb}\cos\theta$ and $-\delta f_{xb}\sin\theta$ in the x and z channels of the reference frame respectively. These errors then propagate as position errors which increase as the square of time, $\delta f_{xb}\cos\theta\ t^2/2$ and $-\delta f_{xb}\sin\theta\ t^2/2$, the result of the double integration process required to compute position estimates. Initial attitude errors propagate in a similar manner, whilst a bias on the output of the gyroscope causes a position error which increases with time cubed owing to the additional integration process required to determine body attitude.

The form of the position errors caused by the various error contributions shown in Figure 11.1 is tabulated in Table 11.1:

Table 11.1 Propagation of errors in 2-D strapdown inertial navigation system

Error source		Position error	
		x axis	z axis
Initial position errors	δx_0	δx_0	–
	δz_0	–	δz_0
Initial velocity errors	δv_{x0}	$\delta v_0 t$	–
	δv_{z0}	–	$\delta v_{z0} t$
Initial attitude error	$\delta\theta_0$	$\delta\theta_0 f_{zi}\dfrac{t^2}{2}$	$-\delta\theta_0 f_{xi}\dfrac{t^2}{2}$
Accelerometer biases	δf_{xb}	$\delta f_{xb}\cos\theta\dfrac{t^2}{2}$	$\delta f_{xb}\sin\theta\dfrac{t^2}{2}$
	δf_{zb}	$\delta f_{zb}\sin\theta\dfrac{t^2}{2}$	$\delta f_{zb}\cos\theta\dfrac{t^2}{2}$
Gyroscope bias	$\delta\omega_{yb}$	$\delta\omega_{yb} f_{zi}\dfrac{t^3}{6}$	$-\delta\omega_{yb} f_{xi}\dfrac{t^3}{6}$

The outline analysis given above has drawn attention to the way in which different sources of error propagate in an inertial navigation system. It is evident that there is inherent coupling of errors between the navigation channels caused by the process of resolving the specific force measurements into the designated reference frame. Hence, a simple rigorous calculation of errors is not usually practical.

11.2.2 Navigation in a rotating reference frame

Consideration is now given to the particular situation in which it is required to navigate in the vicinity of the Earth. Navigation is assumed to take place in the local geographic reference frame as considered in Chapter 3. The revised two-dimensional system mechanisation is as described in Figures 3.3 and 3.4.

In this system, the x and z reference axes are coincident with the local horizontal and the local vertical respectively, and the navigation system provides estimates of velocity in each of these directions. The estimated horizontal velocity divided by the radius of the Earth, which constitutes the transport rate term referred to in Chapter 3, is fed back and subtracted from the measured body rates for the purpose of calculating body attitude with respect to the local geographic frame. The effect of this feedback is to revise the system error dynamics as described in the following text.

The error dynamics of the local geographic system are analysed here for the condition where true body attitude is zero, i.e. $\theta = 0$. In this case, the coupling between the channels is nominally zero, allowing each channel to be analysed separately. Despite the simplification of the analysis which this approach affords, the propagation of errors in the vertical and horizontal channels of the system can still be illustrated without loss of generality in the form of the results. In addition, it is assumed that the navigation system is mounted in a vehicle which is at rest on the Earth, or one which is travelling at a constant velocity with respect to the Earth. Under such conditions, the only force acting on the vehicle is the specific force needed to overcome the gravitational attraction of the Earth. In this situation, $f_{xg} = 0$ and $f_{zg} = g$.

The error equations, correct to first order, for the vertical and horizontal channels can now be written as shown below:

Table 11.2

Horizontal channel equations	Vertical channel equations
$\delta\dot{\theta} = \delta\omega_{yb} - \delta v_x / R_0$	$\delta\dot{v}_z = \delta f_{zb}$
$\delta\dot{v}_x = g\,\delta\theta + \delta f_{xb}$	$\delta\dot{z} = \delta v_z$
$\delta\dot{x} = \delta v_x$	

Errors in the vertical channel propagate with time in a similar manner to those of the inertial frame indicated earlier. However, in the horizontal channel there is a closed loop as shown by the block diagram representation in Figure 11.2. This loop is oscillatory owing to the presence of two integrators in the closed path.

The single axis navigator shown here is an electronic analogue of a hypothetical simple pendulum of length equal to the radius of the Earth. This is referred to as the Schuler pendulum, the properties of which are described in the following section.

11.2.3 The Schuler pendulum

The direction of the local vertical on the surface of the Earth can be determined using a simple pendulum consisting of a mass suspended

KEY :

$\frac{1}{s}$ = INTEGRATION.

Σ = SUMMING JUNCTION.

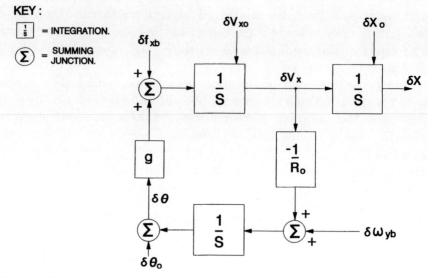

Figure 11.2   Inertial navigation system horizontal channel Schuler loop

from a fixed support point by a string. However, if the support point is moved from rest with an acceleration a, the string which supports the pendulum mass will be deflected with respect to the vertical by an angle θ equal to $\arctan(a/g)$ and will therefore no longer define the direction of the local vertical. Hypothetically, if the length of the support string is increased to equal the radius of the Earth, it will always remain vertical irrespective of the acceleration of the support point relative to the centre of the Earth. This is referred to as the Schuler pendulum after its formulation by Professor Max Schuler. A single-axis navigation system configured in the manner described above, as shown in Figure 11.2, is referred to as a Schuler tuned system since it behaves like a Schuler pendulum, as will now be demonstrated.

The measured specific force is resolved into the reference frame stored within the inertial navigation system. The resolved component of specific force is integrated once to give vehicle velocity and then again to give position. The transport rate is calculated by dividing the indicated velocity by the radius of the Earth (R_0). This signal is used to modify the stored attitude reference as the inertial system moves around the Earth. In the event that the stored attitude reference is in error by an angle $\delta\theta$, the direction of the horizontal indicated by the system will be tilted with respect to the true horizontal by this angle and the resolved accelerometer measurement will include a component of gravity equal to $g\delta\theta$. The closed loop that results is referred to as the Schuler loop. The loop is unstable as a result of the two integrators in the closed path and its dynamic behaviour is governed by the characteristic equation:

$$1 + \frac{g}{s^2 R_0} = 0 \tag{11.2}$$

where s is the Laplace operator.

Eqn. 11.2 may be rewritten as:

$$s^2 + \frac{g}{R_0} = 0$$

or $\quad s^2 + \omega_s^2 = 0 \tag{11.3}$

which is the equation for simple harmonic motion with natural frequency $\omega_s = \sqrt{(g/R_0)} = 0.00124$ rads/s for the Schuler pendulum. This is the Schuler frequency. The period of the Schuler oscillation is given by:

$$T_s = \frac{2\pi}{\omega_s} = 2\pi \sqrt{(R_0 / g)} = 84.4 \text{ minutes} \tag{11.4}$$

This is of the same form as the equation for the period of a simple pendulum of length l:

$$T = 2\pi \sqrt{(l/g)} \tag{11.5}$$

Therefore, the Schuler oscillation can be considered as the motion of a hypothetical pendulum of length equal to the radius of the Earth, R_0. A pendulum tuned to the Schuler frequency will always indicate the vertical on a moving vehicle provided it has been initially aligned to it. It is for this reason that Schuler tuned systems are most commonly employed for inertial navigation in the vicinity of the surface of the Earth.

11.2.4 Propagation of errors in a Schuler tuned system

In the single-axis navigation system, oscillations at the Schuler frequency will be excited in the presence of system errors. The block diagram given in Figure 11.2 shows errors in the initial estimates of attitude, velocity and position, $\delta\theta_0$, δv_0 and δx_0 respectively, and fixed biases in the gyroscopic and accelerometer measurements, $\delta\omega_{yb}$ and δf_{xb}. The propagation of these error terms with time may be derived from the differential equations, as shown in Table 11.3.

Table 11.3 Single axis error propagation

Error source		Position error
Initial position error	(δx_0)	δx_0
Initial velocity error	(δv_0)	$\delta v_x \dfrac{\sin \omega_s t}{\omega_s}$
Initial attitude error	$(\delta \theta_0)$	$\delta \theta_0 R_0 (1 - \cos \omega_s t)$
Fixed acceleration bias	(δf_{xb})	$\delta f_{xb} \left(\dfrac{1 - \cos \omega_s t}{\omega_s^2} \right)$
Fixed gyroscope bias	$(\delta \omega_{yb})$	$\delta \omega_{yb} R_0 \left(t - \dfrac{\sin \omega_s t}{\omega_s} \right)$

11.2.5 Discussion of results

It is apparent from these results that over long periods of time, several Schuler periods or more, the errors in the simple navigation system are bounded as a result of the Schuler tuning. This is true for all sources of error with the exception of the bias of the gyroscope which gives rise to a position error which increases linearly with time, $\delta \omega_{yb} R_0 t$, in addition to an oscillatory component. It is clear, therefore, that the performance of the gyroscope is critical in the achievement of long term system accuracy. This is one of the reasons why so much effort has been expended over the years in perfecting the performance of gyroscopes.

It follows that an approximate indication of inertial navigation system performance can be deduced solely from knowledge of gyroscopic measurement accuracy. For example, a system incorporating 0.01°/hour gyroscopes should be capable of navigating to an accuracy of ~1 km/hour. This relationship is often used to provide a rule of thumb guide to navigation system performance. The physical significance of this effect will be appreciated when it is remembered that the gyroscopes are used to store an attitude reference within the navigation system, and that the stored reference changes at the 'drift' rate of the gyroscope. On the surface of the Earth, at the Equator, 1° of longitude corresponds to 111 km (approximately 60 nautical miles). Hence, one minute of arc is equivalent to a displacement of approximately one nautical mile.

The analyses presented thus far relate to a simplified navigation system operating in a single plane. As will now be shown, the complexity of the error model is increased greatly in a full three-dimensional inertial navigator, particularly in the presence of vehicle manoeuvres and as a result of coupling between the respective channels of the system. The following section describes a generalised set of error equations which may be used to predict inertial navigation system performance.

11.3 General error equations

In this section, the growth of errors in a full three-dimensional navigation system is examined. The equations given here relate to a terrestrial system operating close to the Earth in a local geographic reference frame.

11.3.1 Derivation of error equations

11.3.1.1 Attitude errors

The orientation of the instrument cluster in a strapdown system with respect to the navigation reference frame may be expressed in terms of the direction cosine matrix, C_b^n. The estimated attitude, denoted by $\tilde{C}_b^n$, may be written in terms of the true direction cosine matrix, C_b^n, as follows:

$$\tilde{C}_b^n = BC_b^n \tag{11.6}$$

where B represents the transformation from true reference axes to estimated reference axes, the misalignment of the reference frame stored in the inertial navigation system computer. For small angles of misalignment, the matrix B may be approximated as a skew symmetric matrix which may be written as follows:

$$B = \left[I - \Psi\right] \tag{11.7}$$

where I is a 3×3 identity matrix and Ψ is given by:

$$\Psi = \begin{pmatrix} 0 & -\delta\gamma & \delta\beta \\ \delta\gamma & 0 & -\delta\alpha \\ -\delta\beta & \delta\alpha & 0 \end{pmatrix} \tag{11.8}$$

The elements, $\delta\alpha$ and δb, correspond to the attitude errors with respect to the vertical, the level or tilt errors, whilst $\delta\gamma$ represents the error about vertical, the heading or azimuth error. These terms are analogous to the physical misalignments of the instrument cluster in a stable platform navigation system and may be equated approximately to the roll, pitch and yaw Euler angle errors for small angle misalignments.

The estimated direction cosine matrix may now be written as follows:

$$\tilde{C}_b^n = \left[I - \Psi\right]C_b^n \tag{11.9}$$

which may be rearranged to give:

$$\dot{\Psi} = -\dot{\tilde{C}}_b^n C_b^{nT} - \tilde{C}_b^n \dot{C}_b^{nT} \tag{11.10}$$

Differentiating this equation yields:

$$\dot{\Psi} = -\dot{\tilde{C}}_b^n C_b^{nT} - \tilde{C}_b^n \dot{C}_b^{nT} \tag{11.11}$$

As shown in Chapter 3, the direction cosine matrix, C_b^n, propagates as a function of the absolute body rate (Ω_{ib}^b) and the navigation frame rate (Ω_{in}^n) in accordance with the following equation:

$$\dot{C}_b^n = C_b^n \Omega_{ib}^b - \Omega_{in}^n C_b^n \tag{11.12}$$

Similarly, the time differential of the estimated matrix $\tilde{C}_b^n$

$$\dot{\tilde{C}}_b^n = \tilde{C}_b^n \tilde{\Omega}_{ib}^b - \tilde{\Omega}_{in}^n \tilde{C}_b^n \tag{11.13}$$

where $\tilde{\Omega}_{ib}^b$ and $\tilde{\Omega}_{in}^n$ represent the measured body rate and the estimated turn rate of the navigation reference frame respectively.

Substituting for $\dot{\tilde{C}}_b^n$ and $\dot{C}_b^n$ in eqn. 11.11 gives:

$$\dot{\Psi} = -\tilde{C}_b^n \tilde{\Omega}_{ib}^b C_b^{nT} + \tilde{\Omega}_{in}^n \tilde{C}_b^n C_b^{nT} + \tilde{C}_b^n \Omega_{ib}^b C_b^{nT} - \tilde{C}_b^n C_b^{nT} \tilde{\Omega}_{in}^n$$

$$= -\tilde{C}_b^n \left[\tilde{\Omega}_{ib}^b - \Omega_{ib}^b \right] C_b^{nT} + \tilde{\Omega}_{in}^n \tilde{C}_b^n C_b^{nT} - \tilde{C}_b^n C_b^{nT} \Omega_{in}^n \tag{11.14}$$

Substituting for $\tilde{C}_b^n$ from eqn. 11.9 gives:

$$\dot{\Psi} = -\left[I - \Psi \right] C_b^n \left[\tilde{\Omega}_{ib}^b - \Omega_{ib}^b \right] C_b^{nT}$$

$$+ \tilde{\Omega}_{in}^n \left[I - \Psi \right] C_b^n C_b^{nT} - \left[I - \Psi \right] C_b^n C_b^{nT} \Omega_{in}^n \tag{11.15}$$

writing $\delta\Omega_{in} = \tilde{\Omega}_{in} - \Omega_{in}$ and $\delta\Omega_{ib} = \tilde{\Omega}_{ib} - \Omega_{ib}$ and ignoring error output product terms, we have:

$$\dot{\Psi} \approx \Psi\Omega_{in}^n - \Omega_{in}^n \Psi + \delta\Omega_{in}^n - C_b^n \delta\Omega_{ib}^b C_b^{nT} \tag{11.16}$$

It can be shown from an element by element comparison that the above equation may be expressed in vector form as:

$$\dot{\psi} \approx -\omega_{in}^n \times \psi + \delta\omega_{in}^n - C_b^n \delta\omega_{ib}^b \tag{11.17}$$

where $\psi = \begin{bmatrix} \delta\alpha & \delta\beta & \delta \end{bmatrix}^T$, the misalignment vector and

$$\psi \times = \Psi$$

$$\omega_{in}^n \times = \Omega_{in}^n$$

$$\delta\omega_{in}^n \times = \delta\Omega_{in}^n$$

$$\delta\omega_{ib}^b \times = \delta\Omega_{ib}^b$$

11.3.1.2 Velocity and position errors

The velocity equation may be expressed as:

$$\dot{\mathbf{v}} = \mathbf{C}_b^n \mathbf{f}^b - \left(2\boldsymbol{\omega}_{ie}^n + \boldsymbol{\omega}_{en}^n\right) \times \mathbf{v} + \mathbf{g}_l \tag{11.18}$$

where $\mathbf{f}^b$ represents the specific force in body axes.

Similarly, the estimated velocity may be assumed to propagate in accordance with the following equation in which estimated quantities are again denoted by a tilde:

$$\dot{\tilde{\mathbf{v}}} = \tilde{\mathbf{C}}_b^n \tilde{\mathbf{f}}^b - \left(2\tilde{\boldsymbol{\omega}}_{ie}^n + \tilde{\boldsymbol{\omega}}_{en}^n\right) \times \tilde{\mathbf{v}} + \tilde{\mathbf{g}}_l \tag{11.19}$$

Differencing these two equations we have:

$$\delta\dot{\mathbf{v}} = \dot{\tilde{\mathbf{v}}} - \dot{\mathbf{v}}$$

$$= \tilde{\mathbf{C}}_b^n \tilde{\mathbf{f}}^b - \mathbf{C}_b^n \mathbf{f}^b$$

$$- \left(2\tilde{\boldsymbol{\omega}}_{ie}^n + \tilde{\boldsymbol{\omega}}_{en}^n\right) \times \tilde{\mathbf{v}} + \left(2\boldsymbol{\omega}_{ie}^n + \boldsymbol{\omega}_{en}^n\right) \times \mathbf{v} + \tilde{\mathbf{g}}_l - \mathbf{g}_l \tag{11.20}$$

Substituting for $\tilde{\mathbf{C}}_b^n = \left[\mathbf{I} - \boldsymbol{\Psi}\right]\mathbf{C}_b^n$ and writing $\tilde{\mathbf{f}}^b = \delta\mathbf{f}^b$, $\tilde{\mathbf{v}} - \mathbf{v} = \delta\mathbf{v}$, $\tilde{\boldsymbol{\omega}}_{ie}^n - \boldsymbol{\omega}_{ie}^n = \delta\boldsymbol{\omega}_{ie}^n$ and $\boldsymbol{\omega}_{en}^n - \boldsymbol{\omega}_{en}^n = \delta\boldsymbol{\omega}_{en}^n$, and expanding, ignoring product terms gives :

$$\delta\dot{\mathbf{v}} = -\boldsymbol{\Psi}\mathbf{C}_b^n \mathbf{f}^b + \mathbf{C}_b^n \delta\mathbf{f}^b$$

$$- \left(2\boldsymbol{\omega}_{ie}^n + \boldsymbol{\omega}_{en}^n\right) \times \delta\mathbf{v} - \left(2\delta\boldsymbol{\omega}_{ie}^n + \delta\boldsymbol{\omega}_{en}^n\right) \times \mathbf{v} - \delta\mathbf{g}$$

writing $\mathbf{C}_b^n \mathbf{f}^b = \mathbf{f}^n$ and re-arranging gives:

$$\delta\dot{\mathbf{v}} = \left[\mathbf{f}^n \times\right]\boldsymbol{\psi} + \mathbf{C}_b^n \delta\mathbf{f}^b$$

$$- \left(2\boldsymbol{\omega}_{ie}^n + \boldsymbol{\omega}_{en}^n\right) \times \delta\mathbf{v} - \left(2\delta\boldsymbol{\omega}_{ie}^n + \delta\boldsymbol{\omega}_{en}^n\right) \times \mathbf{v} - \delta\mathbf{g} \tag{11.21}$$

Ignoring errors in the Coriolis terms and in knowledge of the gravity vector, this equation reduces to:

$$\delta\dot{\mathbf{v}} = \left[\mathbf{f}^n \times\right]\boldsymbol{\psi} + \mathbf{C}_b^n \delta\mathbf{f}^b \tag{11.22}$$

Finally, the position errors, $\delta\mathbf{p}$, may be expressed as follows

$$\delta\dot{\mathbf{p}} = \delta\mathbf{v} \tag{11.23}$$

The velocity and position errors are predominantly functions of the specific force, $\mathbf{f}^n$, to which the inertial navigation system is subjected, the attitude errors, $\boldsymbol{\psi}$, and inaccuracies in the measurements of specific force provided by the accelerometers, $\delta\mathbf{f}^b$. In addition, errors arise in a local vertical terrestrial navigator through errors in the Coriolis terms, imperfect knowledge of the local gravity vector and incorrect assumptions regarding the shape of the Earth.

Equations of the above form may be used to describe the propagation of errors in each of the strapdown mechanisations discussed in Section 3.3. For example, for a system operating in local geographic axes, ω_{in} represents the sum of the Earth's rate and transport rate terms, although it becomes zero for a system operating in space fixed co-ordinates.

11.3.1.3 State space form

Eqns. 11.17, 11.21 and 11.23 may be combined to form a single matrix error equation as follows:

$$\delta \dot{x} = \mathbf{F}\,\delta x + \mathbf{G}\,u \tag{11.24}$$

where $\delta x = \begin{bmatrix} \delta\alpha & \delta\beta & \delta\gamma & \delta v_N & \delta v_E & \delta v_D & \delta L & \delta l & \delta h \end{bmatrix}^T$ $\qquad$ (11.25)

$$u = \begin{bmatrix} \delta\varphi_x & \delta\varphi_y & \delta\varphi_z & \delta f_x & \delta f_y & \delta f_z \end{bmatrix}^T \tag{11.26}$$

$$\mathbf{G} = \begin{pmatrix}
-c_{11} & -c_{12} & -c_{13} & 0 & 0 & 0 \\
-c_{21} & -c_{22} & -c_{23} & 0 & 0 & 0 \\
-c_{31} & -c_{32} & -c_{33} & 0 & 0 & 0 \\
0 & 0 & 0 & c_{11} & c_{12} & c_{13} \\
0 & 0 & 0 & c_{21} & c_{22} & c_{23} \\
0 & 0 & 0 & c_{31} & c_{32} & c_{32}
\end{pmatrix} \tag{11.27}$$

11.3.2 Discussion

The set of coupled differential equations given in the last section define the propagation of errors in a local geographic inertial navigation system. A simplified block diagram representation of the error model is given in Figure 11.3. This model is applicable to a navigation system installed in a vehicle which is moving over the surface of a spherical earth. The diagram shows the Schuler loops and a number of the cross coupling terms which give rise to longer term oscillations described below.

The errors which propagate in an inertial navigation system over long periods of time are characterised by three distinct frequencies:

(i) The Schuler oscillation, $\omega_s = \sqrt{(g/R_0)}$ which manifests itself as an oscillation in each horizontal channel. The period of this oscillation is approximately 84.4 minutes as described in the Section 11.2.3;

(ii) The Foucault oscillation, $\omega_f = \Omega \sin L$. This maintains itself as a modulation of the Schuler oscillation, the modulation in the two horizontal channels being 90° apart in phase. The Foucault oscillation has a period of $2\pi/\Omega \sin L$ where Ω is the angular

$$
\mathbf{F} =
\begin{pmatrix}
0 & -\Omega\sin L-\dfrac{v_E}{R}\tan L & \dfrac{v_N}{R} & 0 & -\dfrac{1}{R} & 0 & -\Omega\sin L & 0 & -\dfrac{v_E^2}{R^2} \\[2ex]
\Omega\sin L+\dfrac{v_E}{R}\tan L & 0 & \Omega\cos L+\dfrac{v_E}{R} & \dfrac{1}{R} & 0 & 0 & 0 & 0 & \dfrac{v_N}{R^2} \\[2ex]
-\dfrac{v_N}{R} & -\Omega\cos L-\dfrac{v_E}{R} & 0 & 0 & -\dfrac{\tan L}{R} & 0 & -\Omega\cos L-\dfrac{v_E}{R\cos^2 L} & 0 & \dfrac{v_E\tan L}{R^2} \\[2ex]
0 & -f_D & f_E & \dfrac{v_D}{R} & -2\!\left(\Omega\sin L+\dfrac{v_E}{R}\tan L\right) & \dfrac{v_N}{R} & -v_E\!\left(2\Omega\cos L+\dfrac{v_E}{R\cos^2 L}\right) & 0 & \dfrac{1}{R^2}\!\left(v_E^2\tan L-v_N v_D\right) \\[2ex]
f_D & 0 & -f_N & 2\Omega\sin L+\dfrac{v_E}{R}\tan L & \dfrac{1}{R}\!\left(v_N\tan L+v_D\right) & 2\Omega\cos L+\dfrac{v_E}{R} & 2\Omega(v_N\cos L-v_D\sin L)+\dfrac{v_N v_E}{R\cos^2 L} & 0 & -\dfrac{v_E}{R^2}\!\left(v_N\tan L+v_D\right) \\[2ex]
-f_E & f_N & 0 & -\dfrac{2v_N}{R} & -2\!\left(\Omega\cos L-\dfrac{v_E}{R}\right) & 0 & 2\Omega v_E\sin L & 0 & \dfrac{1}{R^2}\!\left(v_N^2+v_E^2\right) \\[2ex]
0 & 0 & 0 & \dfrac{1}{R} & 0 & 0 & 0 & 0 & -\dfrac{v_N}{R^2} \\[2ex]
0 & 0 & 0 & 0 & \dfrac{1}{R\cos L} & 0 & \dfrac{v_E\tan L}{R\cos L} & 0 & -\dfrac{v_E}{R^2\cos L} \\[2ex]
0 & 0 & 0 & 0 & 0 & -1 & 0 & 0 & 0
\end{pmatrix}
\tag{11.20}
$$

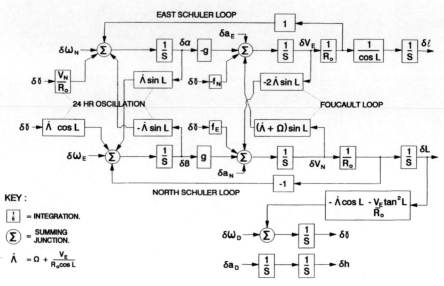

Figure 11.3   Error block diagram

frequency of the Earth's rotation and L is the latitude of the system. The period of this oscillation is about 30 hours for moderate latitudes;

(iii) A 24 hour oscillation, ω_g, which is directly related to the period of rotation of the Earth, showing itself mainly as a latitude/azimuthal oscillation.

As described earlier, the dynamics of the horizontal channels of an inertial navigation system are analogous to the motion of a simple pendulum having a length equal to the Earth's radius. The Schuler motion of a freely swinging pendulum of length R_0 suspended above a rotating Earth will be modulated at the Foucault frequency, which corresponds to the vertical component of the Earth's rate. The Foucault oscillation is named after the French physicist who used a freely swinging pendulum to demonstrate the rotation of the Earth. In a moving system, this frequency is modified by motion of the navigation system about the Earth. These effects are illustrated with the aid of the error propagation examples given below.

Examples

In Figure 11.4, plots are given illustrating the propagation of navigation errors with time over a 36 hour period. The system is assumed to be stationary and located on the surface of the Earth at a latitude of 45°. The following error sources are included:

initial alignment errors with respect to the vertical	0.1 mrad
initial heading error	1.0 mrad
gyroscopic bias	0.01 °/hour
accelerometer bias	0.1 mg

The distribution of the errors is assumed to be Gaussian and the above figures represent 1σ values, as described in Appendix B. The resulting attitude and position errors given in the Figure are also Gaussian 1σ values, i.e. there is a 68% probability of each error lying within the limits indicated.

The presence of the Schuler, Foucault and Earth's rate oscillations are clearly apparent in Figure 11.4. At a latitude of 45° considered here, the period of the Foucault oscillation is approximately 34 hours. The Schuler components of the attitude errors are shown to be modulated at this frequency, whilst the affect on the propagation of position errors is second order.

In general, the full error model described in the preceding section is only required to assess the performance of inertial navigation systems operating for long periods of time, such as several days. For many applications, including aircraft and missile systems, flight times are typically of the order of hours or minutes, rather than days. In such cases, some simplifications can be made in the error models used to assess system performance, since the terms which give rise to the Foucault and 24 hour oscillations can often be disregarded. This is illustrated in Figure 11.5 where the growth of navigation errors over a four hour period is shown. It can be seen that the Schuler frequency components are dominant in this situation.

For navigation over a few hours or less, much of the coupling between the north, east and vertical channels of the inertial navigation system can be ignored allowing each channel to be treated largely in isolation. The analysis of such systems becomes more tractable, as illustrated in the following section.

11.4 Analytical assessment

A full analytical solution of the error eqn. 11.24 given in Section 11.3 is extremely onerous mathematically and it is common practice to

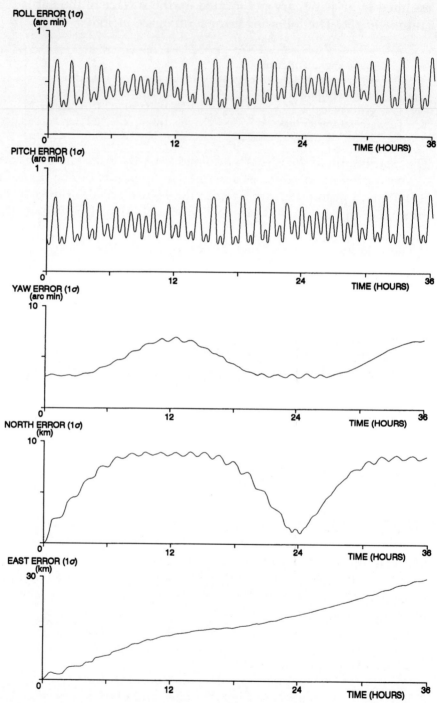

Figure 11.4   Simulated navigational accuracy (36 hour period)

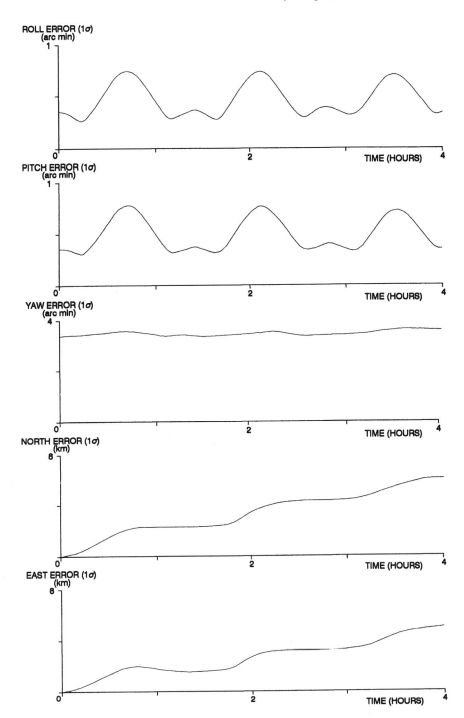

Figure 11.5   Simulated navigational accuracy (4 hour period)

solve it using a computer. Methods by which this may be accomplished using a computer model are the subject of Section 11.5. However, for periods of navigation up to a few hours, the effects of the Foucault and 24 hour oscillations may safely be ignored for many applications and the propagation of errors in the north, east and vertical channels can be considered separately. Under such conditions, a simplified analysis similar to that described in Section 11.2 may be undertaken. To illustrate the analytical methods which may be applied, the propagation of navigation errors in the north channel alone is examined.

11.4.1 Single-channel error model

For a strapdown inertial navigation system mounted in a vehicle travelling at constant speed and at constant height above the Earth, the error dynamics for the north channel may be expressed in terms of the following set of coupled differential equations, in accordance with the error equations given in the preceding section:

$$\delta\dot{\beta} = \left(\Omega\cos L + v_E / R\right)\delta\gamma - \delta v_n / R_0 - \delta B_{gl}$$

$$\delta\dot{\gamma} = -\delta B_{gD}$$

$$\delta\dot{v}_N = g\,\delta\beta + \delta B_{aN}$$

$$\delta\dot{x}_N = \delta v_N \tag{11.29}$$

where δB_{gE} and δB_{gD} represent the effective gyroscopic biases acting about the east and vertical axes respectively, and δB_{aN} is the net accelerometer bias acting in the north direction. These terms may be expressed in terms of the gyroscopic measurement errors $(\delta B_{gx},\ \delta B_{gy},\ \delta B_{gz})$ and the accelerometer errors $(\delta B_{ax},\ \delta B_{ay},\ \delta B_{az})$ as follows:

$$\delta B_{gE} = c_{21}\,\delta B_{gx} + c_{22}\,\delta B_{gy} + c_{23}\,\delta B_{gz}$$

$$\delta B_{gD} = c_{31}\,\delta B_{gx} + c_{32}\,\delta B_{gy} + c_{33}\,\delta B_{gz}$$

$$\delta B_{gN} = c_{11}\,\delta B_{ax} + c_{12}\,\delta B_{ay} + c_{13}\,\delta B_{az} \tag{11.30}$$

If, as assumed here, the gyroscopic and accelerometer errors may be represented as fixed biases, the instrument bias dynamics may be represented by the following set of trivial differential equations:

$$\delta\dot{B}_{gE} = 0$$

$$\delta\dot{B}_{gD} = 0$$

$$\delta\dot{B}_{aN} = 0 \tag{11.31}$$

These equations may be expressed in matrix form as follows:

$$\delta\dot{\mathbf{x}} = \mathbf{F}\,\delta\mathbf{x} \tag{11.32}$$

where $\delta\mathbf{x} = [\delta\beta \ \ \delta\gamma \ \ \delta v_N \ \ \delta x_N \ \ \delta B_{gE} \ \ \delta B_{gD} \ \ \delta B_{aN}]^T$ (11.33)

$$\text{and} \quad \mathbf{F} = \begin{pmatrix} 0 & \dot{\Lambda}\cos L & -\dfrac{1}{R} & 0 & -1 & 0 & 0 \\ 0 & 0 & 0 & 0 & 0 & -1 & 0 \\ g & 0 & 0 & 0 & 0 & 0 & 1 \\ 0 & 0 & 1 & 0 & 0 & 0 & 0 \\ 0 & 0 & 0 & 0 & 0 & 0 & 0 \\ 0 & 0 & 0 & 0 & 0 & 0 & 0 \\ 0 & 0 & 0 & 0 & 0 & 0 & 0 \end{pmatrix} \tag{11.34}$$

in which $\dot{\Lambda} = \Omega + v_E / R\cos L$

It is noted that position is given here in terms of distance (x_N) rather than latitude (L). A block diagram representation showing the instrument errors and the initial condition errors is given in Figure 11.6.

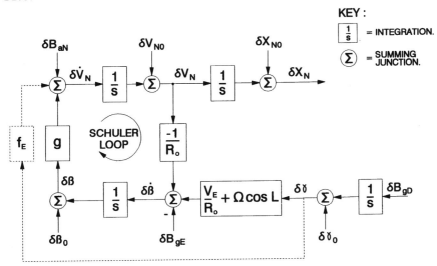

Figure 11.6 Simplified block diagram of the north channel of an inertial navigation system —medium term navigation

The diagram shows the various error sources and the Schuler loop. It should be noted that in the presence of a vehicle acceleration, the azimuth alignment error is coupled directly into the horizontal accelerometer, as indicated by the additional signal path shown dotted in the Figure. It will also be noted that this representation is equivalent to the error model used for the simplified inertial navigation system described earlier.

For very short term navigation, i.e. when the navigation time is a small fraction of a Schuler period, the Schuler feedback has relatively little effect on the growth of errors and the single channel error model can be reduced further to the form given in Figure 11.7.

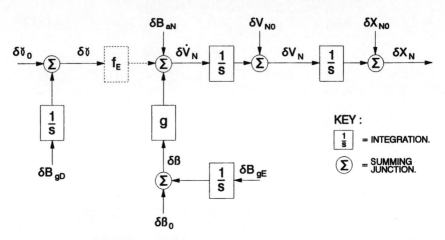

Figure 11.7   Simplified block diagram of the north channel of an inertial navigation system—short term navigation

Expressions for the north position errors which are applicable for medium and short term applications may be derived using state transition matrix methods, as described in the following section.

11.4.2 Derivation of single channel error propagation equations

The solution to eqn. 11.32 may be expressed in terms of the state transition matrix, $\Phi(t) = e^{Ft}$ as:

$$\delta x(t) = \Phi(t - t_0)\,\delta x(t_0) \tag{11.35}$$

where $\Phi(0) = I$ and $x(t_0)$ defines the initial states of the system.

The transition matrix is obtained using:

$$\Phi(t) = L^{-1}(sI - F)^{-1} \tag{11.36}$$

in which s is the Laplace operator and L^{-1} denotes the inverse Laplace transform.

The state transition matrix may be written as:

$$\Phi = \begin{pmatrix}
\cos_s t & \dfrac{\dot{\Lambda}\sin\omega_s t}{\omega_s} & -\dfrac{\sin\omega_s t}{\omega_s R_0} & 0 & -\dfrac{\sin\omega_s t}{\omega_s} & \dot{\Lambda}\left(\dfrac{1-\cos\omega_s t}{\omega_s^2}\right) & -\dfrac{(1-\cos\omega_s t)}{g} \\[2mm]
0 & 1 & 0 & 0 & 0 & -t & 0 \\[2mm]
g\omega\dfrac{\sin\omega_s t}{\omega_s} & \dot{\Lambda}g\left(\dfrac{1-\cos\omega_s t}{\omega_s^2}\right) & \cos\omega_s t & 0 & -g\left(\dfrac{1-\cos\omega_s t}{\omega_s^2}\right) & -\dot{\Lambda}R_0\left(t-\dfrac{\sin\omega_s t}{\omega_s}\right) & \dfrac{\sin\omega_s t}{\omega_s} \\[2mm]
R_0(1-\cos\omega_s t) & \dot{\Lambda}R_0\left(t-\dfrac{\sin\omega_s t}{\omega_s}\right) & \dfrac{\sin\omega_s t}{\omega_s} & 1 & -R_0\left(t-\dfrac{\sin\omega_s t}{\omega_s}\right) & -\dot{\Lambda}R_0\left\{\dfrac{t^2}{2}-\left(\dfrac{1-\cos\omega_s t}{\omega_s^2}\right)\right\} & \left(\dfrac{1-\cos\omega_s t}{\omega_s^2}\right) \\[2mm]
0 & 0 & 0 & 0 & 1 & 0 & 0 \\[2mm]
0 & 0 & 0 & 0 & 0 & 1 & 0 \\[2mm]
0 & 0 & 0 & 0 & 0 & 0 & 1
\end{pmatrix}$$

$$(11.37)$$

where $\omega_s = \sqrt{(g/R_0)}$, the frequency of the Schuler oscillation. The terms in a particular row of the transition matrix describe the dynamic effect of each error term on a particular error state. For example, the first term in row four indicates that a tilt error ($\delta\beta_0$) will give rise to a position error which propagates with time as $\delta\beta_0 R_0 (1-\cos\omega_s t)$. Similarly, it may be inferred that:

a constant velocity error (δv_0) gives rise to a position error of:

$$\delta v_0 \frac{\sin\omega_s t}{\omega_s}$$

an effective acceleration bias acting in the north channel (δB_{aN}) gives rise to a position error of:

$$\delta B_{aN}\left(\frac{1-\cos\omega_s t}{\omega_s^2}\right)$$

an effective angular rate bias acting about the east axis (δB_{gE}) gives rise to a position error of:

$$\delta B_{gE} R_0 \left(t - \frac{\sin \omega_s t}{\omega_s} \right)$$

similarly, a heading error ($\delta \gamma_0$) gives rise to a position error of:

$$\delta \gamma_0 \dot{\Lambda} R_0 \left(t - \frac{\sin \omega_s t}{\omega_s} \right)$$

an effective angular rate bias acting about the vertical axis (δB_{gD}) gives rise to a position error of:

$$\delta B_{gD} \dot{\Lambda} R_0 \left\{ \frac{t^2}{2} - \left(\frac{1 - \cos \omega_s t}{\omega_s^2} \right) \right\}$$

Over very short periods of navigation, i.e. navigation over a small fraction of a Schuler period, further simplifications may be made to these expressions. The position error contributions in the medium and short term are summarised in Table 11.4:

Table 11.4 Growth of position errors in the medium and short term

Error source		Position errors	
		medium error	short term
Initial attitude error	$(\delta \beta_0)$	$R_0(1 - \cos \omega_s)\delta \beta_0$	$g\delta \beta_0 \dfrac{t^2}{2}$
Initial attitude error	$(\delta \gamma_0)$	$R_0 \dot{\Lambda} \left(t - \dfrac{\sin \omega_s t}{\omega_s} \right) \delta \gamma_0$	$\dot{\Lambda} g \delta \gamma_0 \dfrac{t^3}{6}$
Initial velocity error	(δv_{NO})	$\dfrac{\sin \omega_s t}{\omega_s} \delta v_{NO}$	$\delta v_{NO} t$
Initial position error	(δx_{NO})	δx_{NO}	δx_{NO}
Gyroscope bias	(δB_{gE})	$R_0 \dot{\Lambda} \left(t - \dfrac{\sin \varphi_s t}{\varphi_s} \right) \delta B_{gE}$	$\delta B_{gE} \dfrac{t^3}{6}$
Gyroscope bias	(δB_{gD})	$-R_0 \dot{\Lambda} \left\{ \dfrac{t^2}{2} - \left(\dfrac{1 - \cos \omega_s t}{\omega_s^2} \right) \right\} \delta B_{gD}$	$\dot{\Lambda} \delta B_{gD} g \dfrac{t^4}{24}$
Accelerometer bias	(δB_{aN})	$\left(\dfrac{1 - \cos \omega_s t}{\omega_s^2} \right) \delta B_{aN}$	$\delta B_{aN} \dfrac{t^2}{2}$

Expressions similar to those given in the Table 11.4 can be derived for the east channel of the inertial navigation system. As with the north channel, the growth of many of the errors is bounded by the effects of the Schuler tuning. However, this is not the case in the vertical channel where the errors increase rapidly with time. For example, a vertical accelerometer bias, B_{az}, will give rise to a position error of $B_{az}t^2/2$. It is for this reason that aircraft navigation systems commonly operate in conjunction with a barometric or radio altimeter in order to restrict the growth of vertical channel errors. The scope for aiding inertial systems in this way is discussed more fully in Chapter 12, particularly in relation to integrated navigation systems. As shown in that chapter, aided systems rely on error models of the form discussed here to predict the growth of inertial navigation system errors with time.

11.4.3 Single-channel error propagation examples

Some example calculations are given here to illustrate how navigation errors may be determined using the single-axis error models described in the previous section. Example calculations for both aircraft and missile applications are presented.

11.4.3.1 Aircraft INS error propagation

The growth of navigation errors under benign flight conditions may be assessed using the medium-term error equations given in Table 11.4. Sample plots are given in Figure 11.8 which show the growth of north position error with time in an aircraft navigation system over a four hour period. For the purposes of this simple example, the aircraft is assumed to be cruising at a constant speed. Alignment accuracies of 0.05 mrad (1σ) in level and 1 mrad (1σ) in azimuth have been assumed, whilst the instrument performance is typical of a high grade airborne inertial navigation system. The gyroscope and accelerometer biases have been set to constant values of 0.01°/hour and 50 μg (1σ) respectively.

The Figure illustrates the form of the position errors resulting from the various error sources whilst the upper curve represents the combined effect of the individual errors. The upper curve has been obtained by summing the individual error components quadratically to give the total navigation error. It will be seen that the gyroscope bias and azimuthal misalignment contributions grow with time whilst the other terms are bounded as a result of Schuler tuning. It should be noted that these are simplified results for a single channel. In a full model, the Foucault effect is noticeable even at the first Schuler period.

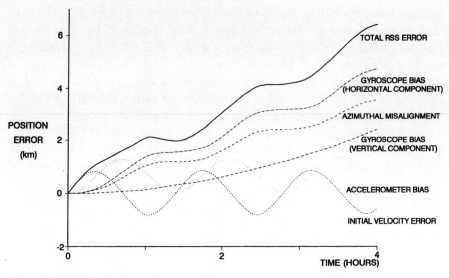

Figure 11.8   Single-channel error propagation

The analysis of inertial navigation system performance for airborne applications rapidly becomes complex when account is taken of realistic vehicle trajectories and manoeuvres, in which case the analyst will usually turn to simulation to aid the design process. However, under many circumstances some useful analysis can still be carried out to obtain an initial indication of system performance.

In addition to the usual effects of alignment errors and sensor biases illustrated in Figure 11.8, a number of error contributions arises as a result of the acceleration experienced by an aircraft during take-off. Both alignment errors and mounting misalignments of the accelerometers will give rise to cross-track velocity errors as the aircraft accelerates during take-off. For example, an azimuthal misalignment of $\delta\gamma_0$ will give rise to a velocity error of $V\delta\gamma_0$, where V is the cruise speed of the aircraft. The velocity error which has built up during take-off then propagates during the subsequent cruise phase of flight in the same way as an initial velocity error. Similarly, gyroscopic mass unbalance introduces a tilt error during take-off which propagates in the same manner as an accelerometer bias during the cruise phase of flight. Sensor scale factor inaccuracy and acceleration dependent biases will give rise to additional navigation errors in the event of an aircraft manoeuvre. The effects of such manoeuvres are most conveniently assessed through simulation. Errors which are dependent on the motion of the host vehicle are discussed separately later in the chapter. As will be shown, many of these errors are of particular concern in strapdown navigation systems.

11.4.3.2 Tactical missile INS error propagation

During short periods of flight, navigation errors propagate as simple functions of time. For example, in the absence of a missile manoeuvre, a gyroscopic bias (B_G) propagates as $gB_Gt^3/6$, as shown in Table 11.4. However, in the presence of missile accelerations and turn rates, other errors exert a considerable influence on the navigation performance, as illustrated in the following example.

Consider a missile which accelerates from rest at 200 m/s^2 (approximately 20 g) for a period of five seconds by which time it reaches a speed of 1000 m/s. Thereafter, the vehicle is assumed to maintain this speed for a further ten seconds. Hence, the total flight duration is 15 seconds and the overall distance travelled is 12.5 km. The onboard inertial system is assumed to contain gyroscopes and accelerometers having 1σ measurement biases of 50°/hour and 10 mg respectively. Table 11.5 below indicates the dominant contributions to cross-track position error for the flight path described, together with approximate mathematical expressions for the error propagation. The values used for the instrument errors and misalignments are typical for this type of application.

Table 11.4 Tactical missile INS error analysis Note: a(t) is the longitudinal acceleration of the missile

Error source	1σ magnitude	Cross-track position error	Cross-track position error at $t = 15$ s (m)
Initial misalignment			
position (δx_0)	1 m	δx_0	1
velocity (δv_0)	1 m/s	$\delta v_0 t$	15
attitude pitch $(\delta\theta_0)$	0.5°	$g\,\delta\theta_0 t^2/2$	10
yaw $(\delta\psi_0)$	0.5°	$\int\int \delta\psi_0 a(t)\, dt\, dt$	109
Accelerometer errors			
fixed bias (B_A)	10 mg	$0.5 B_a t^2$	11
mounting misalignment/ cross-coupling (M_A)	0.25%	$\int\int M_A a(t)\, dt\, dt$	31
Gyroscope errors			
fixed bias (B_G)	50°/hr	$\int\int B_G a(t) t\, dt\, dt$	7
g-dependent bias (B_g)	25°/hr/g	$\int\int\int B_g a(t)^2\, dt\, dt\, dt$	72
anisoelastic bias (B_a)	1°/hr/g²	$\int\int\int B_a a(t)^2 g\, dt\, dt\, dt$	3
		1σ root sum square position error	136 m

It is clear from the above results that the largest contributions to position error are caused by the initial angular misalignment in yaw and gyroscope mass unbalance error, the g-dependent bias.

11.5 Assessment by simulation

11.5.1 Introductory remarks

Whilst the analytical techniques described earlier provide a broad indication of inertial navigation system performance accuracy for various applications as a function of instrument quality and alignment accuracy, such methods are limited for the following reasons:

- they take only limited account of coupling between channels of the inertial navigation system;
- it is difficult to take account of realistic vehicle manoeuvres without the solution to the error equations becoming mathematically intractable;
- it is necessary to make simplifying assumptions about the instrument errors in each channel. In general, the effective angular rate and specific force measurement errors in each of the north, east, and vertical channels are functions of the errors in all three gyroscopes and accelerometers.

A more detailed investigation of inertial system errors and their interactions can be carried out using simulation.

11.5.2 Error modelling

The model of the inertial system that is to be assessed must include all sources of error which are believed to be significant. A full simulation must therefore incorporate alignment errors, representative models of the inertial sensors, including their errors, and any imperfections in the computational processes which are to be implemented.

11.5.2.1 Alignment errors

Unless the alignment process is to be modelled in detail, typical values for alignment errors are summed with the true attitude, velocity and position to define the navigation system estimates of these quantities at the start of navigation. Note that the alignment process itself can result in correlation between initial errors and the sensor errors as described in Section 11.6.1.1.

11.5.2.2 Sensor errors

Generalised sensor error models suitable for simulation purposes are given here based upon the gyroscope and accelerometer error models discussed in Chapters 4, 5 and 6. The errors in the measurements of angular rate provided by a set of gyroscopes ($\delta\omega_x$, $\delta\omega_y$, $\delta\omega_z$), whose sensitive axes are orthogonal, may be expressed mathematically as:

$$\begin{pmatrix} \delta\omega_x \\ \delta\omega_y \\ \delta\omega_z \end{pmatrix} = \mathbf{B}_G + \mathbf{B}_g \begin{pmatrix} a_x \\ a_y \\ a_z \end{pmatrix} + \mathbf{B}_{ac} \begin{pmatrix} a_y a_z \\ a_z a_x \\ a_x a_y \end{pmatrix} + \mathbf{B}_{ai} \begin{pmatrix} \omega_y \omega_z \\ \omega_z \omega_x \\ \omega_x \omega_y \end{pmatrix}$$

$$+ \mathbf{S}_G \begin{pmatrix} \omega_x \\ \omega_y \\ \omega_z \end{pmatrix} + \mathbf{M}_G \begin{pmatrix} \omega_x \\ \omega_y \\ \omega_z \end{pmatrix} + \mathbf{w}_G \tag{11.38}$$

where a_x, a_y, a_z are the accelerations acting along the principle axes of the host vehicle, and ω_x, ω_y, ω_z are the applied angular rates acting about these same axes, as defined by the reference model. The measurements of angular rate generated by the navigation system gyroscopes are generated by summing errors with the true rates.

For example, the measured rate about the x axis, $\tilde{\omega}_x$, may be expressed as :

$$\tilde{\omega}_x = \omega_x + \delta\omega_x$$

In eqn. 11.38 $\mathbf{B}_G$ is a three element vector representing the residual fixed biases which are present, $\mathbf{B}_g$ is a 3×3 matrix representing the g-dependent bias coefficients, $\mathbf{B}_{ac}$ is a 3×3 matrix representing the anisoelastic coefficient, $\mathbf{B}_{ai}$ is a 3×3 matrix representing the anisoinertia coefficients, $\mathbf{S}_G$ is a diagonal matrix representing the scale factor errors, $\mathbf{M}_G$ is a 3×3 skew symmetric matrix representing the mounting misalignments and cross-coupling terms, $\mathbf{w}_G$ is a three element vector representing the in-run random bias errors.

All of these gyroscopic errors are present to a greater or lesser extent in conventional gyroscopes, rate sensors and vibratory devices. However, as described in Chapter 5, the acceleration dependent biases are usually insignificant in optical sensors such as ring laser gyroscopes.

The modelling of in-run random errors has been discussed at some length by King [1]. Of particular concern is the effect of random walk errors which arise in optical gyroscopes and propagate as an angular error which is a function of the square root of time. The propagation of this type of error has been described by Flynn [2].

The errors in the measurements of specific force provided by an accelerometer triad may be expressed as shown below. It is assumed that the sensors are mounted with their sensitive axes nominally aligned with the principal axes of the host vehicle:

$$\begin{pmatrix} \delta f_x \\ \delta f_y \\ \delta f_z \end{pmatrix} = \mathbf{B}_A + \mathbf{B}_v \begin{pmatrix} a_y a_z \\ a_z a_x \\ a_x a_y \end{pmatrix} + \mathbf{S}_A \begin{pmatrix} a_x \\ a_y \\ a_z \end{pmatrix} + \mathbf{M}_A \begin{pmatrix} a_x \\ a_y \\ a_z \end{pmatrix} + \mathbf{w}_A \qquad (11.39)$$

In this equation $\mathbf{B}_A$ is a three element vector representing the fixed biases, $\mathbf{B}_v$ is a 3×3 matrix representing the vibro-pendulous error coefficients, $\mathbf{S}_A$ is a diagonal matrix representing the scale factor errors, $\mathbf{M}_A$ is a 3×3 skew symmetric matrix representing the mounting misalignments and cross-coupling terms and $\mathbf{w}_A$ is a three element vector representing the in run random bias errors.

The accelerometer errors described are particularly relevant for the pendulous force feedback accelerometer which is most commonly used at the present time for many different strapdown system applications.

11.5.2.3 Computational errors

As described in Chapter 10, inaccuracies will arise in a strapdown navigation system computer as a result of:

- bandwidth limitations, i.e. restricted computational frequency;
- truncation of the mathematical functions used in the strapdown attitude and navigation algorithms;
- limitations on the order of numerical integration schemes selected.

The effects of computational imperfections are most effectively assessed in a forward time stepping simulation by implementing the attitude and navigation algorithms in full. Comparison is then made between the navigation estimates generated in this way with those obtained using more precise calculations carried out within the reference model.

The inertial system designer will usually endeavour to ensure that the navigation errors arising from computational inadequacies are small compared with the alignment and instrument error contributions. Where this is so, attention may well be concentrated on the alignment and sensor error contributions, certainly during the early stages of a project. However, any computational imperfections which are expected to give rise to a sustained error should be taken into account at this stage of the design process. For example, a net linear acceleration or angular rate bias may be expected to arise in a system designed to operate in a particular vibratory environment. Such an effect may be modelled as a bias of appropriate magnitude and summed with an instrument bias in an attempt to model the effect approximately.

11.5.3 Simulation techniques

Alternative techniques which may be used for computer assessment of inertial systems are indicated below.

Time stepping simulation

A forward time stepping simulation is commonly adopted for the detailed assessment of an inertial navigation system.

When using a simulation of this type, it is necessary to define some standard or reference against which the performance of the simulated inertial system can be judged, as depicted in Figure 11.9. The reference defines the actual, or true, motion of the vehicle in which the inertial system is required to operate. In defining this reference system, all aspects of the computation must be implemented as precisely as possible in order to ensure that it represents, as closely as possible, the true motion of the vehicle in response to the stimuli which are causing it to move. Thus, it may be considered to be a perfect inertial system in which all alignment, sensor and computational errors are set identically to zero.

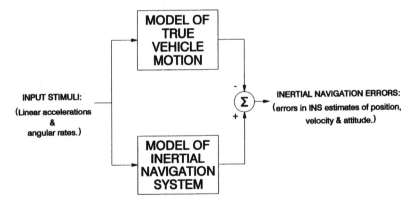

Figure 11.9   Block diagram of strapdown system simulation

The inertial system senses the specific force acceleration and turn rates to which it is subjected, and these same stimuli are used to drive the reference model. An assessment of inertial system performance can then be made by comparing the outputs of the inertial system model with those of the reference system.

Using this approach, a so-called Monte Carlo simulation may be carried out in which each error source in the inertial navigation system is modelled as a random process in the error model. A large number of simulation runs are then undertaken to generate estimates of the variances of the position, velocity and attitude estimates provided by the inertial system.

Covariance simulation

Using this approach, the error model equations given in Section 11.4 are transformed into covariance equations and are solved directly to determine the variances of the outputs as functions of time [3].

This technique avoids the need to perform very large numbers of individual runs to carry out a statistical assessment of system performance. Unlike the time stepping simulation method described in the previous section, a single covariance simulation yields the standard deviations of the position, velocity and attitude errors caused by initial misalignments and instrument errors.

Adjoint simulation

The adjoint technique is a 'computer efficient' method of determining the effect of initial condition errors, deterministic errors and random inputs on the values attained by a number of parameters of a system at a particular time [4, 5, 6]. For example, given an inertial navigation system with alignment and sensor errors, an adjoint simulation may be used to determine the contribution of each error source to the north or east position error at a given time.

The contribution to the north, or the east, position error made by each of the separate error components is indicated by a single simulation run. Using a conventional forward time stepping simulation, separate runs would be required for each error source to extract the same information.

The adjoint method can also be used to assess the performance of an inertial system when operating in a manoeuvring vehicle.

Where the sensitivity of a full strapdown system is to be examined in this way, it will be appreciated that the computational savings over a forward time stepping simulation can become very significant indeed. However, it should be noted that this technique provides no information about the transient behaviour of a system and is not valid for non-linear systems. To examine such effects, a conventional forward time stepping simulation will usually need to be used.

Whichever method is used, it will be vital to establish confidence in the simulation through verification and validation before proceeding with any detailed analysis. Hence, comparisons of the results of the simulation against theoretical results are recommended wherever possible to verify that the simulation is operating correctly. In some circumstances it may be helpful to use a combination of methods in order to validate the complete model.

11.6 Motion dependency of strapdown system performance

The error propagation equations given in Section 11.4 are broadly applicable to all types of inertial navigation systems. In this section, attention is focused on system imperfections which are dependent on vehicle motion, many of which are of particular concern in strapdown inertial navigation systems.

The performance of a strapdown inertial navigation system is very dependent on the motion of the host vehicle. Strapdown inertial sensors are subjected to the full range of heading and attitude changes and turn rates which the vehicle experiences along its flight path. This is in marked contrast to the inertial sensors in a stable platform navigation system which remain nominally fixed in the chosen reference frame and are not subjected to the rotational motion dynamics of the vehicle. For example, aircraft turn rates sensed by strapdown gyroscopes are typically four or five orders of magnitude greater than the inertial rates of the local geographic frame to which instruments in a platform system are subjected.

The need to operate in a relatively harsh dynamic environment whilst being able to measure large changes in vehicle attitude with sufficient accuracy has a major effect on the choice of inertial sensors. For example, gyroscope scale factor accuracy and cross-coupling errors must be specified more precisely in a strapdown system than would be necessary for a platform system of comparable performance. In addition, a number of error effects need to be taken into account which do not have a major impact on the performance of platform systems and hence are not addressed in many earlier texts on the subject of inertial navigation. A number of motion dependent errors are discussed below, including various manoeuvre dependent terms and inaccuracies introduced through cyclic or vibratory motion of the host vehicle.

11.6.1 Manoeuvre dependent error terms

The turn rates and accelerations which act on a vehicle as it manoeuvres excite a number of error sources within an onboard strapdown inertial navigation system. These include gyroscope and accelerometer scale factor errors, cross-coupling effects and sensitivity to non-orthogonality of the sensors' input axes. In addition, for systems which use conventional spinning mass gyroscopes, various acceleration dependent errors are induced as a result of mass unbalance and anisoelasticity, as described in Chapter 4. Therefore, the accuracy of the angular rate and specific force measurements generated by the onboard sensors during a manoeuvre can degrade substantially compared with that achieved under more benign conditions. The resulting measurement errors propagate giving rise to additional inaccuracies in the navigation system estimates of vehicle attitude, velocity and position.

Such effects are of particular significance in the airborne inertial navigation systems used for combat aircraft and agile missile applications. In missile applications, for example, the onboard sensors are often subjected to high levels of acceleration and rapid rates of turn, which give rise to substantial navigation errors. An example is given in Figure 11.10 showing the growth of navigation error in a short range tactical missile navigation system. The missile contains a medium grade strapdown inertial navigation system with gyroscope biases of 10°/hour and accelerometer biases of 10 mg. The growth of attitude and position errors is largely determined by g-dependent gyroscopic biases, sensor scale factor inaccuracy and initial misalignments of the onboard system. Also shown in the figure, for comparison purposes, are the errors which arise in a missile flying straight and level for a similar period of time. The importance of taking account of vehicle manoeuvres when assessing inertial system performance is clear.

Similarly, a combat aircraft may need to perform a variety of manoeuvres during the course of a mission. Examples include jink or S manoeuvres in which the aircraft performs a series of co-ordinated turns for purposes of low level terrain avoidance or tactical evasion and pop-up manoeuvres for ground attack sorties. The effect of these manoeuvres on the overall navigation accuracy achieved during flight is often highly dependent on their precise order and timing [7].

11.6.1.1 De-correlation of error terms

When an inertial system is aligned using gyro-compassing techniques, as described in Chapter 9, residual tilt and heading errors remain

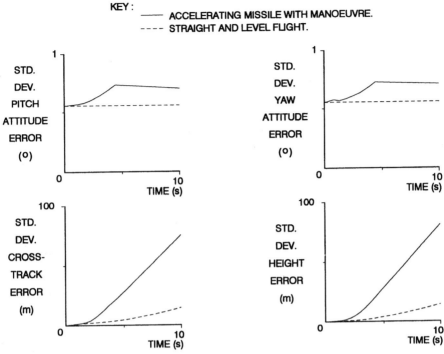

Figure 11.10   Illustration of manoeuvre dependent error propagation

which are caused by gyroscope and accelerometer biases. For example, alignment to the local vertical may be achieved by adjusting the stored direction cosine matrix until the resolved measurements of specific force in the horizontal plane become zero. On achieving this condition, biases in the accelerometer measurements are offset or cancelled by the tilt errors. Consequently, the tilt errors and biases are then said to be correlated.

Following this alignment process, provided the orientation of the inertial sensors with respect to the navigation reference frame in which the tilt errors are defined remains fixed, as occurs normally in a stable platform system, neither the biases nor the tilt errors propagate as navigation errors. However, in a strapdown system, the orientation of the sensors with respect to the navigation reference frame is unlikely to be maintained with the result that these errors will not remain correlated for long. As soon as the vehicle rotates in the reference frame, the instrument biases are no longer cancelled by the tilt errors. In fact, a rotation of 180° about the vertical will result in a reinforcement rather than a cancellation of the errors, because both the tilt error and the bias propagate separately giving rise to errors in the navigation system estimates of velocity and position. It is precisely this effect which gives rise to the Schuler pumping phenomenon discussed below.

11.6.1.2 Schuler pumping

As shown in Section 11.4, for flight times up to a few hours error propagation in the horizontal channels of an inertial navigation system is governed by the behaviour of the so-called Schuler loop. The Schuler loop may be represented as an undamped oscillator with a natural period of 84.4 minutes. It is to be expected that by stimulating this loop with a particular error signal at or near its natural frequency, a large and increasing error would result. For example, errors of this type may arise in an airborne navigation system if the aircraft executes a series of 180° turns at intervals of 42 minutes. This effect is referred to as Schuler pumping [7, 8]. Although not unique to strapdown inertial navigation systems, it is more likely to occur in strapdown systems than other types of inertial system mechanisation. This is because the correlation which exists between certain of the error sources, attitude errors and accelerometer biases for example, is maintained in a platform system but is lost in a strapdown system when the host vehicle changes course.

11.6.2 Vibration dependent error terms

This section is concerned specifically with the effects of vibratory and oscillatory motion on the performance of a strapdown navigation system. The various errors considered below are categorised as follows:

> **instrument rectification errors** arise through the rectification of the applied oscillatory motion by the sensor, and manifests themselves as a bias giving rise to an erroneous measurement of the vehicle motion;
> **system errors** these errors refer to bandwidth limitations and imperfections in the strapdown computation which inhibit the capability of the system to follow both angular and translational oscillatory motion correctly. Coning and sculling motion are of particular significance in this context;
> **pseudo-motion errors** are caused by false instrument outputs which the navigation system interprets incorrectly as true vehicle motion; pseudo-coning motion is a typical example.

Examples of each are described briefly below.

11.6.2.1 Instrument rectification errors

Many of the inertial sensor errors described previously in Section 11.5 are functions of products of the applied angular rates or linear accelerations—anisoelasticity, scale factor linearity and vibro-pendulous errors for example. Vibratory motion will be rectified by such effects resulting in additional biases on the inertial sensor outputs.

As an example, consider the effect of anisoelasticity in a spinning mass gyroscope which will give rise to a measurement bias ($\delta\omega$) which is a function of the linear acceleration acting simultaneously along orthogonal axes:

$$\delta\omega = B_{ae}\, a_x\, a_y \tag{11.40}$$

where a_x and a_y represent components of applied acceleration acting along orthogonal axes x and y respectively, and B_{ae} is the anisoelastic coefficient. In the presence of sustained oscillatory motion of frequency ω and phase difference φ between the two axes of motion, $a_x = A \sin \omega t$ and $a_y = A \sin(\omega t + \varphi)$, the bias becomes:

$$\delta\omega = B_{ae} A^2 \sin \omega t \sin(\omega t + \varphi)$$
$$= 0.5\, B_{ae} A^2 \{\cos \varphi - \cos(2\omega t + \varphi)\} \tag{11.41}$$

The mean value of this expression, $0.5\, B_{ae}A^2 \cos \varphi$, represents a constant bias which will be maximised when the accelerations in the two channels are exactly in phase. Consider a single-axis spinning mass gyroscope which is subjected to a sustained sinusoidal oscillation of amplitude 10 g in a direction at 45° to its spin and input axes, and normal to its output axis. If the magnitude of its anisoelastic coefficient is 0.5°/hour/g², an angular rate bias of 25°/hour would result.

11.6.2.2 System errors

Oscillatory motion can give rise to navigation system errors owing to limited sensor bandwidth, dynamic mismatch between sensors and insufficient computational speed all of which prevent the system from interpreting such motion correctly. The effects of cyclic angular motion by the vehicle, known as coning motion, or combinations of angular and translational motion, known as sculling, can be particularly detrimental to system performance. If the navigation system fails to detect such motion or to process accurately the inertial measurements obtained in the presence of such motion, significant navigation errors can arise.

11.6.2.3 Coning errors

Coning is the conical (or near conical) motion in inertial space of one of the gyroscope input axes, as illustrated in Figure 11.11. Such motion results from the simultaneous application of angular oscillations about two orthogonal axes of the system, where the oscillations are out of phase.

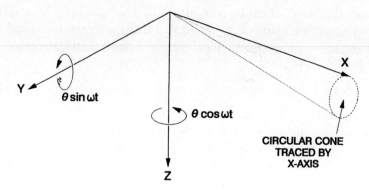

Figure 11.11   Coning motion

Taking the situation where a single gyroscope is subjected to motion such that its input axis follows a closed conical path, it can be shown [9, 10] that σ, the attitude of the gyroscope with respect to its initial position, after a full cycle of the motion, is given by:

$$\sigma = \int_0^T \omega \, dt + \varepsilon \tag{11.42}$$

where ω is the angular rate of the gyroscope about the line of the cone axis (as if no coning motions were taking place), T is the time taken to complete one coning cycle and ε is an additional rotation caused by the movement of the input axis around the closed path. The error term ε is identically equal to the area traced out by the input axis on the surface of a unit sphere centred on the origin of the gyroscope axis. The error ε, is a real effect, correctly measured by the gyroscope. It results from the fact that the input axis of the gyroscope is slightly displaced from its nominal *x* direction, as illustrated in Figure 11.11. The consequence is that when a small rotation is caused by θ sin ω*t* about the *y* axis, the gyroscope senses a small amount of the θ cos ω*t* about the *z* axis and vice versa. These small rotations keep changing size and sign. If, as shown in Figure 11.11, the motions about *y* and *z* are at the same frequency and not in phase, there is a net angle sum.

Thus, coning is purely a geometric effect resulting from the real motion of the gyroscope. If the attitude is computed solely on the basis of this one measurement, in the absence of knowledge of the cyclic rotations which have taken place about the other axes, the value of σ will be in error by ε. It therefore appears that the measurement of angular rate is in error owing to the presence of a bias of ε/T. This bias is referred to as the coning error and would be present even if a perfect gyroscope without any errors was used.

In a strapdown inertial navigation system which contains three single-axis gyroscopes, or an equivalent configuration, mounted such that their input axes are mutually orthogonal, the cyclic motions can be measured accurately and taken account of in the full attitude computation process. However, a coning error will still arise if the bandwidths of these gyroscopes are insufficient to measure or observe the angular motion, or if the attitude computation process is not performed at a sufficiently high rate.

Consider now the case of classical coning motion in which sinusoidal motions which are 90° out of phase are applied about two orthogonal axes. In addition, a constant rate is applied about the third axis in order to ensure that the body returns to its original position at the end of each coning cycle. For small rotations, the instantaneous attitude of the body may be expressed as:

$$\sigma = \begin{bmatrix} -\theta\cos\beta t & \theta\sin\beta t & 0 \end{bmatrix}^T \tag{11.43}$$

where θ is the amplitude of the coning motion and β is the frequency. The body may be returned to its original position at any time by rotating through an angle equal to the magnitude of σ, θ in this case. The associated angular rate is given by:

$$\omega = \begin{bmatrix} \beta\theta\sin\beta t & \cos\beta t & \beta\theta^2/2 \end{bmatrix}^T \tag{11.44}$$

Failure of the inertial navigation system to keep track of the oscillatory components of ω means that the measured motion, denoted ω', and the computed attitude, σ', will be as follows:

$$\omega' = \begin{bmatrix} 0 & 0 & \beta\theta^2/2 \end{bmatrix}^T \tag{11.45}$$

$$\sigma = \begin{bmatrix} 0 & 0 & \beta\theta^2 t/2 \end{bmatrix}^T \tag{11.46}$$

i.e. the computed attitude drifts at a rate $\beta\theta^2/2$ about the coning axis.

Hence, coning motion of 0.1° at a frequency of 50 Hz (~300 rad/s), for instance, can result in a drift in the computed attitude of almost 100°/hour. Clearly this can be a very significant error. The effects of coning motion on navigation system performance are addressed in Reference 8.

A more general development shows that the coning error which arises when the phase shift between the two orthogonal rotations is γ, and their respective amplitudes are θ_x and θ_y, can be expressed as follows:

$$\text{coning error} = \frac{1}{2}\beta\theta_x\theta_y \sin \gamma \qquad (11.47)$$

Clearly the resulting drift error is maximised when γ is 90° and falls to zero when γ is zero.

11.6.2.4 Sculling errors

Sculling is made up of a combination of linear and angular oscillatory motions of equal frequency in orthogonal axes. In the presence of such motion, errors can arise in the strapdown computing task which is concerned with the resolution of the measured specific force vector into the chosen navigation reference frame. An acceleration bias can arise through failure to take account of the rapid changes of attitude occurring between successive specific force vector resolutions.

For example, if a vehicle rotates sinusoidally about its y axis such that $\theta_y = \theta \sin(\omega t + \varphi)$, whilst oscillating linearly along its z axis such that $a_z = A \sin \omega t$, a component of the acceleration $(A_z \sin \theta_y)$ will appear in the x direction if these rotations are not correctly sensed and resolved into the navigation reference frame. For small angle perturbations, the x component of acceleration can be approximated as:

$$a_B = a_z\theta_y \qquad (11.48)$$

Substituting for a_z and θ_y gives:

$$a_B = 0.5 \, A\theta\{\cos \varphi - \cos(2\omega t + \varphi)\} \qquad (11.49)$$

Therefore, a steady acceleration component of $0.5A\theta \cos \varphi$ occurs in the x direction. It is stressed that this error term can arise even when using perfect accelerometers, being purely a function of the inaccuracy in the resolution process. If, for example, $A = 10$ g, $\theta = 0.1°$ and the phase difference is zero, the resulting acceleration bias is ~9 mg.

11.6.2.5 Size effect errors

The specific force acting on a vehicle is detected by sensing motion along three orthogonal axes, often using a triad of linear accelerometers. In order to navigate, it is necessary to sense the linear accelerations acting at a particular point in the vehicle, at its centre of gravity for example. Whether the inertial navigation system is mounted precisely at the vehicle centre of gravity or, as is more usual, at some offset location, it provides a measure of the motion of that point within the vehicle. This assumes that the inertial system is able to sense all motion accurately, including any centripetal and tangential forces induced by vehicle rotation, and to process accurately the inertial measurements which are generated.

Ideally, it is required that all three accelerometers should be mounted precisely at the same location in the vehicle. This is clearly impossible because the sensors are of finite size and because of design constraints on the positioning of hardware. Centripetal and tangential forces are sensed by the accelerometers because of their physical displacements with respect to the desired position, referred to as the size effect.

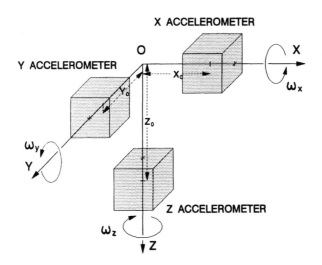

Figure 11.12   Illustration of size effect

Consider the situation depicted in Figure 11.12 where the sensitive axes of the accelerometers intersect at the point O. The x-axis accelerometer is mounted with its sensitive element displaced a distance x_o from O and its sensitive axis pointing along that axis. In the presence of angular rates, ω_y and ω_z about the y and z axes respectively, the x accelerometer will be subject to a centrifugal acceleration:

$$a_x = -(\omega_y^2 + \omega_z^2)x_o \qquad (11.50)$$

Similarly, the y and z sensors will sense accelerations:

$$a_y = -(\omega_x^2 + \omega_z^2)y_o \qquad (11.51)$$

$$a_z = -(\omega_x^2 + \omega_y^2)z_o \qquad (11.52)$$

In the presence of continuous rotations, as occur in a freely rolling missile for example, the effect of these additional accelerations will integrate to zero over a few cycles provided that the resolution of the measurements into the navigation reference frame is implemented accurately. Size effect errors can arise in this situation as a result of imperfections in the strapdown computational algorithms.

Of particular concern here is the effect of oscillatory motions which will be rectified to give steady acceleration errors. For instance, if $\omega_y = \omega\theta_y \sin \omega t$ and $\omega_z = \omega\theta_z \sin(\omega t + \phi)$, then:

$$a_x = -\left\{ \omega^2\theta_y^2 \sin^2 \omega t + \omega^2\theta_z^2 \sin^2(\omega t + \xi) \right\} x_0$$

$$= -\frac{1}{2}\left(\theta_y^2 + \theta_z^2 \right) x_0 + \frac{1}{2}\omega^2\left\{ \theta_y^2 \cos 2\omega t + \theta_z^2 \cos(2\omega t + 2\xi) \right\} x_0 \qquad (11.53)$$

Thus a steady bias acceleration of magnitude $0.5\omega^2(\theta_y^2 + \theta_z^2)x_0$ is introduced as a result of size effect. In the presence of cyclic angular motion of amplitude 0.1° and frequency 50 Hz for example, a bias of ~0.15 mg arises for a 10 cm displacement. This is not related in any way to imperfections in the accelerometer and will arise even when perfect sensors without any errors are used in this configuration.

11.6.2.6 Pseudo-motion errors

The inertial sensors themselves may produce false signals which are correlated with each other. The navigation system will then interpret these signals as indicating the presence of coning or sculling motion, for instance, where no such motion is actually present. The apparent motions are sometimes referred to as pseudo-coning or pseudo-sculling. If these motions are treated as true by the navigation system computer, the performance of the system will be degraded.

For example, pseudo-coning can arise in systems which use spinning mass gyroscopes as a result of the angular acceleration sensitivity of such sensors. In the presence of a cyclic angular rate about a single axis, $\omega\theta \sin \omega t$, an apparent rate may be detected about an orthogonal axis proportional to the applied angular acceleration, $\omega^2\theta \cos \omega t$. The vehicle will therefore appear, according to the measured rates, to exhibit coning motion as characterised by the two cyclic oscillations which are 90° out of phase. For a system using rate integrating gyroscopes, it can be shown [11] that the resulting bias, ω_b, is given by:

$$\omega_b = \frac{I}{2H}\omega^2\theta^2 \qquad (11.54)$$

where H is the angular momentum of the gyroscope and I is the moment of inertia of the float assembly, as described in Chapter 4.

Pseudo-coning can also arise in the absence of any applied motion, i.e. purely as a result of gyroscope imperfections. In a dual-axis sensor, such as the dynamically tuned gyroscope, outputs will arise if there is a misalignment between the rotor and the pick-offs which sense angular displacements about the two input axes of the sensor. These outputs will approximate to sinusoidal functions of time. Since the pick-offs are displaced by 90° about the spin axis, the two outputs will be 90° out of phase thus giving an erroneous indication of coning motion at the gyroscope spin frequency. If the spin frequency is 1000 rad/s, a misalignment of 1 arc minute will result in a coning error of 1°/hour. Of course, for many applications this would be insignificant.

11.7 Summary

The performance accuracy of an inertial navigation system can be expressed in terms of a series of equations. Inaccuracies arise in such a system because of initial alignment errors, imperfections in the performance of the inertial instruments and limitations in the computational process. These errors can be quantified enabling a designer to estimate the performance accuracy of a proposed navigation system. The analysis can be simplified in some circumstances, such as for very short duration flight. In other cases, particularly where there is coupling between channels, a deterministic solution is not possible, and simulation is necessary to provide accurate information on performance.

Inertial sensors are sensitive to various external stimuli. Fortunately these sources of error are frequently well behaved and consequently can be expressed as a deterministic equation, the coefficients of each term representing the various sensitivities to a given stimulus. Care must be taken when processing the various sensor signals in the presence of angular motion, particularly in the presence of coning and skulling motion. In these cases, the bandwidths of the sensors and the speed of the computation must be high enough to sense and record the actual motion, otherwise significant errors can arise even if perfect sensors were to be available.

References

1 KING, A.D.: 'Characterisation of gyro in-run drift'. Proceedings of DGON Sympsium on *Gyro technology* Stuttgart, Germany, 1984

2 FLYNN, D.J.: 'The effect of gyro random walk on the navigation performance of a strapdown inertial navigator'. Proceedings of DGON Sympsium on *Gyro technology*, Stuttgart, Germany, 1982

3 CHIANG-FANG LIN: *'Modern navigation, guidance and control processing'* (Prentice-Hall, 1991)

4 LANING, J.H. and BATTIN, R.H.: *'Random processes in automatic controll'* (McGraw-Hill, New York, 1956)

5 NESLINE, F.W. and ZARCHAN, P.: 'Miss distance dynamics in homing missiles', AIAA Paper 84-1845, August 1984

6 ALPERT, J.: 'Miss distance analysis for command guidance missiles' *J. Guidance and Control Dyn.*, 1988, **11** (6)

7 WATSON, N.F. and CAMPBELL, E.A.F.: 'Cost-effective strapdown INS design and the need for standard flight profiles'. Proceedings of DGON Sympsium on *Gyro technology* , Stuttgart, Germany, 1987

8 FENNER, P.J.: 'Requirements, applications and results of strapdown inertial technology to commercial airplanes' in 'Advances in strapdown inertial systems', AGARD Lecture Series 133, May 1984

9 GOODMAN, L.E. and ROBINSON, A.R.: 'Effect of finite rotations on gyroscopic sensing devices' *Trans. ASME. E, J. App. Mech.*, 1958

10 FLYNN, D.J.: 'A discussion of coning errors exhibited by inertial navigation systems', Royal Aircraft Establishment Technical Memorandum, Rad-Nav 243, 1984

11 'A study of the critical computational problems associated with strapdown inertial navigation systems', NASA CR-96, 1968

Chapter 12
Integrated navigation systems

12.1 Introduction

For many vehicles requiring a navigation capability, there are two basic but conflicting requirements to be considered by the designer, namely those of achieving high accuracy and low costs. This chapter examines the scope for satisfying these requirements by using integrated navigation systems, in which strapdown inertial navigation systems are used in conjunction with other navigation aids. The variety of modern navigation aids now available, coupled with advances in estimation processing techniques and high speed computers, has resulted in greater application of integrated navigation systems in recent years.

As discussed in Chapter 11, the performance of an inertial navigation system is characterised by a time dependent drift in the accuracy of the position estimates it provides. The rate at which navigation errors grow over long periods of time is governed predominantly by the accuracy of the initial alignment, imperfections in the inertial sensors which the system uses and the dynamics of the trajectory followed by the host vehicle. Whilst improved accuracy can be achieved through the use of more accurate sensors, there are limits to the performance which can reasonably be achieved before the cost of the inertial system becomes prohibitively large and incompatible with all except those special applications where there is no easy alternative, such as submarine navigation systems.

An alternative approach, suitable for many applications and one which has received much attention in recent years, is to employ some additional source of navigation information, external from the inertial system, to improve the accuracy of the inertial system.

12.2 Basic principles

In an aided inertial system, one or more of the inertial navigation system output signals is compared with independent measurements of identical quantities derived from an external source. This is illustrated in Figure 12.1. Corrections to the inertial navigation system are then derived as functions of these measurement differences. By judicious combination of this information, it is possible to achieve more accurate navigation than would be achieved using the inertial system in isolation.

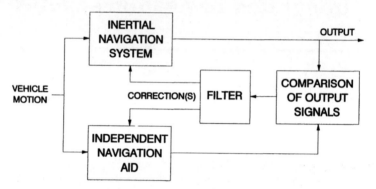

Figure 12.1   Basic principle of integrated navigation system

As a simple example, take the case of an aircraft which is able to detect a radio signal when it flies over a beacon on the ground. Provided the aircraft has precise knowledge of the position of the beacon, an accurate position fix is provided at the instant it flies over the beacon and this fix may be used to update an onboard inertial navigation system. In the event of a number of such position fixes being available, at discrete intervals of time, it is possible to update other quantities within the inertial system which are not directly measurable. For example, it may be possible to update and improve the inertial system estimates of velocity and heading.

   Integrated systems of this type usually make use of two independent sources of information with complementary characteristics, one source providing data with good short-term accuracy and the second providing good long-term stability. For example, a radio beacon can provide accurate position fixes at discrete intervals of time and so bound the long-term drift of an inertial navigation system. Meanwhile, the inertial system provides low noise continuous navigation data between the fixes which are accurate in the short-term and not subject to external interference.

In broad terms, the various types of navigation aid that are available may be categorised under the following headings:

external measurements, obtained by receiving signals or by viewing objects outside the vehicle requiring navigation. Such observations may be provided by radio navigation aids, satellites, star trackers or a ground based radar tracker, for example. In some cases, data may be transmitted to the vehicle during its journey, whilst in others there will be a receiver or viewer to accept the observations. Navigation aids of this type usually provide a position fix which may be expressed either in terms of vehicle latitude and longitude or as co-ordinates with respect to a local reference frame;

onboard measurements are derived using additional sensors carried onboard the vehicle requiring navigation. Navigation aiding of this type may be provided by altimeters, Doppler radar, airspeed indicators, magnetic sensors and radar or electro-optical imaging systems. Such sensors may be used to provide attitude, velocity or position updates, any of which may be used to improve the quality of the inertial navigation system.

Navigation aids which fall into these two categories are described in the two sections which follow. The later sections are concerned with methods of mixing measurement data provided by different navigation sensors or systems to form an integrated navigation system.

12.3 External navigation aids

12.3.1 Radio navigation aids

Radio navigation, based on ground transmitting stations, is perhaps the oldest of the modern navigation aids. The development of radio direction finders for both ships and aircraft allowed bearings to any radio transmission station at a known location to be determined and used for navigation purposes. Given measurements of bearing to two or more ground stations at known locations, the position of the vehicle may be calculated by the process of triangulation.

Many communications and broadcasting stations use low and medium frequencies to obtain large areas of coverage and these stations can be used at long ranges, well beyond the visual line-of-sight. However, radio propagation at these frequencies is affected by atmospheric conditions and care has to be taken when using such broadcasts at night. More accurate measurements can be obtained at higher frequencies, although their range is more restricted.

To overcome some of the problems which arise when using simple direction finding equipment, a number of systems have been developed based on the use of modulated radio beams, the modulation received being dependent on the position of the vehicle in the beam and hence providing navigational information. A widely adopted scheme is very high frequency omnidirectional radio range (VOR).

Very high frequency omni-directional radio range (VOR)

This is a short range navigation aid, primarily for aircraft use. The ground station has an omni-directional aerial which transmits a VHF carrier amplitude modulated by a reference signal. A series of other aerials is situated around the reference aerial. These aerials transmit a constant carrier frequency which is switched between them to simulate a cardioid shaped beam rotating once per cycle of the reference signal, as indicated in Figure 12.2. At a receiver, this gives a frequency modulated carrier modulated at the rotation frequency.

The timing of the rotating beam is adjusted so that for a receiver situated due north of the VOR station the frequency modulation is in phase with the amplitude modulation. The modulation phase at any other location is then equal to the angle from North along which the receiver is situated. Thus, by measuring this phase angle, the aircraft is able to determine the bearing from the station to itself. The reference direction at the VOR station is magnetic north.

The operation of the VOR navigational aid depends on maintaining line-of-sight contact between the aircraft and the ground stations. For this reason, operational range varies with aircraft altitude. At 300 m altitude, for example, the effective range is about 75 km, whilst at 6 km altitude this increases to 350 km. Typical bearing measurement accuracies are about 2°. The error in a position fix derived using VOR increases with the range from the ground stations.

A major improvement in navigational accuracy is possible if the distance to the radio stations can be determined. This can be achieved by transmitting signals at known times and measuring the time of arrival at the receiver. Since the propagation velocity is known, distance can be determined from the measured time delay. This is the basis of virtually all modern radio navigation systems, including GPS which is described later, the differences being in the means of timing and the transmission frequencies.

To measure the time delay between transmission and reception of a radio signal, the transmitter and receiver must have clocks which are

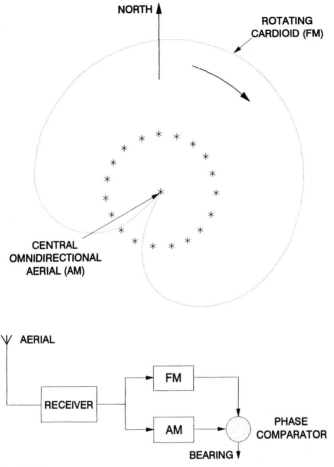

Figure 12.2   VOR system

synchronised to a common time. This is not particularly easy since a three microsecond timing error corresponds to a range error approaching 1 km. Assuming that clock re-synchronisation can be performed only once per hour, this corresponds to a drift in the measurement of time of one part in 10^9. Such accuracies were not achievable before the invention of atomic clocks, and other methods of measuring range were sought to overcome this difficulty. A number of ranging systems are described next.

Distance measuring equipment (DME)

Many VOR stations are equipped with a microwave transponder which provides range information in response to a signal emanating from the aircraft. The aircraft transmits pairs of pulses with a unique separation and repetition rate. A ground station receives the signals and retransmits them after a fixed time delay. The aircraft receives the re-transmitted signals and measures the time delay between transmission and reception, deducts the fixed delay in the ground station, and so determines the two way range to and from the ground station. Position can then be determine by measuring the range to two or more DME stations. DME observations are accurate typically to within 300 m. The principle of the method is shown in Figure 12.3.

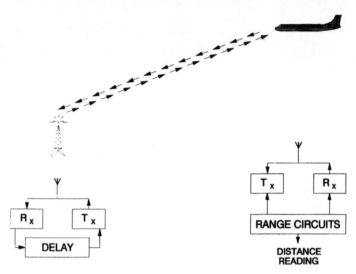

Figure 12.3   DME

Tactical air navigation system (TACAN)

The tactical air navigation system provides the same type of measurements as those obtained using VOR and DME, as described above, but with increased precision through the use of ultra high frequency (UHF) transmissions in the 1 GHz frequency band. Typical bearing accuracy is in the region of ±0.5° with ranging errors usually better than ±1% of the distance between the aircraft and the beacon. The maximum range is altitude dependent because of the characteristics of the propagation of UHF radio waves.

Hyperbolic systems

The need for accurate absolute time in the receiver is also eliminated if signals are transmitted in synchronism from two or more ground stations and the time interval between their arrival at the receiver is measured. In systems based on this principle, a master station transmits a signal which is received in the aircraft or ship and also at slave transmitter stations on the ground. The slave stations lock their clocks to the master signal, with allowance for the propagation time for the fixed distance between master and slave. The corrected clocks are then used to generate the slave station transmissions.

The receiver's clock can be locked to the received master signal and used to measure the time interval to the receipt of the slave signals. The long-term stability is now governed by the stability of the master station clock, and the short-term stability of crystal clocks is perfectly adequate for the time difference measurements. The time interval between signals received from two stations gives a measure of the difference in range to the stations. A given reading indicates a position for the receiver which lies on a hyperbola with the two stations at its foci, as indicated in Figure 12.4. By measuring the time intervals obtained between three stations, two hyperbolae and hence a fix are obtained.

Two methods of measuring range differences are used by terrestrial hyperbolic navigation radio aids, namely:

- phase measurements from continuous waves;
- direct time measurements from pulse transmissions.

Decca Navigation System

This is a typical system which uses continuous waves and phase measurements to determine hyperbolic position lines. Transmissions are in the low frequency band from 70 kHz to 130 kHz giving a usable range of around 250 km. Four stations, a master and three slaves, form a chain of transmitting stations. The slave transmissions are phase locked and harmonically related to the transmissions from the master station.

A complete phase cycle of 360° at the comparison frequencies represents a distance of between 500 and 800 metres on the baseline between the stations. Hence measuring phase to an accuracy of 10° gives a resolution of around five metres. However, the resulting position is ambiguous since the same phase measurement repeats every cycle or 500 metres. Special facilities, which essentially involve making phase comparisons at synthetically produced lower frequencies, have to be introduced to resolve the ambiguity.

When originally introduced during World War II, the system required maps with special overlays printed on them to obtain latitude and longitude information. This limitation has been removed through

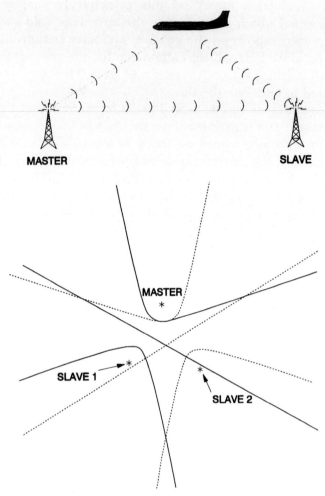

Figure 12.4   Hyperbolic navigation

the use of computers to produce systems which provide latitude and longitude information automatically.

Omega

This is a long range hyperbolic navigation aid operating at very low frequency. The system is based on eight ground transmitting stations distributed around the Earth, each having a nominal range of 8000 nautical miles. Thus, an aircraft or ship located anywhere around the Earth can expect to receive signals from at least three stations, and is able to deduce its position from the phase of the received signals, in the manner outlined above. Typically, a position fix can be obtained to an accuracy of a few nautical miles.

Although the Omega system was developed originally for maritime use, it has been extensively used as an airborne navigation aid, in particular on the trans-oceanic routes.

Long Range Navigation (LORAN)

A low frequency electronic position fixing system using pulsed radio waves with a frequency of 100 kHz, LORAN is a long range aid (1500 km or beyond) obtained by using pulse transmissions rather than continuous waves. It is more precise than Omega but does not have the worldwide coverage of that system.

The LORAN C system has a master transmitting station and two or more slave stations forming a chain. There are many chains located in the northern hemisphere. The system operates by measuring the difference in time of arrival of pulses from the master and its slave stations. The ground wave from the transmitters travels distances of up to 2000 km, the exact range depending on the power of the transmitter, receiver sensitivity and atmospheric attenuation. Position accuracies are dependent on the range of the observer from the chain and vary from a few tens of metres at short ranges, typically less than 500 km, to 100 metres or more at longer ranges of around 2000 km. The transmissions also produce sky waves which can interfere with the ground waves and cause distortion of the received signals and errors in the estimates of position provided by the system.

As with the Omega system, LORAN was developed originally for ships but has found wide use as an airborne navigation aid and, more recently, as an aid for land-vehicle location.

12.3.2 Satellite navigation aids—GPS

Radio positioning, similar to that described in the previous section, can be achieved using satellite transmissions. The global positioning system (GPS), formally known as Navstar, is just such a system which is now reaching full operational status providing a worldwide navigation capability [1–4].

GPS is designed to provide highly accurate, three-dimensional position and velocity data to users anywhere on or near to the Earth. The system is available to an unlimited number of users, each equipped with an antenna and a receiver. GPS comprises a constellation of 24 satellites[1] in near circular orbit around the Earth, as shown in Figure 12.5, and a ground control system. The satellites orbit the Earth about every 12 hours and are arranged in six orbital planes which are inclined at 55° to each another, at an altitude of 20 200 km. The spacing of the satellites is arranged so that, generally, at least four satellites will be in view to a user at any instant of time. Each satellite transmits two navigation signals centred at frequencies of 1.575 GHz and 1.227 GHz respectively. Both signals are derived from an atomic frequency standard.

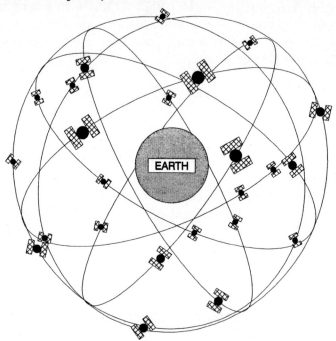

Figure 12.5   GPS satellite constellation

Position is calculated by taking measurements of distance from the satellites. The distance measurements are made by measuring how long it takes for a radio signal to travel from each satellite to the GPS receiver. It is assumed that the satellites and the GPS receiver are generating the same coded signal at exactly the same time. Distance is determined by comparing the arrival time of the satellite signal with that expected by the receiver. The timing signals used by GPS are pseudo-random sequences which enable each satellite to be identified

[1]The constellation has 21 operational satellites and three fully functional spare satellites. Reliability analysis shows that this arrangement provides a 98% probability of system availability.

unambiguously and also allow access to the system to be controlled. Two codes are transmitted by each satellite; a coarse/acquisition (C/A) code and a precision (P) code. As its name suggests, the P code yields the full navigational precision of the system, but is only available to selected users.

In order to make precise distance measurements, accurate timing of the satellite and receiver signals is clearly essential. The satellite signals are accurate because they have atomic clocks onboard. Less accurate clocks are used in the receivers and their timing errors corrected by taking measurements of range to four satellites, four ranges being required to determine the four unknowns; three spatial co-ordinates and time.

It is also necessary to know the precise location in space of each satellite which is being monitored if position is to be determined accurately. The orbits of the satellites are very predictable. However, minor variations that do occur are monitored regularly by ground tracking stations. These data are passed to the satellites enabling them to broadcast information about their exact orbital location in addition to the timing signals discussed earlier.

By using the techniques outlined above it is possible to obtain very accurate distance measurements. By measuring the Doppler shift of the radio frequency carrier, the range rate to each satellite can be calculated in the receiver. Using this information, the vehicle's velocity can be determined since that of the satellite is known.

Measurement errors arise from a variety of causes. The Earth's ionosphere and atmosphere cause delays in the GPS signals which can give rise to errors in the measurement of position, although their effects can be compensated to some extent by modelling. Other sources of error are satellite and receiver clock imperfections and multi-path reception. Further, at certain times, the geometrical arrangement of the satellites being monitored can magnify the errors in the system. However, GPS can enable a vehicle to establish its position anywhere in the world and at any time with an accuracy of a few tens of metres and its velocity to better than 1 m/s.

A substantial increase in the accuracy of the estimates of position can be obtained by the use of a technique known as differential GPS which requires the use of a receiver at a surveyed location. That receiver unit will be able to compare its GPS estimate of position with the known position from the survey, and thus compensate for range errors produced by the GPS. A correction signal can then be transmitted to other receivers in the immediate vicinity allowing errors in their measurements to be reduced dramatically.

The equivalent system produced by the former Soviet Union is known as the Global navigation satellite system (GLONASS). This system also has a constellation of 24 satellites but arranged in three orbital planes [4–5].

12.3.3 Star trackers

The stars may be regarded as fixed points which can be used as references for the purposes of celestial navigation. The geographical position of an observer on or close to the Earth can be determined at any time given knowledge of:

(i)   the positions of two or more celestial objects in relation to the observer;

(ii)   the exact time of the observation.

The basis of celestial navigation is that if the altitude (the angle between the line of sight and the horizontal) of a celestial object is measured, then the observer's position must lie on a specific circle on the Earth. This circle is centred on the point on the Earth's surface which lies directly below the object. If the time of the observation is known, then this point can be found from pre-computed astronomical tables. Given sightings of two objects, then two circles of position are defined and the observer must be located at their intersection, as indicated in Figure 12.6a.

To enable the technique of position estimation from star sightings, or celestial observation, to be used in aircraft, automatic star trackers have been developed. A star tracker is basically a telescopic device having a detector and a scanning mechanism. Sightings of stars may be achieved using a star tracker to provide measurements of the azimuth and elevation angles of a star with respect to a known reference frame within the vehicle. Typically, this would be a space stabilised reference frame, defined by a stable platform on which the star tracker is mounted. For navigation purposes, knowledge of the direction of the local vertical is needed in order to relate the measurements to an Earth reference frame.

Alternatively, measurements may be made with respect to a body fixed frame and used to update a strapdown inertial navigation system in the manner outlined below. Such observations can be compared with stored knowledge of the observed star's position to derive a position fix, or an estimate of vehicle heading.

For navigation purposes, stars may be considered to be positioned on the inner surface of a geocentric sphere of infinite radius, often referred to as the celestial sphere. The projection of the Earth's lines of latitude and longitude on to this sphere establishes a grid in which the position of a celestial object may be defined. The position north or south of the Equator is referred to as the declination of a star, whilst its longitudinal position is expressed in terms of a sidereal hour angle, as indicated in Figure 12.6b. Hence, the direction of a star with respect to an inertial frame, which has its origin at the centre of the Earth, may be expressed in terms of its declination and sidereal hour angle.

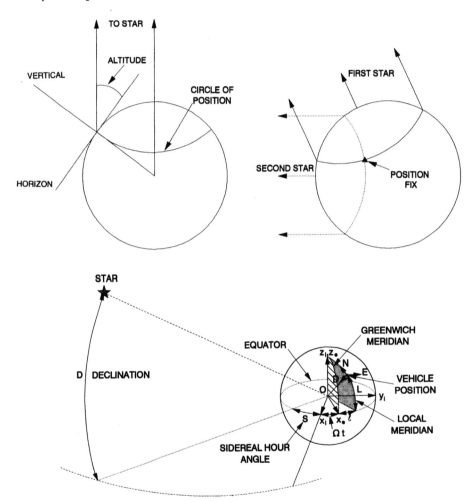

Figure 12.6   *a Star tracker position fixing*
            b Star tracker geometry

A star tracker fixed in the body of a vehicle would provide measurements of the star's bearing and elevation with respect to the body frame. These measurements may then be compared with predictions of these same quantities derived from knowledge of the declination and sidereal hour angle of a star. These quantities may also be expressed in body axes given knowledge of the vehicle's latitude and longitude and its orientation with respect to the local geographic frame. The resulting measurement differences may then be used to update the onboard inertial navigation system in a manner similar to that discussed in Section 12.6.

Star trackers combined with an inertial navigation system are believed to be capable of measurement accuracies of a few arc seconds in a space environment. When a star tracker is used close to the surface of the Earth, corrections to the measurements have to be made for the refraction effects induced by the Earth's atmosphere. Typically, an accuracy of better than 10 arc seconds may be achieved, which corresponds to a position error of around 300 metres on the surface of the Earth.

12.3.4 Surface radar trackers

A ground radar station may be used to provide line-of-sight observations of an aircraft or missile during flight. These observations usually take the form of measurements of a vehicle's range, elevation and bearing as indicated in Figure 12.7. The measurements are derived with respect to a local reference frame, usually the local vertical geographic frame at the location of the radar tracker. The measurement data may be transmitted to the vehicle for in-flight aiding of an onboard inertial navigation system.

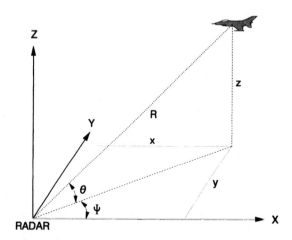

Figure 12.7   Ground radar measurements

The measurements of a vehicle's range (R), azimuth (ψ) and elevation (θ) may be expressed in terms of the Cartesian position co-ordinates of the aircraft (x, y, z) as follows:

$$R = \sqrt{(x^2 + y^2 + z^2)}$$

$$\psi = \tan^{-1}\{y/x\}$$

$$\theta = \tan^{-1}\{z / \sqrt{(x^2 + y^2)}\}$$

These measurements may be used to update the onboard inertial navigation system by comparing them with predictions of the same quantities obtained from information provided by the onboard system. A design example, based on such a system, is presented in Section 12.6.2.

12.4 Onboard measurements

12.4.1 Doppler radar

Doppler radar, which provides a means of measuring a vehicle's velocity, is often used to provide navigation aiding for airborne systems and in some cases, in conjunction with an attitude and heading reference system, as the primary source of navigation data. A Doppler radar operates by transmitting a narrow beam of microwave energy to the ground and measures the frequency shift that occurs in the reflected signal as a result of the relative motion between the aircraft and the ground. In the situation where the aircraft velocity is V and the radar beam slants down towards the ground at an angle θ, the frequency shift is:

$$\frac{2V}{\lambda} \cos \theta$$

where λ is the wavelength of the transmission. For a typical system operating in the frequency range 13.25 GHz to 13.4 GHz ($\lambda \approx 2.2$ cm), the frequency shift is approximately 47 Hz per knot. Given knowledge of the wavelength of the transmission and the slant angle, an estimate of the velocity of the aircraft can be determined from the measured frequency shift. Because the aircraft is able to move in three dimensions, the minimum number of beams needed to establish aircraft velocity is three. The beams are often directed forwards and to the rear of the aircraft as illustrated in Figure 12.8.

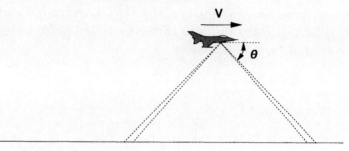

Figure 12.8   Doppler radar beam geometry

Modern Doppler systems generate the beams using a planar array, the aerial being attached rigidly to the body of the aircraft. The reflected signals from each of the beams are processed separately, enabling estimates of aircraft velocity components to be calculated in a computer. Such estimates are derived in a co-ordinate frame which is fixed with respect to the aerial. In order to carry out the navigation function, it is necessary to resolve these velocity estimates into the chosen navigation reference frame. An onboard attitude and heading reference system is required for this purpose. Alternatively, the Doppler sensor may be integrated with an onboard inertial navigation system. In this case, the Doppler velocity measurements would be compared with estimates of those same quantities generated by the inertial system.

Typically, a Doppler sensor operating over land is able to provide measurements of velocity to an accuracy of about 0.25%, or better. Performance is degraded during flights over water owing to poor reflectivity, scattering of the reflected signal giving rise to a bias in the measurement, wave motion, tidal motion and water currents. However, this navigation aid offers good long term stability and a chance to bound the position and velocity estimates provided by an inertial navigation system.

12.4.2 Magnetic measurements

The Earth has a magnetic field similar to that of a bar magnet with poles located close to its axis of rotation. This means that the direction of the horizontal component of the Earth's magnetic field lies close to true north, and the magnetic north, as determined by a magnetic field sensor, or compass, can be used as a working reference. Unfortunately, the angle between true north and magnetic north is not constant. It varies with position on the Earth and slowly with time, although it is possible to compensate for both of these effects. The direction of the

Earth's magnetic field at any point on the Earth is defined in terms of its orientation with respect to true north, known as the angle of variation, and its angle with respect to the horizontal, the angle of dip, as indicated in Figure 12.9. The vehicle in which the magnetic sensor is mounted will almost always have a magnetic field which cannot be distinguished from that of the Earth and consequently it is necessary to compensate for this effect as well as the others mentioned above.

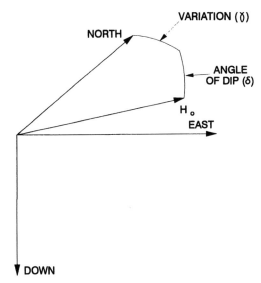

Figure 12.9   Components of Earth's magnetic field

The magnetic compass is one of the oldest navigation aids known to man, having been used for hundreds of years to provide a directional reference for steering and dead reckoning navigation. A practical device which may be used to sense the Earth's magnetic field and to provide a measure of the attitude of a moving vehicle is the fluxgate magnetometer, as described in Section 4.7.2. In the absence of local magnetic disturbances, this device senses the components of the Earth's magnetic field (H_0) acting along its sensitive axis. A three-axis device may be mounted in a vehicle to sense the components of the Earth's field along its principal body axes, (H_x, H_y, H_z). Expressing body attitude, with respect to the local geographic frame, as a direction cosine matrix, $\mathbf{C}_n^b$, the relationship between the magnetic measurements and body attitude may be written as follows:

$$\begin{pmatrix} H_x \\ H_y \\ H_z \end{pmatrix} = \mathbf{C}_n^b \begin{pmatrix} H_0 \cos \delta \cos \gamma \\ H_0 \sin \delta \sin \gamma \\ H_0 \sin \delta \end{pmatrix}$$

where δ is the angle of dip and γ is the angle of variation. Given knowledge of the angle of dip, and the variation between true and magnetic north, estimates of vehicle attitude can be deduced from the measurements provided by the magnetometer. Rotations about the local magnetic vector can not be detected. For this reason, such a device must operate in conjunction with a vertical reference system to determine body attitude in full.

Other possibilities for navigation aiding have been suggested which involve using measurements of the Earth's magnetic field in different ways [6]. For example, it would be possible in principle to obtain position fixes either by comparing field measurements in the local geographic frame with stored maps of magnetic variation and dip angle, or by attempting to match magnetic anomalies. The former method would require precise knowledge of the directions of true north and the local vertical to provide accurate fixes. The latter scheme, which is analogous to the terrain matching technique discussed in Section 12.4.4, would clearly be reliant on the availability of sufficiently detailed magnetic anomaly maps and on the stability of these anomalies. In regions with a large number of significant and stable anomalies, this system has the potential for good position fix accuracy.

12.4.3 Altimeters

Barometric altimeters are invariably used for height measurement in aircraft. As supplementary navigation aids, they are widely used for restricting the growth of errors in the vertical channel of an inertial navigation system. In a Schuler tuned inertial navigation system, although the propagation of errors in the horizontal channels is bounded, the velocity and position errors in the vertical channel are not bounded and can become very large within a relatively short period of time unless there is an independent means of checking the growth of such errors. For example, the effect of a net acceleration bias (B) acting in the vertical direction gives rise to a position error $Bt^2/2$. Hence, a bias of only 10 micro-g would result in a height error in excess of 2.5 km over a two hour period.

A barometric altimeter, relying on atmospheric pressure readings, provides an indirect measure of height above a nominal sea level, typically to an accuracy of 0.1%. Most airborne inertial systems requiring a three-dimensional navigation capability operate with barometric aiding in order to bound the growth of vertical channel errors.

A radio altimeter provides a direct measure of height above ground which is equally important for many applications. Such measurements may be used in conjunction with a stored map of the terrain over which an aircraft is flying to provide position updates for an inertial navigation system. The subject of terrain referenced navigation is addressed separately below.

12.4.4 Terrain referenced navigation

Terrain referenced navigation uses a radar altimeter, an onboard baro-inertial navigation system and a stored contour map of the area over which the aircraft or missile is flying. The radar altimeter measurements of height above ground, in combination with the estimates of height above sea level, provided by the inertial navigation system, allow a reconstruction of the ground profile beneath the flight path to take place within a computer onboard the vehicle. The resulting terrain profile is then compared with the stored map data to achieve a fit, from which the position of the vehicle may be identified.

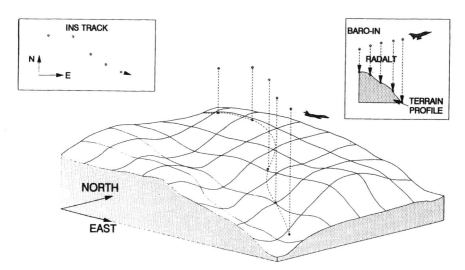

Figure 12.10   Terrain referenced navigation

A position fix is possible because of the random nature of the Earth's surface, which tends to give each section of terrain a shape unique to its location. The accuracy of the position fix which can be obtained is generally a function of the roughness of the terrain, a more precise fix being achievable when the contour variation is greatest.

Various schemes for extracting vehicle position during flight have been devised. The scheme outlined above relies on taking a succession of height measurements, which may be used retrospectively, to determine the location of the vehicle. An alternative method involves the comparison of estimates of terrain height, derived by differencing the inertial system indicated height with the radar altimeter measurement, with estimates of terrain height extracted from the stored contour map. The map heights are derived at the co-ordinate location of the vehicle indicated by its onboard inertial navigation system. Updates of vehicle position may be derived given knowledge of the terrain slope beneath the vehicle. In this way, each altimeter measurement is processed separately, and then used to update the onboard navigation system on an almost continuous basis.

Operation of this scheme is clearly dependent on the availability of a good quality terrain data base for the area over which the vehicle is to be flown. Typically, radial position accuracies of a few tens of metres can be achieved, the precise accuracy varying with the roughness of the terrain beneath the aircraft. Clearly, navigation accuracy degrades when over flying large flat and featureless areas of terrain, and particularly over water, where the accuracy of navigation becomes solely dependent on the performance of the unaided inertial system. Further degradations in performance can result from tree foliage and snow cover which can affect radar altimeter accuracy.

12.4.5 Scene matching

Scene matching area correlation (SMAC) techniques may be used to provide highly accurate position fixes based on images of the ground beneath an aircraft or missile. The operation of this system is equivalent to the technique used by a human navigator, navigating from recognition of remembered landmarks or ground features.

The fundamental principle is illustrated in Figure 12.11. An imaging system, typically an infrared line-scan device, builds up a picture of the terrain beneath the aircraft as it moves forward. When a fix is required, a portion of the infrared line-scan image is stored to form a scene of the terrain beneath the aircraft. By this process, the image is converted into an array of pixels, each pixel having a numerical value indicating its brightness of that part of the image. The captured scene is processed in order to remove noise and to enhance those features which are likely to provide navigation information, road junctions and railway lines, for example. In the next stage of analysis, a correlation algorithm is used to search for recognisable patterns, which appear in a pre-stored data base of ground features. Having found a match between a feature in the scene and one in the data base, geometrical calculations based on the aircraft's attitude and height above ground enable its position to be calculated at the time the scene was captured. The various stages in the scene matching process are illustrated in Figure 12.11.

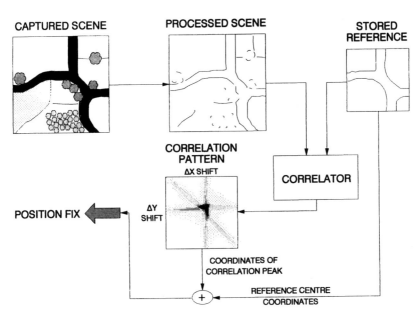

Figure 12.11   The principle of scene matching

12.5 System integration

The remainder of this chapter is concerned with techniques which may be used to combine inertial measurement data with information provided by one or more of the navigation aids discussed in the preceding section. In general, the available measurements are corrupted with noise. Consequently, some form of on-line filtering technique is required to achieve good navigation performance.

Early systems used complementary filtering techniques of the form described below for a baro-inertial system and illustrated in Figure 12.12.

The difference between the inertial system estimate of height and the pure barometric height measurement is fed back via the gains K_1 and

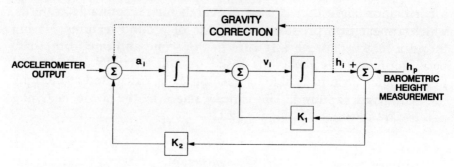

Figure 12.12   Baro-inertial height measurement

K_2, shown in the figure, to correct the velocity and height estimates. Values for the gains are selected to allow the baro-inertial system to follow the long term variations in the barometric measurements, whilst filtering out any higher frequency fluctuations. Typically, $K_1 = 2/T$ and $K_2 = 1/T^2$, where the value of T may be 30 seconds [7]. In the integrated system, a bias (B_z) on the inertial estimate of vertical acceleration no longer propagates as a height error with time squared, but settles to a steady state value of $T^2 B_z$. Hence, a bias of 100 micro-g gives rise to a height error of approximately one metre. The major limitation of this technique is that any longer term errors in the barometric altimeter, resulting from weather conditions and the position of the device, persist in the integrated system.

As with all filtering techniques, the objective is to make use of the available knowledge about the long-term behaviour of a signal, contaminated with noise, in order to derive a better estimate of the signal than could be obtained using a single measurement. A practical form of filter, which is applicable for on-line estimation, relies on generating a mathematical model of the process producing the signal, and adjusting the parameters of the model to minimise the mean square deviation between the signal and the output of the model. A best estimate of the signal is derived based on knowledge of the expected errors in the model and the measured signal. This technique is known as Kalman filtering. Kalman filtering has become a well established technique for the mixing of navigation data in integrated systems. It is particularly suitable for on-line estimation, being a recursive technique which lends itself to implementation in a computer.

The principles of Kalman filtering are described in Appendix A and its application is illustrated in the following section.

12.6 Application of Kalman filtering to aided inertial navigation systems

12.6.1 Introduction

As described in Appendix A, Kalman filtering involves the combination of two estimates of a variable to form a weighted mean, the weighting factors being chosen to yield the most probable estimate. One estimate is provided by updating a previous best estimate, in accordance with the known equations of motion, whilst the other is obtained from a measurement. In an integrated navigation system the first estimate is provided directly by the inertial navigation system, i.e. in filtering terms the inertial system constitutes the model of the physical process which produces the measurement. The second estimate, the measurement, is provided by the navigation aid. The same technique can be applied irrespective of the source of measurement information.

A generalised block diagram representation is shown in Figure 12.13, and a design example is described in the following section.

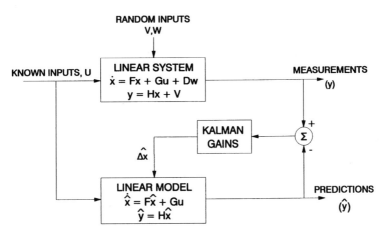

Figure 12.13   Kalman filter for linear systems

Since the measurements provided by navigation aids are often non-linear combinations of the inertial navigation system estimates, and also as the inertial system equations themselves are non-linear, it is customary to use an extended Kalman filter formulation for an aided inertial navigation system.

12.6.2 Design example of aiding

In this section, a scheme for aiding a missile onboard inertial navigation system is described which relies upon tracking the missile during flight using a sensor on the launch platform. Suitable measurements may be provided by a multi-function radar or an infrared tracker in combination with a laser range finder. In either case, it is assumed that measurements of a missile's range, elevation and bearing with respect to the chosen navigation frame will be provided. These measurements may be passed via an uplink transmitter to the missile and used to aid the onboard navigation system.

The transmitted measurements are combined, using a Kalman filter, with the measurements provided by the missile's inertial navigation system. This not only allows improved estimates of the missile's position to be derived from noisy measurement data, but also provides a means of estimating errors in the states of the navigation system which are not directly measurable, the velocity and attitude estimates for example. The form of the filter is as shown in Figure 12.14.

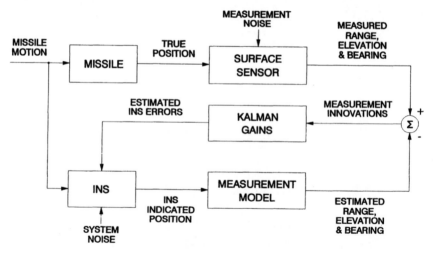

Figure 12.14   INS aiding using ground tracker measurements

In order to provide estimates of the three attitude errors, the three velocity errors and the three position errors in the onboard inertial navigation system, a nine state Kalman filter is required. The associated system and measurement equations are described in the sections which follow.

12.6.2.1 The system equations

To formulate an extended Kalman filter to update the onboard navigation system, it is necessary to develop a linear dynamic model of the errors which are to be estimated. For the purposes of this design example, a simplified version of the error model given in Chapter 11 may be used. The error model may be expressed in matrix form as:

$$\delta \dot{x} = F \, \delta x + D w \qquad (12.1)$$

The vector δx represents the error state of the system. For the purposes of this illustration, δx consists of the three attitude errors $(\delta \alpha \; \delta \beta \; \delta \gamma)$, three velocity errors, $(\delta v_x \; \delta v_y \; \delta v_z)$ and three position errors $(\delta x \; \delta y \; \delta z)$.

In order to allow a discrete Kalman filter to be constructed, it is necessary to express the system error eqn. 12.1 in discrete form. If δx_k represents the inertial navigation system error states at time t_k, and δx_{k+1} the error states at time t_{k+1}, we may write:

$$\delta x_{k+1} = \Phi_k \, \delta x_k + w_k \qquad (12.2)$$

where Φ_k is the system transition matrix at time t_k, which may be expressed in terms of the system matrix F as follows:

$$\Phi_k = \exp[F(t_{k+1} - t_k)] \qquad (12.3)$$

12.6.2.2 The measurement equations

Measurements of the missile's position with respect to the radar are assumed to be available at discrete intervals of time throughout flight. The radar provides measurements in polar co-ordinates, i.e. measurements of range (R), elevation (θ) and bearing (ψ). The polar quantities may be expressed in terms of the Cartesian co-ordinates (x, y, z) as follows:

$$R^2 = x^2 + y^2 + z^2$$

$$\theta = \arctan\left\{ \frac{z}{\sqrt{(x^2 + y^2)}} \right\}$$

$$\psi = \arctan\left\{ \frac{y}{x} \right\} \qquad (12.4)$$

Writing $\tilde{z} = \begin{bmatrix} R & \theta & \psi \end{bmatrix}^T$, the radar measurements, denoted by z, may be expressed as :

$$\tilde{z} = z + v \qquad (12.5)$$

where **v** represents the error in the measurements. **v** is assumed to be a zero-mean, Gaussian white noise process.

Estimates of the radar measurements, **z**, may be obtained from the inertial navigation system estimates of position (x, y, z) as follows:

$$\hat{z} = \begin{pmatrix} \hat{R} \\ \hat{\theta} \\ \hat{\psi} \end{pmatrix} = \begin{pmatrix} \sqrt{\left(\hat{x}^2 + \hat{y}^2 + \hat{z}^2\right)} \\ \arctan\left\{\dfrac{\hat{z}}{\sqrt{\left(\hat{x}^2 + \hat{y}^2\right)}}\right\} \\ \arctan\left\{\dfrac{\hat{y}}{\hat{x}}\right\} \end{pmatrix} \tag{12.6}$$

The difference between the radar measurements $(\tilde{z})$ and the estimates $(\hat{z})$ of those quantities is referred to as the filter measurement innovation (δz) and is generated as follows:

$$\delta z = \tilde{z} + \hat{z}$$
$$= H \delta x \tag{12.7}$$

where

$$H = \begin{pmatrix} 0 & 0 & 0 & 0 & 0 & 0 & \dfrac{x}{R} & \dfrac{y}{R} & \dfrac{z}{R} \\ 0 & 0 & 0 & 0 & 0 & 0 & \dfrac{xz}{R^2\sqrt{\left(x^2 + y^2\right)}} & \dfrac{yz}{R^2\sqrt{\left(x^2 + y^2\right)}} & \dfrac{\sqrt{\left(x^2 + y^2\right)}}{R^2} \\ 0 & 0 & 0 & 0 & 0 & 0 & \dfrac{y}{\left(x^2 + y^2\right)} & \dfrac{x}{\left(x^2 + y^2\right)} & 0 \end{pmatrix} \tag{12.8}$$

12.6.2.3 The Kalman filter

Eqns. 12.2 and 12.7 are the system and measurement equations needed to construct a Kalman filter. The equations for the Kalman filter, given in Appendix A, take the following form for the radar-aided inertial system considered here.

Filter prediction step

Following each measurement update, the inertial navigation system is corrected using the current best estimates of the errors in position, velocity and attitude. Therefore, after an update, the best estimate of each of the inertial system errors becomes identically zero and the state prediction equation reduces to:

$$\delta \mathbf{x}_{k+1/k} = 0 \tag{12.9}$$

The covariance matrix is predicted forward in time using the expression:

$$\mathbf{P}_{k+1/k} = \mathbf{\Phi}_k \mathbf{P}_{k/k} \mathbf{\Phi}_k^T + \Delta \mathbf{Q}' \Delta^T \tag{12.10}$$

where $\mathbf{\Phi}_k$ is the transition matrix given by eqn. 12.3. $\mathbf{P}_{k+1/k}$ denotes the expected value of the covariance matrix at time t_{k+1} predicted at time t_k. It is set up initially as a diagonal matrix, the individual elements being chosen according to the expected variances of the errors in the initial attitude, velocity and position passed to the missile navigation inertial system prior to launch. $\mathbf{Q}'$, the system noise matrix, is set up according to the expected level of noise on the inertial measurements of linear acceleration and angular rate.

Filter update

The estimates of the errors in the inertial navigation system states are derived using:

$$\delta \mathbf{x}_{k+1/k+1} = \mathbf{K}_{k+1} \, \delta \mathbf{z}_{k+1} \tag{12.11}$$

and the covariance matrix is updated according to:

$$\mathbf{P}_{k+1/k+1} = [\mathbf{I} - \mathbf{K}_{k+1} \mathbf{H}_{k+1}] \mathbf{P}_{k+1/k} \tag{12.12}$$

$$\text{where} \quad \mathbf{K}_{k+1} = \mathbf{P}_{k+1/k+1} \mathbf{H}_{k+1}^T [\mathbf{H}_{k+1} \mathbf{P}_{k+1/k} \mathbf{H}_{k+1}^T + \mathbf{R}']^{-1} \tag{12.13}$$

$\mathbf{H}$ is defined by eqn. 12.8 and $\mathbf{R}'$, the measurement noise is a 3×3 diagonal matrix, the elements of which are selected in accordance with the anticipated level of radar measurement noise.

Inertial navigation system correction

The inertial navigation states, $\hat{\mathbf{x}}$, are corrected immediately after each measurement update using the current best estimates of the errors. The correction equations are given below.

Velocity and position correction

Velocity and position may be corrected by simply subtracting the estimate error from the inertial system estimates of these quantities using:

$$\mathbf{x}_c = \hat{\mathbf{x}} - \delta \mathbf{x} \tag{12.14}$$

where $\mathbf{x}_c$ is the corrected state.

Attitude correction

As described earlier, the computed direction cosine matrix may be expressed in terms of the true matrix using:

$$\hat{C} = \left[I - \Psi\right]C$$

The corrected direction cosine matrix, C_c, may therefore be expressed as follows:

$$C_c = \left[I - Y\right]C \qquad (12.15)$$

where $\Psi = \psi \times$

and $\psi = \left[\delta\alpha \quad \delta\beta \quad \delta\gamma\right]^T$.

By writing C_c and C in component form as functions of the corrected and estimated quaternion parameters, denoted $[a_c \; b_c \; c_c \; d_c]$ and $[a \; b \; c \; d]$ respectively, and equating terms, it can be shown that the estimated quaternion parameters may be corrected directly using:

$$a_c = a + 0.5(\delta\alpha b + \delta\beta c + \delta\gamma d)$$

$$b_c = b + 0.5(-\delta\alpha a + \delta\beta d - \delta\gamma c)$$

$$c_c = c + 0.5(-\delta\alpha d - \delta\beta a + \delta\gamma b)$$

$$d_c = d + 0.5(\delta\alpha c - \delta\beta b - \delta\gamma a) \qquad (12.16)$$

12.6.2.4 Results

The results presented in Figures 12.15 and 12.16 illustrate the effectiveness of the measurements provided by a surface sensor in improving the performance of a missile's onboard inertial navigation system when using the nine state Kalman filter, described in the previous section. Results are presented in graphical form showing the standard deviations of attitude and position errors, with and without aiding, over a typical short range missile flight of ten seconds duration. Over this period of time, the missile follows a boost/coast trajectory, accelerating at 20 g for the first 4 seconds of flight and slowing under the influence of aerodynamic drag thereafter.

Figure 12.15 shows navigation performance when a high grade inertial navigation system is used. The fixed biases for the gyroscopes and accelerometers used here were 0.01°/hour (1σ) and 100 micro-g (1σ) respectively. For the purposes of this analysis, the standard deviations of the initial condition errors in the missile system were chosen to be as follows:

initial attitude errors	10 mrad
initial velocity errors	1 m/s
initial position errors	1 m

Sensor measurement accuracies were set to values of 3 mrad (1σ) in elevation and bearing and 10 m (1σ) in range, and a data update rate of 1 Hz was assumed. The dramatic improvement in navigation performance with aiding is illustrated clearly in Figure 12.15. The position errors are corrected very rapidly and remain within 20 m for the remainder of the flight. The attitude errors settle to less than 0.2° after three measurements have been received. Because the instrument errors are small in this case, the system error model on which the Kalman filter is based provides a representative model of the actual system. This allows filter convergence to take place and accurate estimates of the inertial system errors to be derived.

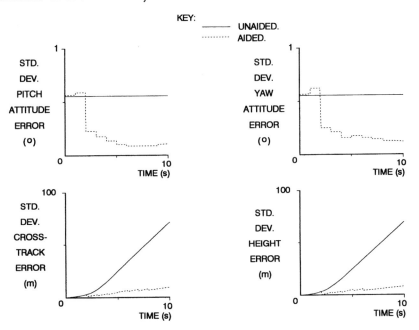

Figure 12.15   Simulation results from aiding—high grade INS

Where the instrument errors become large, failure to model them correctly in the Kalman filter means the error model does not provide an accurate representation of what is happening in the actual system. Under such conditions, there is said to be a mismatch between the filter error model and the actual system. The sensor errors now introduce additional contributions to the measurement differences which will be interpreted incorrectly as alignment errors. As a result, the Kalman filter estimates of attitude, velocity and position will be in error.

This effect is illustrated in Figure 12.16. The simulation described above was repeated under identical conditions with the exception that the high grade inertial navigation system has been replaced by a system of more modest performance. The sensors have been replaced by gyroscopes and accelerometers having biases of 30°/hour and 10 milli-g respectively, typical of the grade of sub-inertial quality instruments often specified for use in tactical missile systems. Although a substantial improvement in navigation performance is still achieved, when compared with the unaided system, the resulting accuracy is reduced and the rate of convergence increased in comparison with the previous set of results shown in Figure 12.15. The build up of attitude errors which occurs between the measurement updates is indicative of the mismatch that now exists in the Kalman filter. A major source of error in this particular case is gyroscopic mass unbalance which introduces a rate bias that varies with the applied acceleration. This accounts largely for the shape of the attitude error curves shown in that Figure.

In theory, there may be scope to improve the performance of the aided system by including additional filter states to model explicitly the dominant sensor errors [8]. By adopting this approach, the Kalman filter can be used to achieve some degree of in-flight sensor calibration, so leading to even greater enhancements of overall navigation performance. However, such a system will usually require time for the affects of modelling the sensor errors to become apparent.

In conclusion, the Kalman filter described earlier allows the effect of initial alignment errors to be reduced dramatically, and so provides sufficiently accurate inertial data for many short range missile applications, without recourse to much higher quality and more expensive inertial sensors. It is noted that, for some applications, the in-flight aiding scheme described above may well allow some relaxation in the accuracy of the pre-flight alignment required. Pre-flight alignment methods are discussed in Chapter 9.

12.7 Summary

There are many sources of navigation data which may be combined with inertial system estimates to provide enhanced navigation

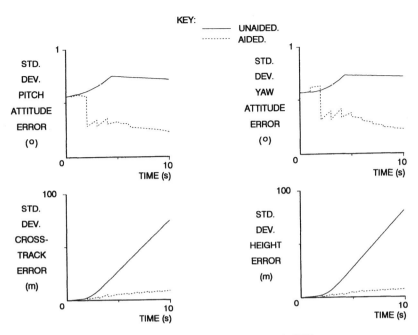

Figure 12.16   Simulation results from aiding—low grade INS

performance. These include external measurements derived from equipment outside the vehicle, such as radio navigation aids and satellites, and measurements derived from additional sensors onboard the vehicle such as various types of altimeters and Doppler radar. The various navigation aids often provide attitude, velocity or position updates, any of which may be used to bound the drift errors arising in an inertial navigation system, and so improve its performance.

The resulting integrated navigation systems often permit substantial improvements in navigation accuracy compared with that which may be achieved purely using inertial systems, even when the inertial navigation system uses very accurate sensors. Although very accurate navigation performance may be achieved through the use of higher quality inertial sensors and more precise alignment techniques, the use of integrated systems, such as those described here, often provides a more cost effective solution.

Techniques for mixing inertial and other measurement data have been described culminating in the description of an algorithm specifically for aided inertial systems. A Kalman filter algorithm may be used for the integration of different measurement data with inertial measurements. The design example described has shown how external measurement data can be combined with the inertial system information to bound the growth of navigation errors. As a result, it may be possible to allow some relaxation in pre-flight alignment accuracy and in the precision of the inertial sensors. Such techniques can be extended to achieve a measure of sensor calibration as part of the aiding process.

References

1 VAN DRIEL, N.: 'A review of the status and capabilities of Navstar-GPS' NLR TP 92042 U, National Aerospace Laboratory, The Netherlands, October, 1992
2 FITZSIMMONS, B.: 'Satellite navigation', *Aerospace*, Royal Aeronautical Society Journal, 1994
3 'Navstar-GPS user equipment—introduction (Public release version)' (Navstar-GPS Joint Program Office, Los Angeles Air Force Base)
4 KAPLAN, E.E. (Ed.): 'Understanding GPS-principles and applications' (Artech House, Boston, 1996)
5 KAYSER, D. and SCHANZER, G.: 'Effects of the specific military aspects of satellite navigation on the civil use of GPS/GLONASS'. Proceeedings of AGARD conference on *Dual usage in military and commercial technology in guidance and control*, March 1995 (556)
6 HINE, A.: *'Magnetic compasses and magnetometers'* (Adam Hilger Ltd., London, 1968)
7 STEILER, B. and WINTER, H.: 'AGARD flight test instrumentation series volume 15 on gyroscopic instruments and their application to flight testing', AGARDograph 160, September 1982
8 TITTERTON, D.H. and WESTON, J.L.: 'In-flight alignment and calibration of a tactical missile INS'. Proceedings of DGON Symposium on *Gyro technology*, Stuttgart, Germany, 1990

Chapter 13
Design example

13.1 Introduction

The objective of this final chapter is to apply the various aspects of strapdown inertial navigation, discussed in the preceding chapters, to a design example. Before any design of an inertial navigation system can be undertaken, it is necessary to make a careful study of the requirements to be placed upon the navigation or stabilisation system in order that the resulting design may be capable of meeting that requirement in a cost effective manner.

The definition of the performance requirement of a navigation or stabilisation system will usually define the accuracy to which it must perform. For example, in the case of a navigation system, it will state the accuracy in position, velocity and attitude to be achieved for a given vehicle and the period of time for which that navigation must take place. It is also vitally important to define the conditions existing at the start of navigation, and the total environment in which the system must operate. These factors, taken together, will influence each of the following aspects of the design solution:

- the system configuration required;
- the navigation system error budget;
- the methods by which the system can be aligned immediately prior to the start of navigation;
- the inertial sensors which are most appropriate for the task;
- the design of the system computational algorithms;
- system testing, calibration and compensation requirements.

These aspects are discussed in the context of a generic requirement for a strapdown inertial navigation system which is needed to achieve midcourse guidance of a surface launched tactical missile. It is assumed that preliminary studies have been conducted at the weapon system definition stage resulting in a set of missile system requirements and the formulation of a navigation system performance specification. The navigation system requirement is first presented followed by a series of analytical steps leading to the definition of a navigation system capable of satisfying the requirement.

13.2 Background to the requirement

A navigation system is required to be designed for installation in a guided missile that will be launched from a ship. This missile is part of a weapon system that has been defined to provide naval vessels with protection against a variety of airborne threat weapons that could attack from any direction or elevation of approach. In the case of a multiple attack, it has been assumed that the attack has been co-ordinated so that the threat weapons could arrive at the ship simultaneously either from the same direction or from different directions.

The proposed guided missile is to be launched vertically, directly from a storage silo (magazine) on the ship, and then undertake a turn manoeuvre in the pitch plane during the first few seconds of its flight. This vertical launch and turnover manoeuvre from a silo gives the weapon system complete flexibility of direction in which it can fly and a high rate of fire enabling a large number of attacking weapons (targets) to be intercepted. Additionally, the missile will be required to be as autonomous as possible after launch in order to achieve this high rate of fire.

The weapon system has a target tracking device, such as a multi-function radar, that is able to track targets as they approach the ship. A tracking algorithm predicts the future path of each approaching target. This information is used to predict where the missile launched from the ship would intercept an approaching enemy weapon. Hence, during the turnover manoeuvre, the ship launched missile is turned so that its flight path is directed towards this predicted interception point.

It is proposed in the missile design that the missile should use inertial guidance during the launch, turnover and post turnover or mid-course phases of flight. However, when the missile is within a given range of the target, a target seeker onboard the missile will be activated to acquire and track the target. At this stage, the missile will use the information provided by the seeker to steer itself to intercept the target.

During the phases of flight that use inertial guidance techniques, the missile is to use knowledge of its position, velocity and attitude provided by the onboard inertial navigation system, together with target position and velocity provided by a tracking device on the ship and transmitted to the missile during flight using a radio data link. In order that the homing head may acquire the target to allow terminal guidance to be initiated, the homing head must be pointed in the right direction. This may be accomplished during the mid-course phase of flight using position and attitude information provided by the onboard navigation system in combination with the measurements of target position provided by the shipboard sensor.

In addition, throughout the flight of the missile, the navigation system is required to supply measurements of the components of missile linear acceleration and turn rate in body axes for missile control purposes. These quantities are required to provide the missile autopilot with suitable feedback signals

13.3 The navigation system requirement

Navigation data required

The requirement is for a navigation system to be installed in a ship launched guided missile to provide position, velocity and attitude data in a suitable reference frame together with measurements of linear acceleration and angular rate in body axes. Ideally, the navigation system should be capable of operating autonomously throughout flight with a view to minimising the quantity of data which need to be passed to the missile when in flight.

Operating and storage environment

Each missile is to be launched perpendicular to the deck of a ship operating in various sea states up to and including rough sea conditions. Typically, the flight profile of the missile will involve a boost phase, during which the missile executes its turn manoeuvre, followed by a period of sustained motor thrust to maintain near constant speed to interception. The missile is assumed to accelerate from rest at up to 500 m/s² (approximately 50 g) for a period of three seconds by which time it reaches a speed of 1000 m/s. Thereafter, the vehicle is assumed to maintain this speed for a further 12 seconds. Example trajectory and speed profiles are shown in Figures 13.1 and 13.2 respectively.

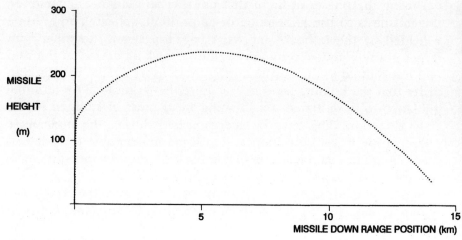

Figure 13.1   Missile trajectory profile

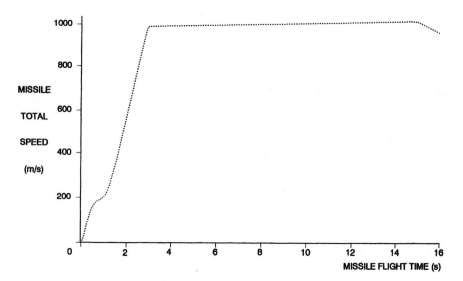

Figure 13.2   Missile speed profile

Over the total flight duration of 15 seconds, the distance travelled is approximately 12.5 km. The precise details of the flight path which the missile may be required to follow will vary depending on the target trajectory. However, it is assumed that the system may be subjected to pitch and yaw turn rates of up to 5 rad/s during flight. The missile is to be roll position stabilised throughout flight, but may experience transient roll rates as high as 5 rad/s.

During flight, the onboard navigation system may be subjected to linear vibration having a power spectral density of 0.05 g²/Hz over a frequency band from 10 Hz to 1 kHz, such motion arising as a result of motor induced vibration and missile body bending induced by changes in fin commands. In addition, the system must be capable of withstanding shock accelerations of 100 g of 0.5 millisecond duration, and it is required to operate in temperatures ranging from –20°C to +50°C.

In addition to the requirements given above, it is customary to specify the environmental conditions which the unit must be able to withstand during storage. For military systems, the storage requirements are often more extreme than those specified for the operational system. However, for this design example, the emphasis will be on the in-flight conditions.

Performance

Estimates of missile position, velocity and attitude are required to an accuracy which will enable the missile to be guided from its launch point to a point at which there is a high probability of its homing head acquiring the designated target. However, previous analysis of the missile system reveals that the critical parameters are those which determine the accuracy with which the homing head can be pointed in the direction of the target, namely cross track position, height, pitch attitude and yaw attitude. In order that a successful transition or handover from inertial mid course to terminal homing guidance may take place, the navigation system must be capable of providing cross-track position and height information with respect to the chosen reference frame to an accuracy of 50 metres (1σ) together with pitch and yaw attitude accurate to 1° (1σ).

Similar accuracies are required in along track position and roll, and estimates of the velocity components in all three channels are required to an accuracy of 10 m/s (1σ). These accuracies or better must be maintained over a flight time of up to 15 seconds.

The measurements of body acceleration and angular rate are required for missile control purposes to an accuracy of 0.1 g and 1°/s respectively. The system must be able to detect variations in these quantities over a bandwidth of 50 Hz.

System reaction time

It is required that the missile is able to be launched a short time after the decision to fire has been taken, say within two seconds. It is therefore important that the missile sensors are able to provide meaningful outputs within a second of being switched on and that the system can be initiated, or aligned, within a similar period.

Physical characteristics

In addition to the performance and environmental requirements, physical constraints are usually imposed on the size and weight of the complete unit. In this application, it is required that the volume of the navigation system should not exceed three litres and that the weight should be less than 3 kg.

Table 13.1 summarises the performance requirements of the onboard navigation system. This table also shows the environment in which it is expected to operate, physical constraints and its required reaction time.

Table 13.1 Summary of navigation system specification

1σ navigational accuracy requirements[1]:	
position —all axes	50 m
velocity—all axes	10 m/s
attitude—yaw, pitch and roll	1°
linear acceleration—all axes[2]	0.1 g
angular rate—all axes[2]	1°/s
environmental requirements:	
maximum lateral acceleration—pitch and yaw	50 g
maximum longitudinal acceleration	50 g
maximum turn rate—all axes	5 rad/s
vibration power spectral density	0.05 g²/Hz
	in bandwidth 10 Hz–1 kHz
shock	1000 g, 0.5 ms
operating temperature range	−20°C to +50°C
rate of change of temperature	5°C per minute
maximum altitude	15 km
physical characteristics:	
mass	< 3 kg
size	system must be capable of being accommodated within a cylinder of length = 10 cm diameter = 20 cm (~3 litres in volume)
System reaction time	< 2s

[1]Many of these are independent and may each contribute to the design.
[2]Autopilot requirements.

13.4 Why choose strapdown inertial navigation?

In order to satisfy the requirement for an autonomous onboard missile navigation capability, an inertial system is considered to be the most suitable for the application considered here. Although both strapdown or stable platform inertial systems may be designed to satisfy the navigation performance requirements, a strapdown system is ideally suited to this type of application in view of the physical constraints of size and weight imposed by the system specification.

In addition, there is a requirement for body mounted gyroscopes and accelerometers to be used to provide the measurements of linear acceleration and angular rate in missile body axes needed for control purposes. By using a strapdown system the same sensors may be used to provide the measurements needed for control purposes and to implement the navigation function. If a platform navigation system were to be chosen, separate strapdown gyroscopes and accelerometers would be needed for missile control purposes. Additionally, a four gimbal platform [1] would be required for the system to operate satisfactorily over the full range of dynamic conditions demanded by the system specification. Because of the extra cost incurred with this approach, the strapdown system option is definitely preferred.

The relative merits of strapdown and platform technology for the application considered here are summarised in Table 13.2, the ticks and crosses indicating conformance or non-conformance respectively with the requirement.

Table 13.2 Comparison of strapdown and stable platform options

	Strapdown	Platform
Size	✓	✗
Weight	✓	✗
Performance	✓	✓
Environment	✓	✓
Outputs	✓	✗[1]

[1]Separate body mounted gyroscopes and accelerometers required.

13.5 Navigation system design and analysis process

13.5.1 Introduction

In this section, a systematic approach to the design of a strapdown navigation system is outlined. In general, a sequence of design stages of the type described below will need to be followed for any strapdown system application, although the specific approach adopted and the areas of design emphasis will often vary substantially from one application to another.

For the system considered here, which is to be installed in a tactical missile, particular emphasis must be placed on the dynamic conditions arising during flight. The navigation system will be subject to in-flight manoeuvres, high levels of acceleration and an exacting vibratory environment, all of which will greatly influence the choice of inertial sensors and the accuracy of the measurements which they can be expected to provide during flight. In addition, the requirement for the missile to be launched from a moving platform will affect the accuracy to which the navigation can be initialised or aligned prior to missile launch. Throughout the design process it is important that a reasonable balance is struck between all potential sources of total system error. For example, there is nothing to be gained by relaxing the requirements placed on the inertial sensors at the expense of an alignment specification which is unrealisable or difficult to meet.

As part of the design process, consideration must first be given to the most appropriate system mechanisation for this application. This is followed by an error budget analysis to assess the magnitudes of alignment, instrument and computational errors which can be accepted. Assessments of error budget requirements may be conducted at various levels from relatively simple hand calculations using the single-plane error models discussed in Chapter 11 to more rigorous analysis using simulation which allows the effects of dynamic motion to be assessed more precisely. For the type of application being considered here, in which dynamic effects are expected to be of major significance, simplified calculations are likely to be of limited use and the designer will rapidly need to resort to very complex calculations or, more usually, simulation for the assignment of error budgets. In general, several iterations of the error budget analysis process will need to be carried out before a set of error parameter values which are both practical and realisable can be defined.

Having defined instrument performance characteristics and alignment accuracies, and also identified any potential computational difficulties, the tasks of defining suitable sensors and computational algorithms can begin. Throughout this process, further iterations of the error budget calculations may need to be undertaken in the light of the types of sensors considered before converging on a satisfactory design.

13.5.2 Choice of system mechanisation

During the inertial mid-course phase of missile flight, guidance commands are to be generated by combining the estimates of missile position and velocity provided by the onboard inertial navigation system with estimates of target position and velocity generated by a tracking device on the ship. In order that this may be accomplished satisfactorily, both the target and missile estimates must be generated in the same reference frame.

It may be assumed that the ship is fitted with an attitude and heading reference system, or possibly a full ship's inertial navigation system, which defines a co-ordinate reference frame which is nominally aligned with the directions of true north and the local vertical. Further, it is assumed that all shipboard equipment is harmonised with this reference system. Hence, the tracking device can provide estimates of target position and velocity in this reference frame. Similarly, the missile navigation system may be nominally aligned with this same frame before the missile is launched. Navigation will then take place in this reference frame, the origin of which will be the location of ship's attitude and heading reference system (or navigation system) at the instant when missile navigation begins.

Therefore, the missile navigation system will provide estimates of its position, velocity and attitude with respect to an Earth fixed reference frame defined at the time when the missile system starts to navigate, at or shortly before the missile is launched. In order that the target position and velocity estimates may be specified in this same frame, it will be necessary to correct the measurements provided by the tracker, which are generated in a reference frame which translates with the motion of the ship, to take account of any ship motion which takes place during missile flight. This approach is considered preferable to navigating the missile in a reference frame which moves with the ship to avoid the need to transmit motion of the ship data to the missile during flight. An Earth frame strapdown navigation system mechanisation of the type which may be adopted for this application is described in Chapter 3.

An orthogonal inertial sensor arrangement is considered appropriate for this application, there being no requirement for sensor redundancy or the measurement of particularly high turn rates about any axis which could lead to a skewed-sensor configuration being considered. The sensitive axes of the respective inertial sensors should be mounted coincident with the principal body axes of the missile, hence providing direct measures of the lateral accelerations and the turn rates required for autopilot feedback purposes.

13.5.3 Error budget calculations

Having defined the inertial system in broad terms, it is now possible to specify the alignment accuracy required and the performance of the gyroscopes and accelerometers needed to achieve the desired navigation accuracy at the end of the inertial phase of flight. As described in Chapter 11, the overall performance of the inertial navigation system depends upon the values of a large number of error parameters and on the way in which they propagate during the flight of the missile.

Each error parameter may be characterised as a random variable, with a probability distribution having a zero-mean and a known standard deviation [2], which varies from system to system and from flight to flight. The effect of each error source may be quantified with the aid of simulation over a given flight profile and its contribution to the total error budget determined in the manner described below.

Assuming conventional (mechanical) sensors are to be used, the dominant sensor error contributions in a tactical missile application of the type considered here are:

- fixed bias uncertainty;
- g-dependent biases (gyroscope only);
- anisoelastic biases (gyroscope only);
- scale factor errors;
- sensor mounting misalignments/cross-coupling.

It is assumed in the analysis which follows that variations with temperature in these various error terms will be compensated, so that the overall effect can be ignored.

In addition, for a system of this type which must be aligned on a moving ship, alignment errors are expected to contribute significantly to the overall error budget. For alternative sensor technologies, the sensor error contributions which need to be considered will need to be revised, as discussed later.

Various approaches may be used to allocate acceptable error contributions between the various processes which contribute to the overall error. A simple approach is to divide the total error equally between the contributing processes. Preliminary values may be determined in this way for a typical flight profile. However, this approach is unlikely to yield a practical set of error parameter values which bear close relation to state-of-the-art technology and alternative design procedures are therefore sought.

A common alternative method is to assess the sensitivity of the total error budget to each of the contributing errors sources. From this sensitivity analysis, combined with an assessment of the ease with which a given level of performance can be achieved, it is possible to make an informed allocation of the various component error sources. This approach relies heavily upon the designer's knowledge and experience of what is technically feasible and practical. As a result of this process, the size of many of the error parameters may be reduced, where this may be achieved without incurring an excessive cost penalty. By reducing the size of some error parameters, it may be possible to accommodate some of the larger or significant error sources, which are difficult or costly to reduce, in the error budget.

An illustration of this design process is given below.

Preliminary error allocation process

In Table 13.3, a list is given of 30 errors which are expected to contribute significantly to the overall error budget in the missile application considered here. Initially, values are assigned to each of the error parameters assuming that each contributes equally, in the root sum squared (RSS) sense [2], to the overall cross-track position and height errors which propagate over a 15 second period of flight. In this analysis, particular attention is focused on maintaining the cross-track position and height errors within the performance limits specified since this is crucial for the achievement of a successful inertial mid-course guidance phase. Although it is also vital to meet the attitude (pitch and yaw) accuracy specification, these requirements are more easily satisfied as illustrated in the error budget Tables described below.

Therefore, assuming the standard deviation of the total navigation errors resulting from sensor imperfection and alignment errors can be 50 m, then the contribution from each individual error source is allowed to be $50/\sqrt{30} \approx 9$ m (1σ), as defined in Table 13.3.

Table 13.3 Sensor error budget based on equal position error contributions

Error source	1σ value	cross-track position (m)	height error (m)	pitch error (°)	yaw error (°)
Alignment errors					
pitch attitude errors	0.04° (0.8 mrad)	0	9	0.04	0
yaw attitude error	0.04° (0.8 mrad)	9	0	0	0.04
x-track velocity error	0.6 m/s	9	0	–	–
vertical velocity error	0.6 m/s	0	9	–	–
x-track position error	9 m	9	0	–	–
height error	9 m	0	9	–	–
Gyroscope fixed biases					
B_{Gx}	437°/hr	9	0	0	0.09
B_{Gy}	100°/hr	0	9	0.41	0
B_{Gz}	120°/hr	9	0	0	0.48
Gyroscope g-dependent biases					
B_{gxx}	11.2°/hr/g	9	0	0	0.05
B_{gxz}	115°/hr/g	9	0	0	0.07
B_{gyx}	2.1°/hr/g	0	9	0.07	0
B_{gyz}	21.3°/hr/g	0	9	0.01	0
B_{gzx}	2.9°/hr/g	9	0	0	0.08
B_{gzz}	21.5°/hr/g	9	0	0	0.01
Gyroscope anisoelastic biases					
B_{axx}	2.9°/hr/g²	9	0	0	0.06
B_{ayx}	0.57°/hr/g²	0	9	0.05	0
B_{azx}	0.59°/hr/g²	9	0	0	0.05
Gyroscope scale factor errors					
S_{gy}	0.05%	0	9	0.04	0
Gyroscope cross-coupling					
M_{Gxy}	0.06%	9	0	0	0.04
M_{Gzy}	0.08%	9	0	0	0.04
Accelerometer fixed biases					
B_{ax}	115 mg	0	9	–	–
B_{ay}	8.2 mg	9	0	–	–
B_{az}	8.8 mg	0	9	–	–
Accelerometer scale factor errors					
S_{Ax}	0.42 %	0	9	–	–
S_{Az}	1.06 %	0	9	–	–
Accelerator cross-coupling					
M_{Axz}	3.07 %	0	9	–	–
M_{Ayx}	0.06 %	9	0	–	–
M_{Ayz}	1.01 %	9	0	–	–
M_{Azx}	0.07 %	0	9	–	–
Total RSS errors (1σ)		36 m	33.7 m	0.42°	0.51°

The results given in Table 13.3, which are purely for illustrative purposes, have been derived for a missile which manoeuvres in the pitch plane alone. Under these simplified conditions, the cross-track and height channels of the navigation system remain largely decoupled from each another and the majority of the error sources give rise to navigation errors in one channel only. As a result, the total errors in position and attitude are shown to be well within the requirements specified. In general, the missile will be called upon to manoeuvre in the pitch and yaw planes simultaneously, in which case each error source will contribute to the navigation error in each channel. Further, some additional g-dependent biases and scale factor errors may become significant in the more general situation, owing to the manoeuvring of the missile. However, the simplified analysis given here is adequate for illustrating the fundamentals of the technique.

Refinement of the error budget allocation

Referring to the error terms given in Table 13.3, the angular alignment accuracies and some of the sensor bias values suggested will be very difficult to achieve in practice. Therefore, the design engineer must now bring his or her experience of system design and sensor technology performance to bear on the selection of suitable error coefficient values, bearing in mind those parameters to which the overall errors are most sensitive.

There is usually some limited scope for relaxation of some contributions to the error budget at the expense of other parameters. For example, in a design based upon conventional gyroscopes the fixed bias contributions to the error budget may be allowed to increase whilst reducing the g-dependent biases to levels which can more easily be achieved in practice. Analysis reveals that system performance in this type of application is particularly sensitive to the values of the g-dependent bias coefficients. It can also be seen that certain of the cross-coupling terms associated with both gyroscopes and accelerometers will need to be small in order to achieve the required performance.

At all times, it is of course essential to ensure that the contribution of any one error term does not exceed the total error budget. It will usually be necessary to carry out several iterations of this parameter selection process before arriving at a reasonable set of values. An example set of error parameter values is given in Table 13.4 along with their respective contributions to the total position error budget.

Table 13.4 Sensor error budget based on minimum risk strategy

Error source	1σ value	cross-track position (m)	height error (m)	pitch error (°)	yaw error (°)
Alignment errors					
pitch attitude errors	0.16° (3 mrad)	0	36	0.16	0
yaw attitude error	0.16° (3 mrad)	36	0	0	0.16
x-track velocity error	0.6 m/s	9	0	–	–
vertical velocity error	0.6 m/s	0	9	–	–
x-track position error	1 m	1	0	–	–
height error	1 m	0	1	–	–
Gyroscope fixed biases					
B_{Gx}	50°/hr	1	0	0	0.01
B_{gy}	50°/hr	0	4.5	0.21	0
B_{gz}	50°/hr	3.75	0	0	0.2
Gyroscope g-dependent biases					
B_{gxx}	5°/hr/g	4	0	0	0.02
B_{gxz}	5°/hr/g	0.4	0	0	0.003
B_{gyx}	5°/hr/g	0	21.4	0.17	0
B_{gyz}	5°/hr/g	0	2.1	0.002	0
B_{gzx}	5°/hr/g	15.5	0	0	0.14
B_{gzz}	5°/hr/g	3.8	0	0	0.004
Gyroscope anisoelastic biases					
B_{axx}	0.5°/hr/g²	1.6	0	0	0.01
B_{ayx}	0.5°/hr/g²	0	7.9	0.05	0
B_{azx}	0.5°/hr/g²	7.7	0	0	0.04
Gyroscope scale factor errors					
S_{gy}	0.05%	0	9	0.04	0
Gyroscope cross-coupling					
M_{Gxy}	0.1%	15	0	0	0.07
M_{Gzy}	0.1%	11.25	0	0	0.05
Accelerometer fixed biases					
B_{ax}	10 mg	0	0.8	–	–
B_{ay}	10 mg	11	0	–	–
B_{az}	10 mg	0	10.2	–	–
Accelerometer scale factor errors					
S_{Ax}	0.3%	0	6.4	–	–
S_{Az}	0.3%	0	2.5	–	–
Accelerator cross-coupling					
M_{Axz}	0.1%	0	0.3	–	–
M_{Ayx}	0.1%	15	0	–	–
M_{Ayz}	0.1%	0.9	0	–	–
M_{Azx}	0.1%	0	12.9		
Total RSS errors (1σ)		49.2 m	48.2 m	0.32°	0.32°

Table 13.4 shows clearly the dominant sources of error to be the angular alignment errors together with certain of the g-dependent gyroscope biases and accelerometer cross-coupling, both of which give rise to large position errors in the presence of missile longitudinal acceleration. In addition, gyroscope cross-coupling contributes substantially to the overall error budget in the presence of the pitch turn manoeuvre which the missile undergoes during its boosted phase of flight.

Using the alignment and sensor errors given in Table 13.4, the total along track position error, roll error and velocity errors, which are not given in the table, have been calculated to be as follows:

RSS along-track position error= 41 m
RSS roll error   = 0.3°
RSS along-track velocity error = 0.7 m/s
RSS cross-track velocity error = 3.7 m/s
RSS vertical velocity error   = 3.8 m/s

It can be seen that each lies within the limits defined by the specification.

Having generated detailed sensor specifications, it is essential to evaluate the performance of the resulting system over a representative set of missile trajectories. The effect that many of the errors have on overall navigation performance is often highly dependent on the precise motion to which the system is subjected during flight. It may become necessary to refine further some of the error coefficient values at this stage of the design in order to specify the system adequately.

Whilst following the type of procedure described above, the designer may well wish to combine certain errors which propagate in a similar manner and have similar effects on navigation system performance. Some examples are given below.

- Gyroscope anisoelasticity—in the presence of cyclic motion, biases arise on the output of a conventional gyroscope owing to the effects of unequal bearing compliance;
- Accelerometer vibro-pendulous error—additional biases arise on the outputs of pendulous accelerometers in the presence of vibration.
- Coning and sculling motion—further angular rate and linear acceleration biases can arise if the inertial sensors are subjected to coning and sculling motion respectively;

The gyroscope and accelerometer biases used in the error budget analysis may need to be increased to take account of such effects, all of which are discussed at some length in Chapter 11.

13.5.4 System alignment

A vital factor in the performance of an inertial navigation system is the accuracy to which it can be aligned, or initialised, prior to the start of navigation. If the alignment contribution to the total error budget is to be as suggested in Section 13.5.3, then a 1σ angular alignment accuracy of approximately 0.16° (10 arc minutes) is required together with initial velocity and position estimates accurate to 0.6 m/s and 1 m (1σ) respectively.

Although the achievement of sufficiently small angular misalignments is of particular concern, significant errors can also arise in the velocity estimates used to initialise the navigation system as a result of lever-arm motion between the ship's reference and the missile. For a tactical missile system which is to be launched from a moving platform, the achievement of satisfactory alignment accuracy is particularly difficult. In addition to the alignment accuracy required, a major factor influencing the choice of the alignment scheme is the time available to accomplish such an alignment.

Following from the discussion of shipboard alignment methods in Chapter 9, it is postulated that an alignment to this order of accuracy may best be achieved within the short period of time available (~1 second) by the rapid transfer of alignment data from an inertial navigation system mounted on the missile launch silo on the ship using the so-called one-shot alignment method. The alignment of this unit would need to be maintained independently using shipboard measurements provided by the ship's own inertial reference system, or through the use of satellite data updates.

In view of the potential difficulties of achieving an accurate alignment of the missile system onboard a moving ship, it may well be advisable to relax the alignment requirement further at the expense of tightening instrument performance. However, it can be seen from the error budget that there is very limited scope for further relaxation of the alignment requirement and for the purposes of this design exercise it is assumed that the system can be aligned to the accuracy suggested.

13.5.5 Choice of inertial instruments

Having defined a set of performance characteristics which must be satisfied by the inertial sensors in order to meet the overall system specification, the task of selecting suitable gyroscopes and accelerometers can be undertaken.

With reference to Chapters 4, 5 and 6 on inertial sensor technology, there is a wide range of gyroscopes and accelerometers capable of satisfying the performance requirements under the flight conditions described above. Previous experience with other strapdown navigation systems would suggest that possible candidates for the body mounted gyroscopes could be:

- dynamically tuned gyroscope (DTG) or flex gyroscope
- rate integrating gyroscope (RIG)
- fibre optic gyroscope (FOG)
- ring laser gyroscope (RLG)
- vibratory gyroscope.

In the case of the accelerometers, the leading candidate is likely to be a form of pendulous force feedback accelerometer, although other possibilities are as follows:

- silicon accelerometer;
- surface acoustic wave (SAW) accelerometer

The compliance of these sensors with the various parameters selected from the requirements in the system specification and the error budget analysis are indicated in the Tables 13.5 and 13.6.

Table 13.5 Gyroscope options

	Mechanical		Optical		Other
	RIG	DTG	RLG[1]	FOG	Vibratory
Measurement range	✓	✓	✓	✓	✓
Measurement accuracy	✓	✓	✓	✓	X^2
Environmental performance	✓	✓	✓	✓[3]	✓[4]
Size and weight	✓	✓	✓	✓	✓
Reaction time	✓[5]	✓[5]	✓	✓	✓
Risk[6] (technical maturity)	low	low	low/ medium	low/ medium	low/ medium

[1]Even a small RLG is likely to have performance well in excess of this requirement.
[2]This type of sensor could possibly be compensated to give performance close to the desired accuracy.
[3]Care would possibly be required in the design and packaging of the system to make this sensor sufficiently insensitive to various aspects of the environment.
[4]Elaborate temperature compensation of this type of sensor is likely to be necessary and may be prohibitively complex. However, it would be very rugged.
[5]The ability of mechanical gyroscopes to achieve stable operation shortly after switch on has been examined in the past. It has been demonstrated that sub-inertial sensors of the type required for this type of application can be run-up in about one second by applying extra power to the spin motor at switch-on, using so-called over-volting techniques. Some characterisation of sensor error transients may be needed to achieve meaningful outputs within this period of time.
[6]Risk is a multi-faceted problem, encompassing technical, financial and temporal aspects. In this case, only the technical maturity and pedigree of the gyroscopes are considered.

Table 13.6 Accelerometer options

	Pendulous force feedback	SAW	Silicon
Measurement range	✓	✓	✓
Measurement accuracy	✓	✓[1]	✓[1]
Environmental performance	✓[2]	✓[2]	✓[2]
Size and weight	✓	✓	✓
Reaction time	✓	✓	✓
Risk[3] (technical maturity)	Low	Low/Medium	Medium

[1]Careful compensation is likely to be necessary.
[2]Careful packaging design is likely to be necessary for some types.
[3]Risk is a multi-faceted problem, encompassing technical, financial and temporal aspects. In this case, only the technical maturity and pedigree of the accelerometers are considered.

Clearly, a number of different sensors could provide the desired angular rate and acceleration data to the required accuracy. Based on the information given in Table 13.5, it is concluded that there are three leading gyroscope candidates; the two mechanical gyroscopes and the fibre optic sensor, whilst Table 13.6 suggests that any of the three accelerometer types considered may be chosen.

It should be noted that the data provided by the ring laser gyroscopes are likely to be far more accurate than required for this application and, therefore, it is less likely to be chosen. However, the use of such sensors could possibly enable the performance requirements to be relaxed on other processes which contribute to the navigation system error budget, such as the alignment accuracy or accelerometer performance. Alternatively, the performance requirement of another sub-system within the missile may be relaxed. This would be the subject of a trade-off study for the whole system, and is beyond the intended scope of this design example.

In the event that the designer decides to choose sensors based on the maturity of the technology and minimum risk, then he or she is most likely to select the force feedback accelerometer and either the dynamically tuned gyroscope or the single-axis rate integrating gyroscope. A further attraction of the dynamically tuned gyroscope is the fact that only two sensors are required to provide the three axes of angular rate measurements required, potentially offering good value for money.

Alternatively, the more modern sensors may be chosen, i.e. either the SAW or the silicon accelerometer and the fibre optic gyroscope, on the grounds that they could offer low total cost in the future through the ease of manufacture of solid state technology.

The final choice may well be decided by cost, including purchase price and projected total life cycle costs. The purchase price is a complex issue involving availability of components, total order size, production rate and all aspects of calibration and compensation.

13.5.6 Computational requirements

It is important that the data provided by the inertial sensors are processed correctly in order to achieve the required navigation system performance. The onboard computer is required to calculate the missile attitude in the chosen reference frame, resolve the accelerometer measurements into that reference frame and then integrate these quantities to provide estimates of missile velocity and position. In addition estimates of roll angle are to be extracted to provide the feedback signal needed to stabilise the position of the missile about the roll axis.

As discussed in Section 13.5.2, the Earth frame strapdown system configuration described in Chapter 3 is considered appropriate for this application. However, in this particular case, the Earth's rate terms in the attitude computation process could be ignored. Since angular rate biases of 50°/hour can be tolerated, the effects of neglecting Earth's rate (15°/hour) will not cause system performance to degrade significantly. However, the Coriolis correction terms will need to be included in the navigation equation owing, to the high missile velocity (1000 m/s). It is noted that the removal of the Earth's rate terms will not reduce the overall computational burden greatly.

The crucial aspects of the computation are:

- data rate;
- processing speed;
- truncation levels of mathematical functions used in the computation.

Data rate and processing speed are determined by the frequencies of the angular and translational motion to which the navigation system is to be subjected during flight. For the missile application considered here, it is expected that the bulk of the computation, including the attitude update algorithm, the resolution of the accelerometer measurements and the solution of the navigation equation would need to be carried out at a frequency in the region of 100–200 Hz.

It is important to be aware of any cyclic motion which may arise, such as coning or sculling type motions, since they can have an adverse effect on the performance of the system. These types of motion can arise during flight as a result forces exerted by aerodynamic control surfaces which can excite natural body bending modes of the missile airframe. The degradation in performance introduced by cyclic motion can often be reduced by the use of multiple speed algorithms of the type described in Chapter 10, in which fast but relatively simple algorithms are used to implement high speed correction algorithms. This approach may need to be adopted in a missile application where significant levels of vibration can arise. For these fast calculations, computational frequencies of around 400 Hz would normally be sufficient. However, these calculation may need to be carried out more rapidly since the specification indicates that vibratory motion at frequencies up 1 kHz may be present. Detailed calculation is necessary in each specific case to ensure that the processing speed exceeds the frequency of the vibratory motion by a factor of at least two.

Truncation levels of mathematical functions must be selected to allow sufficient accuracy of computation to be achieved without imposing an excessive burden on the system computer. In general, there are trade-off considerations to be made between the speed of computation and the amount of truncation which can be permitted.

In the past, the choice of wordlength has also been a crucial factor in the design of navigation system computers. However, modern systems tend to use floating point processors in which wordlength restrictions do not usually impose a serious limitation.

There is often a temptation to assume that computational errors can be contained within acceptably small limits and that their contribution to the overall error budget is small compared with the sensor and alignment errors. Although this may often be the case, care should be taken in the choice of algorithms, particularly for agile missile applications of the type considered here. As a result of the rapid advances made in computer technology in recent years, the achievement of sufficient computer speed is certainly far less of a problem now than it has been in the past.

13.5.7 Electrical and mechanical interfaces

It is essential to check that the electrical interfaces are compatible on either side of any mechanical interface. Additionally, any mechanical interfaces on the inertial system must be mutually compatible with the fixing points on the missile structure. In particular, any mechanical interface with any other unit must not compromise the structural loading and stiffness requirements. For the purposes of this design study, it will be assumed that these design characteristics can be achieved.

13.6 Testing, calibration and compensation requirements

It is most important to evaluate the prototype navigation system to ensure that it fulfils the design requirement. Suitable test plans for the navigation system, and the individual sensors within it, may be devised in accordance with the overall system specification. A series of laboratory tests will be required including static, rate, centrifuge and temperature tests as discussed in Chapter 7.

During the sensor and system evaluation, various systematic errors may become evident. Compensation techniques can be readily applied to such systematic errors. In some cases, compensation enables the derived specification of basic performance of the inertial sensors to be relaxed further.

A typical systematic error that is likely to require correction is temperature dependence of the scale factors of the inertial sensors. Hence, it may be necessary to incorporate a temperature sensor in the inertial measurement unit. A typical temperature compensation routine is shown below.

Consider a sensor with a scale factor which varies systematically with temperature. This dependence can be determined from the testing procedure described in Chapter 7, Section 7.5 for gyroscopes and Section 7.6 for accelerometers. The slope of the graph in Figure 7.8 gives the scale factor temperature coefficient (SFTC) for a gyroscope. A similar parameter can be deduced from Figure 7.20 for an accelerometer.

The scale factor of a sensor can be corrected for changes induced by variations in the temperature of the sensor, provided that the temperature (T_c) at which the sensor was calibrated is known and the actual temperature (T_a) of the sensor, when the measurement was made, is also known. Then the scale factor at temperature T_a is calculated as follows:

$$SF_a = SF_c + SFTC(T_a - T_c) \cdot 10^{-6}$$

where SF_a is the scale factor at temperature T_a, SF_c is the calibrated scale factor (at temperature T_c) and $SFTC$ is the scale factor temperature coefficient expressed in parts per million per Kelvin (or centigrade degree).

Similarly, any systematic variation in the bias of the sensor can also be compensated.

13.7 Performance enhancement by aiding

If additional navigation measurements could be provided during flight, it may be possible to relax further the sensor performance and the alignment accuracy requirements.

In a tactical missile application of the type considered here, missile position fixes may be provided by tracking the missile during flight and transmitting the data to the missile during flight. Given that such information may be provided at intervals during the flight with sufficient accuracy and passed to the missile, it may be used to correct the onboard inertial navigation system. As a result of this process, there may be scope to relax the specifications of the inertial sensors and to avoid the need for pre-flight alignment to the accuracy proposed for the unaided system.

The reader is referred to Chapter 12 in which a system of this type is described in some detail.

13.8 Concluding remarks

This chapter has provided an outline of the stages needed to produce a design for an inertial navigation system for a generic short range tactical missile application. By analysis of the allocated system error budget in the system specification it is possible to derive the performance specification for the inertial sensors and to define the computing processes which must be implemented.

In the example chosen, a design has been formulated that can fulfil the technical requirements set out in the specification. Therefore, the first stage of the design is complete. This would have to be confirmed by laboratory and field testing. However, if a suitable design could not be achieved, even with aiding and superior performance sensors, then it would be necessary to negotiate concessions in the technical requirements specification, with possible implications on the overall performance of the system.

References

1 BRITTING, K.: *'Inertial Navigation System Analysis'* (Wiley Interscience, New York, 1971)
2 TOPPING, J.: *'Errors of observation and their treatment'* (Chapman and Hall, 1975)

Glossary of principal terms

Inertial navigation systems rely on complex technology, even if the implementation is simple or mundane. Thus it will be inevitable that many technology specific terms and jargon will be in common usage. The more common aspects are covered here together with those that are used in the text of this book. The terms described appear in alphabetical order.

Accelerometers

These devices are used to measure the translational motion of the vehicle in which they are located. In its most basic form, this sensor consists of a proof mass attached, via a spring, to the case of the instrument. The device operates by detecting a displacement of the mass in the presence of an acceleration of the vehicle. An accelerometer actually provides a measure of specific force, the non-gravitational force per unit mass to which it is subjected.

Aiding

This technique refers to the use of external measurement data to enhance the performance of an inertial navigation system. The source of the aiding may be external to the vehicle or derived from an additional onboard sensor. An example of the former type is the use of position information provided by a tracking radar to update periodically an inertial navigation system in a short range tactical missile. Alternatively, an onboard sensor such as a Doppler radar or a barometric altimeter may provide information to aid the inertial system. As with all such techniques which combine measurement data with different characteristics, sophisticated Kalman filtering techniques are often used, and are necessary to produce the best results.

Note that data from the inertial navigation system may be used to aid another sensor, such as a satellite navigation system.

Algorithms

An algorithm is a mechanistic procedure for solving a problem in a finite number of steps. Such a procedure may take the form of a series of mathematical instructions which may be programmed into a computer.

Alignment

This is the process of determining the initial orientation of the measurement axes of the inertial navigation system to the chosen reference frame, i.e. the process of determining the angles between the measurement axes and the axes of the reference frame. An alignment must be carried out prior to the start of navigation. The accuracy with which this can be achieved is of fundamental importance as it can severely influence the performance of an inertial navigation system, or indeed any system which operates by the so-called dead reckoning technique.

Attitude and heading reference system

Such a system will provide, as its name suggests, attitude and heading data for the vehicle in which it is installed. This system is very similar to an inertial navigation system, but does not provide velocity and position of the host vehicle. Generally the sensors are of lower accuracy and consequently less expensive. Such a system is often combined with another sensor such as a Doppler radar to form a full navigation system. The Doppler radar provides measurements of velocity over the ground which can be resolved into the navigation frame defined by the attitude and heading reference system. Integration of these velocity components yields the position of the vehicle.

Bias

Bias refers to the offset in the measurement provided by an inertial sensor. For example, a gyroscope provides a measurement of turn rate about a given axis. The output of the gyroscope may take the form of a voltage or current proportional to the applied turn rate plus a bias term caused by the various imperfections within the sensor.

Calibration

This is the process of establishing the precise value of the electrical signals produced by the inertial sensors. For example, a change of one millivolt in the electrical signal generated may indicate a rotation of a given rate or a specific force acceleration of a given value.

Compensation

All inertial sensors that are subject to systematic errors can have these errors removed by a correction technique called compensation. For example, if a gyroscope has a known tendency to drift when its temperature changes, the effects of this drift can be corrected provided:

- it is a systematic error;
- the temperature at which the measurement being made is known;
- the temperature of the calibration is known;
- the contribution that this change in temperature has to the measured electrical signal is known.

Given this information, the signal produced by the gyroscope can be corrected to give a measure of the true rotation or rotation rate. Similar logic applies to other forms of inertial sensor, and other systematic errors.

Dead reckoning systems

These types of navigation system rely on the continuous updating of the position data derived from inputs of velocity components or speed and heading generated from a known start position. A simple example of this technique is a system which uses a compass heading in combination with a device, such as an odometer, to measure the distance travelled over the ground. Inertial navigation systems also fall into this category. The navigation accuracy that can be achieved using a dead reckoning system is largely influenced by the accuracy to which the start position, the velocity and the heading are known.

Direction cosine

A direction cosine is the cosine of the angle between two vectors. The orientation of a vector with respect to a given co-ordinate frame may be expressed in terms of three direction cosines defining the projection of the vector on to the axes of the frame. The attitude of a body frame with respect to a given reference frame may be expressed as a direction cosine matrix. This is a 3×3 matrix, the columns of which represent unit vectors in body axes projected along the reference axes.

Drift

Drift refers to the rate at which the error in a sensor or system accumulates with time. For example, a mechanical displacement gyroscope will provide a measure of the attitude of the body in which it is installed with respect to the spin axis of the gyroscope. The orientation of the spin axis will change with time as a result of unwanted torques acting on the rotor. As a result, a drift will be present on the measurement of attitude provided by the sensor which may be expressed in units of degrees per hour.

In an inertial navigation system, the estimates of position that it provides will drift with time as a result of various errors within the system. Hence, it is customary to quantify navigation system performance in terms of a drift figure, often expressed in nautical miles per hour.

Euler angles

The attitude of a body with respect to a given reference frame may be specified in terms of three successive rotations about different axes. Hence, a transformation from the reference frame to the body frame can be carried out as follows:

(i)   rotate through angle ψ about reference z axis;

(ii) rotate through angle θ about new y axis;

(iii) rotate through angle φ about new x axis;

where ψ, θ and φ are referred to as the Euler rotation angles. It is, of course, possible to use the same angles in reverse order to define the transformation from body axes to reference axes.

The three angles correspond to the angles which would be measured between a set of mechanical gimbals supporting a stable element or platform. The axes of the stable element represent the reference frame, whilst the body of the host vehicle is attached via a bearing to the outer gimbal. The order of the rotations is important and corresponds to the gimbal order.

Force Feedback

A technique used to return a sensitive element such as an accelerometer's proof mass or a gyroscope's rotor to its null position with respect to the case of the instrument. Null positions can be located and measured far more accurately than measurement of displacements. In some sensors, an electromagnet is used to move the sensitive element to this zero position, the current flowing in the coils of the electromagnet usually being proportional to the quantity being measured.

Gravitational model

This is the mathematical representation of the gravitational attraction to which a navigation system is subjected. Such a representation is required so that the specific force measurements provided by the accelerometers can be converted to true acceleration data. This is important for the navigation of vehicles in close proximity to the Earth as well as vehicles moving through interplantary space. For navigation in the vicinity of the Earth, the Earth's gravitation can be defined to various levels of detail and complexity depending on the accuracy required from the navigation system. For very accurate navigation, it is necessary to take account of local variations or anomalies in the gravitational vector from the generalised mathematical model.

Gyroscopes

These are the sensors that are usually used to measure rotational motion in an inertial system. They commonly take the form of mechanical devices which rely on the inertial properties of a spinning mass for their operation. For example, in a free (two-axis) gyroscope, the spin axis tends to remain fixed in space allowing rotational motion to be sensed about two orthogonal axes (which are nominally at right angles to the spin axes and to each other). Such devices produce measurements of turn angle with respect to inertial space. Single-axis mechanical gyroscopes are usually designed to provide measurements of turn rate with respect to inertial space.

There are a number of other devices that perform a similar function which rely on different physical phenomena for their operation. For example, optical gyroscopes which provide a measure of angle or of angular rate by detecting some physical difference between two counter-propagating beams of light, using the Sagnac effect, and vibratory sensors which measure the Coriolis forces acting on masses which undergo linear vibration whilst being rotated.

Guidance

This process refers to the direct control of the vehicle to constrain it to follow a pre-defined trajectory. For instance, consider an aircraft flying at constant speed that is required to follow a straight track between two way points. It can reasonably be assumed that the aircraft has knowledge of where it should be at any time during the flight. Guidance commands can therefore be generated by differencing the desired and measured aircraft position. These commands are fed to the aerodynamic control surfaces causing the aircraft to move so as to null the detected path difference, thus ensuring that it flies along the desired track. This form of guidance is fairly common in strategic ballistic missiles and some long range tactical guided weapons.

Harmonisation

This term refers to the relative orientation of pieces of equipment on a ship or aircraft.

Inertial guidance

Guidance using knowledge of position, velocity and attitude derived from an inertial navigation system. The distinction is drawn between this process and inertial navigation.

Inertial measurement unit

A sub-set of the full inertial navigation system, the structure that contains the inertial sensors together with their support electronics and power supplies. The inertial measurement unit often contains the electronics required to operate the sensors and a microprocessor to compensate some of the biases in the measurement signals generated by the sensors.

Inertial navigation

The process of establishing the position, velocity, attitude and heading of a vehicle using information derived from inertial sensors. Such systems are widely used for the navigation of aircraft, guided missiles, space vehicles, ships, submarines and land vehicles.

Integrated navigation

This is the process of combining data or information from two or more navigation systems, usually with complementary error characteristics, in order to produce a system with performance characteristics which surpass those of the component systems acting in isolation. Typical examples of such synergy are the use of position fixes derived from ground transmitters or satellite information, terrain reference systems and scene matching techniques to augment information derived from an inertial navigation system.

Performance

The accuracy with which an inertial navigation system can navigate a vehicle during a journey is governed by the accuracy of the data supplied to it at the commencement of navigation (i.e. the alignment accuracy), the quality of the inertial sensors it uses and the precision with which the navigation computation task is carried out. The errors grow with time and it is therefore customary to characterise inertial navigation system performance in terms of a drift in its navigational accuracy with respect to time. The drift is often specified in units of nautical miles per hour, although the error growth is not a linear function of time, in general.

Full inertial performance usually refers to sensors which give navigational accuracy of less than one nautical mile per hour. Such systems typically require gyroscopes with drift rates of 0.01°/hour or better and accelerometers which can provide measurements to an accuracy of 100 micro-g. The other term sometimes used when describing system performance is sub-inertial which relates to tactical applications. Such systems are used for short duration navigation, or possibly to provide an attitude reference. Typically, sub-inertial systems use gyroscopes and accelerometers with measurement biases of the order of 1°/hour and 1 milli-g respectively, which if used in a full inertial navigation system would produce an error growth of approximately one nautical mile per minute.

Position fixing systems

Such systems operate by determining their position with respect to a known location, and rely on the observation of an object or effect at that location or a transmitted signal emanating from it. This form of system can determine its position when activated at any time during a journey provided that the object or effect is observable. The earliest forms of navigation used this technique of navigation by observation and recognition. There are many techniques, both active and passive, that can be used by the observer to fix his/her position. Examples are signals from fixed radio beacons such as DECCA, OMEGA or LORAN and signals transmitted from satellites orbiting the Earth as in the case of GPS or GLONASS. Terrain referenced systems also fall into this category. The received signals are used to derive position updates that are independent of the previous estimate of position. As a rule, such schemes only provide navigation data at discrete time intervals and not on a continuous basis as in dead reckoning systems, although given a knowledge of prior events, some extrapolation is possible.

Quaternion

A four parameter attitude representation. The quaternion attitude representation allows a transformation from one co-ordinate frame to another to be effected by a single rotation about a vector defined in the reference frame. The four elements of the quaternion are functions of the orientation of this vector and the magnitude of the rotation about it.

Reference frame

This refers to the set of axes to which the measurements and estimated quantities generated within an inertial system are referenced. The reference frame may be either a co-ordinate frame defined by three mutually orthogonal axes or a polar co-ordinate system. Various frames of reference are considered in the text of this book. An inertial reference frame is one that is fixed in space, i.e. fixed in relation to the fixed stars. For the purposes of navigation in the vicinity of the Earth, use is made of a geographic axis set which takes the polar form to give latitude, longitude and height above the Earth. In addition a local vertical geographic frame is often used. This is a Cartesian reference frame most commonly defined by the mutually orthogonal directions of true north, east and the local vertical (down).

Resolution

The mathematical process of calculating the various components of a vector quantity in a given co-ordinate reference frame when the vector is given in components referred to in another reference frame. In a strapdown inertial navigation system, it is necessary to resolve the measurements of specific force, provided by the accelerometers in a body axis reference frame, into the chosen navigation reference frame. The resolved specific force measurements can then be integrated to enable velocity and position with respect to the reference frame to be determined. The attitude of the body may be defined using various mathematical representations including Euler angles, direction cosines and the four parameter quaternion form.

The term resolution is also used in the context of inertial sensors, and refers to the smallest change in a measurement quantity which can be detected and distinguished by such a sensor. For example, a particular type of gyroscope may be able to detect or resolve angular movements of one arc second.

Scale factor

The scale factor of an inertial sensor is the relationship between the output signal and the quantity which it is measuring. For instance, the scale factor of a gyroscope which provides an analogue output may be expressed as so many millivolts per degree per second.

Stabilised platform

The original applications of the inertial navigation technology used stable platform techniques which are still in common use, particularly for ships and submarines when accurate navigation is required, unaided and over long time intervals. At the core of this system is a structure, often called the platform or inertial element, on which inertial sensors are mounted. This platform is isolated from the rotational motion of the vehicle using a number of gimbals arranged to provide at least three degrees of rotational freedom and so minimise the coupling between the vehicle and the platform.

Movement of these gimbals is controlled by torque motors which are activated by information provided by the gyroscopes. Additional gimbals may be used for some specific applications, particularly for very agile vehicles to prevent gimbal lock occurring.

The first practical realisation, called the geometric system, had at least five gimbals to provide the basic navigation data of latitude, longitude, roll, pitch and yaw in direct analogue form, these angles being available directly from the angular displacement of the various gimbals. This mechanisation instrumented two reference frames, an inertial non-rotating co-ordinate system and a local navigational co-ordinate system.

This geometric system was superseded by the so-called semi-analytic system which established one of the two reference frames instrumented by the geometric system. Generally, three gimbals are used in this mechanisation, to minimise coupling of the vehicle motion into the stable platform, with the co-ordinates of latitude and longitude being calculated in a computer. As in the geometric system, roll, pitch and yaw are deduced from the relative positions of the gimbals. A variety of different mechanisations of this type have been used. Some are used to establish an inertially non-rotating reference frame, as in space stabilised systems. An alternative system commonly used for navigation in the vicinity of the Earth establishes a local level frame so that the input axes of two of the gyroscopes and two of the accelerometers are constrained to remain with their input axes in the horizontal plane, thus avoiding the explicit calculation of the gravity field vector.

Strapdown systems

In strapdown systems, the inertial sensors are fastened directly to the vehicle and hence are not isolated from its angular motion. Such systems are sometimes referred to as analytic systems. The signals produced by the inertial sensors are resolved mathematically in a computer prior to the usual calculation of navigation information. This use of a computer to establish and resolve the inertial data reduces the mechanical complexity of the inertial navigation system, thus frequently reducing the cost and size of the system and increasing its reliability.

Sub-inertial systems

see *Performance*

Kalman filtering

The combination of independent estimates

Combining two independent estimates of a variable to form a weighted mean value is central to the process of Kalman filtering [1]. The following development assumes only a knowledge by the reader of elementary statistical principles. A full mathematical derivation of the Kalman filter is beyond the intended scope of this book and the reader interested in a mathematical treatise on the subject is referred to the excellent text by Jazwinski [2].

The single dimension case

Consider the situation in which two independent estimates, x_1 and x_2, are provided of a quantity x, where σ_1^2 and σ_2^2 are their respective variances. It is required to combine the two estimates to form a weighted mean, corresponding to the best, or minimum variance, estimate, $\hat{x}$. In general, the weighted mean may be expressed as:

$$\hat{x} = w_1 x_1 + w_2 x_2 \tag{A.1}$$

where w_1 and w_2 are the weighting factors and $w_1 + w_2 = 1$. The expected or mean value of $\hat{x}$, written as $E(\hat{x})$, is given by:

$$E(\hat{x}) = w_1 E(x_1) + w_2 E(x_2) \tag{A.2}$$

The variance of a quantity x is defined as $E[\{x - E(x)\}^2]$. Hence, the variance of $\hat{x}$, denoted σ^2, may be written as:

$$\sigma^2 = E\left\{\left(w_1 x_1 + w_2 x_2 - w_1 E(x_1) - w_2 E(x_2)\right)^2\right\}$$

$$= E\left\{w_1^2\left(x_1 - E(x_1)\right)^2 + w_2^2\left(x_2 - E(x_2)\right)^2\right.$$

$$\left. -2w_1 w_2\left(x_1 - E(x_1)\right)\left(x_2 - E(x_2)\right)\right\} \tag{A.3}$$

Since x_1 and x_2 are independent, $(x_1 - E(x_1))$ and $(x_2 - E(x_2))$ are uncorrelated, $E\{(x_1 - E(x_1))(x_2 - E(x_2))\} = 0$. Hence, σ^2 can be expressed as follows:

$$\sigma^2 = w_1^2 E\left\{\left(x_1 - E(x_1)\right)^2\right\} + w_2^2 E\left\{\left(x_2 - E(x_2)\right)^2\right\} = w_2 \sigma_1^2 + w_2 \sigma_2^2 \tag{A.4}$$

Writing $w_2 = w$ and $w_1 = 1 - w$, the variance s^2 may be expressed as:

$$\sigma^2 = (1 - w)^2 \sigma_1^2 + w^2 \sigma_2^2 \tag{A.5}$$

The value of w, which minimises σ^2, is obtained by differentiating the above equation with respect to w. Hence,

$$\frac{d}{dw}\sigma^2 = -2(1-w)\sigma_1^2 + 2w\sigma_2^2 = 0$$

which yields the optimum weighting factor as:

$$w = \frac{\sigma_1^2}{\sigma_1^2 + \sigma_2^2} \tag{A.6}$$

Substitution in eqns. A.1 and A.5 gives $\hat{x}$ and its variance σ^2:

$$\hat{x} = \frac{\sigma_2^2 x_1 + \sigma_1^2 x_2}{\sigma_1^2 + \sigma_2^2} \tag{A.7}$$

$$\sigma^2 = \frac{\sigma_1^2 \sigma_2^2}{\sigma_1^2 + \sigma_2^2} \tag{A.8}$$

By following this process, the two independent estimates, x_1 and x_2, have been combined to form a weighted mean value, in which the weighting factor has been selected to yield a mean with minimum variance and, hence, maximum probability. In a Kalman filter, one such estimate is usually provided by updating a previous best estimate in accordance with the known equations of motion, whilst the other is obtained from a measurement. If x_2 is taken to be a measurement that is used to improve an updated estimate x_1, the above equations can be expressed in the following form:

$$\hat{x} = x_1 - w(x_1 - x_2) \tag{A.9}$$

$$\sigma^2 = \sigma_1^2(1 - w) \tag{A.10}$$

which shows how the estimate (x_1) and its variance (σ_1) are improved by the measurement (x_2). This derivation is now generalised to the multi-dimensional form necessary for a full Kalman filter implementation.

The multi-dimensional case

Consider now the situation in which $\mathbf{x}_1$ and $\mathbf{x}_2$ are n-element vectors representing two independent estimates of an n-dimensional vector quantity $\mathbf{x}$. The variances of $\mathbf{x}_1$ and $\mathbf{x}_2$ are represented by the two $n \times n$ matrices, $\mathbf{P}_1$ and $\mathbf{P}_2$ respectively.

The weighted mean of $\mathbf{x}_1$ and $\mathbf{x}_2$ may be expressed in the same form as presented earlier for the single dimension case:

$$\hat{\mathbf{x}} = (\mathbf{I} - \mathbf{W})\mathbf{x}_1 + \mathbf{W}\mathbf{x}_2$$
$$= \mathbf{x}_1 + \mathbf{W}(\mathbf{x}_1 + \mathbf{x}_2) \tag{A.11}$$

where $\mathbf{W}$ is an $n \times n$ weighting matrix and $\mathbf{I}$ is a unit matrix. The best estimate of $\mathbf{x}$, denoted $\hat{\mathbf{x}}$, will be provided by the above equation when $\mathbf{W}$ is selected to minimise the variance of $\hat{\mathbf{x}}$.

In most practical cases the dimensions of the two estimates are not equal and one of them is often a function of the individual elements of $\mathbf{x}$. For example, a set of m measurements may be provided, denoted $\mathbf{y}_2$, where $\mathbf{y}_2$ is related only to some of the elements of $\mathbf{x}$. In this situation, the relationship between $\mathbf{y}_2$ and $\mathbf{x}_2$ may be expressed as follows:

$$\mathbf{y}_2 = \mathbf{H}\mathbf{x}_2 \tag{A.12}$$

where $\mathbf{H}$ is an $m \times n$ matrix.

It is, therefore, necessary to form an optimum estimate of $\mathbf{x}$ from one estimate $\mathbf{x}_1$ of variance $\mathbf{P}_1$ and a second estimate $\mathbf{y}_2$ $(=\mathbf{H}\,\mathbf{x}_2)$, with variance denoted here by the symbol $\mathbf{R}$. If we now let the weighting matrix $\mathbf{W} = \mathbf{K}\,\mathbf{H}$, where $\mathbf{K}$ is another arbitrary weighting matrix, then:

$$\hat{\mathbf{x}} = \mathbf{x}_1 - \mathbf{K}\mathbf{H}(\mathbf{x}_1 - \mathbf{x}_2)$$
$$= \mathbf{x}_1 - \mathbf{K}(\mathbf{H}\mathbf{x}_1 - \mathbf{y}_2)$$
$$= (\mathbf{I} - \mathbf{K}\mathbf{H})\mathbf{x}_1 + \mathbf{K}\mathbf{y}_2 \tag{A.13}$$

By definition, the variance $(\mathbf{P})$ of $\hat{\mathbf{x}}$ is given by:

$$\mathbf{P} = E\left\{\left[\hat{\mathbf{x}} - E(\hat{\mathbf{x}})\right]\left[\hat{\mathbf{x}} - E(\hat{\mathbf{x}})\right]^T\right\} \tag{A.14}$$

Similar expressions can be written for the variances $\mathbf{P}_1$ and $\mathbf{R}$. Substituting for $\hat{\mathbf{x}}$ from eqn. A.13 yields:

$$\mathbf{P} = E\{(\mathbf{I} - \mathbf{KH})\mathbf{x}_1 + \mathbf{K}\mathbf{y}_2 - (\mathbf{I} - \mathbf{KH})\,E\{\mathbf{x}_1\} - \mathbf{K}\,E\{\mathbf{y}_2\}]$$

$$-[(\mathbf{I} - \mathbf{KH})\mathbf{x}_1 + \mathbf{K}\mathbf{y}_2 - (\mathbf{I} - \mathbf{KH})E\{\mathbf{x}_1\} - \mathbf{K}\,E\{\mathbf{y}_2\}]^T\}$$

but since $\mathbf{x}_1$ and $\mathbf{y}_2$ are uncorrelated, this reduces to:

$$\mathbf{P} = (\mathbf{I} - \mathbf{KH})\,E\{[\mathbf{x}_1 - E(\mathbf{x}_1)][\mathbf{x}_1 - E(\mathbf{x}_1)]^T\}(\mathbf{I} - \mathbf{KH})^T$$

$$+ \mathbf{K}E\{[\mathbf{y}_2 - E(\mathbf{y}_2)][\mathbf{y}_2 - E(\mathbf{y}_2)]^T\}\mathbf{K}^T$$

$$= (\mathbf{I} - \mathbf{KH})\mathbf{P}_1(\mathbf{I} - \mathbf{KH})^T + \mathbf{KRK}^T \tag{A.15}$$

We now need to find the value of $\mathbf{K}$ which minimises $\mathbf{P}$ in the sense that the diagonal elements of $\mathbf{P}$, the variances of $\mathbf{x}$, are minimised.

It is shown in Reference 1 that this is achieved when:

$$\mathbf{K} = \mathbf{P}_1\mathbf{H}^T[\mathbf{H}\mathbf{P}_1\mathbf{H}^T + \mathbf{R}]^{-1} \tag{A.16}$$

Under such conditions, the best estimate of $\mathbf{x}$ is given by:

$$\hat{\mathbf{x}} = \mathbf{x}_1 - \mathbf{K}[\mathbf{H}\mathbf{x}_1 - \mathbf{y}_2] \tag{A.17}$$

and its variance is given by:

$$\mathbf{P} = \mathbf{P}_1 - \mathbf{KHP}_1 \tag{A.18}$$

where $\mathbf{K}$ takes the value given by eqn. A.16. The weighting process, defined by eqns. A.16 to A.18, is implemented in a Kalman filter.

The Kalman filter

In this section we begin by considering the application of Kalman filtering to linear systems, before moving on to show the extensions which are necessary to overcome any system non-linearity. It is noted that the algorithms presented below are applicable to systems which are linear and time varying, not simply constrained to constant parameter systems. The explicit dependence of some of the parameters with time has been omitted to aid the clarity of the development.

Linear systems

The dynamical behaviour of a linear system may be represented by a set of first order differential equations of the form:

$$\frac{d}{dt}\mathbf{x} = \mathbf{Fx} + \mathbf{Gu} + \mathbf{Dw} \tag{A.19}$$

where the elements of the n vector, $\mathbf{x}(t)$, are called the states of the system, $\mathbf{u}(t)$ is a p vector of deterministic inputs and $\mathbf{w}(t)$ is the system noise. $\mathbf{F}$ is an $n \times n$ matrix, known as the system matrix and $\mathbf{G}$ is an $n \times p$ system input matrix. $\mathbf{F}$, $\mathbf{G}$ and $\mathbf{D}$ are constant or time varying matrices. The noise $\mathbf{w}(t)$ has zero-mean and is normally distributed (Gaussian), with a power spectral density of $\mathbf{Q}$.

Let us now assume that there are m measurements of the system, which are a linear combination of the states, but are corrupted with noise. This can be expressed in terms of the system states by the following equation:

$$\mathbf{y} = \mathbf{H}\,\mathbf{x} + \mathbf{v} \tag{A.20}$$

Here, the m vector, $\mathbf{y}(t)$, is called the measurement vector and $\mathbf{H}$ is an $m \times n$ measurement matrix. $\mathbf{v}(t)$ represents the measurement noise, which also has zero-mean and is normally distributed, with power spectral density $\mathbf{R}$.

The Kalman filter for the system described here seeks to provide the best estimates of the states, $\mathbf{x}$, using:

- the measurements, $\mathbf{y}$;
- model of the system provided by the matrices $\mathbf{F}$, $\mathbf{G}$, $\mathbf{H}$ and $\mathbf{D}$;
- knowledge of the system and measurement statistics given in the matrices $\mathbf{Q}$ and $\mathbf{R}$.

The deterministic or measurable inputs are processed by both the system and the model of the system, as shown in the block diagram representation in Figure A.1.

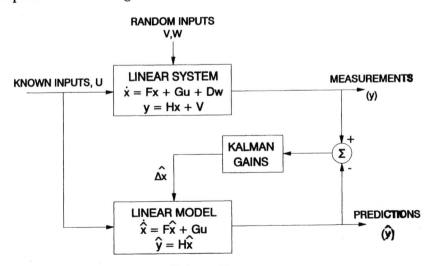

Figure A.1   Block diagram of a Kalman filter

The measurements of the true system are compared with predictions of those measurements, derived from the latest best estimates of the states provided by the system model. The differences between the true and predicted measurements are fed back through a weighting matrix, the Kalman gain matrix, to correct the estimated states of the model.

The Kalman gains are selected to provide best estimates of the states in a least squares sense. It can be shown that this is equivalent to the best estimate in the maximum likelihood sense in the linear, Gaussian noise system described above. It should be noted that it is because there is a feedback of a noisy signal in the Kalman filter that the system must be linear and the noise Gaussian, or normally, distributed. It is only because the distribution of the sum of two normally distributed signals is itself normally distributed, and also because a normally distributed signal remains so after passing through a linear system, that a least squares or maximum likelihood estimation procedure can be applied repeatedly.

Although it is usual for the system to be described mathematically in the continuous differential equation form given above, the measurements are in practice provided at discrete intervals of time. To cope with this, and to provide a computationally efficient filtering algorithm, it is customary to express the continuous equations in the form of difference equations as shown below:

$$\mathbf{x}_{k+1} = \mathbf{\Phi}_k \mathbf{x}_k + \mathbf{\Gamma}_k \mathbf{u}_k + \mathbf{\Delta}_k \mathbf{w}_k \tag{A.21}$$

with measurements:

$$\mathbf{y}_{k+1} = \mathbf{H}_{k+1} \mathbf{x}_{k+1} + \mathbf{v}_{k+1} \tag{A.22}$$

where $\mathbf{x}_k$ is the state at time t_k
$\mathbf{u}_k$ is the input at time t_k
$\mathbf{w}_k$ is the system noise at time t_k
$\mathbf{v}_{k+1}$ is the measurement noise at time t_{k+1}
$\mathbf{\Phi}_\kappa$ is the state transition matrix from time t_k to time t_{k+1}
$\mathbf{H}_{k+1}$ is the measurement matrix calculated at time t_{k+1}
$\mathbf{\Gamma}_k$ and $\mathbf{\Delta}_k$ are appropriate input matrices

The noise is zero-mean, but now discrete, and will be characterised by the covariance matrices $\mathbf{Q}_k$ and $\mathbf{R}_k$ respectively.

These equations are used to formulate a recursive filtering algorithm. In such a formulation it is necessary to consider two distinct sets of equations. The first set is concerned with the prediction of the state of the system based on the previous best estimate, whilst the second involves the updating of the predicted best estimate by combining the prediction with a new measurement.

The prediction process

The best estimate of the state at time t_k is denoted here by $\mathbf{x}_{k/k}$. Since the system noise, $\mathbf{w}_k$, has zero-mean, the best prediction of the state at time t_{k+1} is given by:

$$\mathbf{x}_{k+1/k} = \Phi_k \mathbf{x}_{k/k} \qquad (A.23)$$

whilst the expected value of the covariance at time t_{k+1} predicted at time t_k, is given by:

$$\mathbf{P}_{k+1/k} = \Phi_k \mathbf{P}_{k/k} \Phi_k^T + \Delta_k \mathbf{Q}_k \Delta_k^T \qquad (A.24)$$

The measurement update

On the arrival of a new measurement $\mathbf{y}_{k+1}$, at time t_{k+1}, it is compared with the prediction of that measurement derived from the system model. The measurement is then used to update the prediction to generate a best estimate, following the procedure outlined in the previous section. Hence, the best estimate of the state at time t_{k+1} is given by:

$$\mathbf{x}_{k+1/k+1} = \mathbf{x}_{k+1/k} - \mathbf{K}_{k+1}[\mathbf{H}_{k+1}\mathbf{x}_{k+1/k} - \mathbf{y}_{k+1}] \qquad (A.25)$$

and its covariance by:

$$\mathbf{P}_{k+1/k+1} = \mathbf{P}_{k+1/k} - \mathbf{K}_{k+1}\mathbf{H}_{k+1}\mathbf{P}_{k+1/k} \qquad (A.26)$$

where the Kalman gain matrix is given by:

$$\mathbf{K}_{k+1} = \mathbf{P}_{k+1/k} \mathbf{H}_{k+1}^T [\mathbf{H}_{k+1}\mathbf{P}_{k+1/k} \mathbf{H}_{k+1}^T + \mathbf{R}_{k+1}] \qquad (A.27)$$

where $\mathbf{H}^T$ denotes the transpose of the measurement matrix, $\mathbf{H}$.

The system states may therefore be updated each time a measurement is received by implementing eqns. A.25 to A.27.

Non-linear systems—the extended Kalman filter

So far we have considered only linear dynamical systems with zero-mean, Gaussian noise type disturbances. For such systems, the Kalman filter is optimal in the least squares sense or maximum likelihood sense. If the system is not linear or if the noise is not Gaussian, the Kalman filter is no longer optimal. In these cases, the only way to regain the optimality of the filtering is to design an algorithm specifically for the system under consideration. This, however, is not usually feasible in practice, as the filter would be of infinite dimension. Therefore, it is normal to accept that the performance will be sub-optimal, and use the Kalman filter in such a way as to make the performance as close to optimal as possible. This may involve, for example, predicting the system and its covariance matrix over relatively short time intervals during which the conditions for linearity hold.

Consider a continuous non-linear dynamical system described by the equations:

$$\frac{d}{dt}\mathbf{x} = \mathbf{f}(\mathbf{x}, t)\mathbf{x} + \mathbf{g}(\mathbf{x}, t)\mathbf{u} + \mathbf{d}(\mathbf{x}, t)\mathbf{w} \tag{A.28}$$

with discrete measurements given by :

$$\mathbf{y} = \mathbf{h}(\mathbf{x}, t)\mathbf{x} + \mathbf{v} \tag{A.29}$$

For simplicity, the explicit dependence on time will be dropped, i.e. $\mathbf{f}(\mathbf{x}, t)$ will be written as $\mathbf{f}(\mathbf{x})$. This system may be approximated by linearising it about a nominal set of states commonly referred to as a nominal trajectory.

This approximation will only be valid for short periods of time after which the system will have to be re-linearised. The linearisation most often used is the truncation of the Taylor series. Thus, for the function $\mathbf{f}(\mathbf{x})$, the Taylor series about a nominal trajectory $\tilde{\mathbf{x}}$ is given by:

$$\mathbf{f}(\mathbf{x}) = \mathbf{f}(\tilde{\mathbf{x}}) + \frac{d}{dt}\mathbf{f}\Big|_{\tilde{\mathbf{x}}}(\mathbf{x} - \tilde{\mathbf{x}}) + \frac{d^2}{dt^2}\mathbf{f}\Big|_{\tilde{\mathbf{x}}}\frac{(\mathbf{x} - \tilde{\mathbf{x}})^2}{2} + \dots \tag{A.30}$$

and similarly for other non-linear functions. Defining the nominal trajectory as:

$$\frac{d}{dt}\tilde{\mathbf{x}} = \mathbf{f}(\tilde{\mathbf{x}})\tilde{\mathbf{x}} + \mathbf{g}(\tilde{\mathbf{x}})\mathbf{u} \tag{A.31}$$

$$\tilde{\mathbf{y}} = \mathbf{h}(\tilde{\mathbf{x}})\tilde{\mathbf{x}} \tag{A.32}$$

and subtracting from the original equations, we obtain differential equations governing the deviations from the nominal trajectory:

$$\frac{d}{dt}\delta\mathbf{x} = \frac{d}{dt}\mathbf{f}\Big|_{\tilde{\mathbf{x}}}\delta\mathbf{x} + \frac{d}{dt}\mathbf{g}\Big|_{\tilde{\mathbf{x}}}\delta\mathbf{u} + \frac{d}{dt}\mathbf{d}\Big|_{\tilde{\mathbf{x}}}\mathbf{w} \tag{A.33}$$

$$\delta\mathbf{y} = \frac{d}{dt}\mathbf{h}\Big|_{\tilde{\mathbf{x}}}\delta\mathbf{x} + \mathbf{v} \tag{A.34}$$

If we define :

$$F = \frac{d}{dt} f\Big|_{\tilde{x}}$$

$$G = \frac{d}{dt} g\Big|_{\tilde{x}}$$

$$D = \frac{d}{dt} d\Big|_{\tilde{x}}$$

$$H = \frac{d}{dt} h\Big|_{\tilde{x}} \qquad (A.35)$$

then the discrete Kalman filter equations may be used as given in the previous section. At each measurement interval, the following steps must now be taken:

1 Linearise the equations about the nominal trajectory. The nominal trajectory is usually taken to be the latest estimate of the states.
2 Calculate the transition matrix and other matrices of the discrete equivalent to the linearised system.
3 Integrate the state prediction equations. The actual estimated states may be used here, as there is no difference between this and integrating the differential equations for the nominal trajectory and the deviations from it separately and then adding the result.
4 Implement the Kalman filter equations. This will provide best estimates of the deviations from the nominal. The corrections, given by the Kalman gains multiplied by the measurement differences, can again be added directly to the predicted state estimates.
5 Continue to the next time interval, i.e. return to step 1.

There are situations in which the measurement update rate may be relatively low, in which cases the prediction stage of the filter will need to be run with a smaller time interval. If this is not done, the non-linearities begin to dominate the deviations of the state estimates from the nominal.

Although this form of the Kalman filter, the so-called extended Kalman filter, is used for systems which are explicitly non-linear, it is also used when there is a requirement to identify certain unknown parameters in the system. In this situation, the unknown parameters are defined as states of the system and the state vector is augmented to include them.

References

1 BARHAM, P. M. and HUMPHRIES, D. E.: 'Derivation of the Kalman filtering equations from elementary statistical principles' in 'Theory and application of Kalman filters', NATO Agardograph AG139, 1970
2 JAZWINSKI, A. H.: 'Stochastic processes and filtering theory' (Academic Press, 1970)

Inertial navigation system error budgets

The accuracy of an inertial navigation system is often expressed as a position uncertainty after a given period of navigation, or on reaching a given destination. Alternatively, it is expressed in terms of the rate at which the navigation error builds up with time, in nautical miles per hour, for example. The actual form of expression used for this overall accuracy figure depends upon the application. For example, for an interplanetary mission, the accuracy refers to the desired point of closest approach to the destination planet. For navigation in the vicinity of the Earth, it is usually the errors in two dimensions which are of most interest, the along-track and cross-track position errors over the surface of the Earth. These errors are often combined to yield a single number which expresses navigation accuracy after a given navigation time, the circular error probable (CEP) or circular probable error (CPE) as it is sometimes called. Essentially, this defines a circular area within which the navigation system estimates its true position to be, with a certain probability. The 50 % CEP is a frequently quoted figure. When the probability value is not stated, it usually means a 50 % value should be assumed.

Consider now the composition of this navigation performance figure. In practice, navigation errors propagate owing to a large number of error sources which include alignment errors, a variety of inertial sensor errors and errors attributable to computational inaccuracy. In general, each error may be regarded as comprising a repeatable or predictable component and a random or unpredictable component. The former category produces predictable effects which can be compensated for, if so desired, i.e. electrical signals or software corrections can be applied which should offset the effect of such errors. The remaining errors, which arise as a result of random effects within the system and incomplete compensation of systematic errors, will give rise to navigation inaccuracies.

For the assessment of system performance, random errors are treated statistically to derive a mathematical description of each term and to allow the various contributors to the overall system errors to be combined in what is known as the system error budget. It is common practice to assume that the random errors within the components of an inertial navigation system follow a Gaussian or normal distribution where the probability density function, denoted $p_x(x)$, is expressed mathematically as:

$$p_x(x) = \frac{1}{\sigma\sqrt{2\pi}} \exp\left(-(x-\mu)^2 / 2\sigma^2\right)$$

where μ is the mean value of x. A Gausian distribution is depicted in Figure B.1.

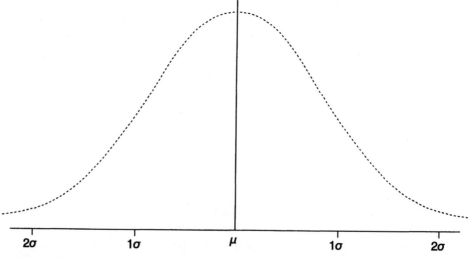

Figure B.1   Gaussian distribution

Based on this assumption, the statistical analysis techniques applicable for normal distributions may be conveniently used for the analysis of inertial systems.

If measurements are made of a bias (x_i) on the measurement provided by a gyroscope or an accelerometer then, over large numbers of sensor samples, it is assumed that the error distribution approximates to this curve. It is common practice to quote a 1σ error or standard deviation for each sensor error. The standard deviation of a normal distribution may be expressed in terms of the mean error (x_m) and the number of samples (n) as follows:

$$\sigma = \frac{\sqrt{\left(\Sigma x_i^2 - nx_m^2\right)}}{\sqrt{n}}$$

The 1σ value represents 68.26% of the area under the curve. Hence, if a 1σ gyroscope bias is quoted as being $1°$/hour, it is expected that over a large number gyroscopes of that type, 68.26 % would have a bias of between $\pm1°$/hour of the mean value. Alternatively, it may be inferred that there is a 68.26 % probability of the bias uncertainty on a given gyroscope being within $\pm1°$/hour. The 2σ and 3σ bias values which are obtained by multiplying the 1σ value by 2 and 3 respectively may also be defined for the sensor, corresponding to 95.46 % and 99.73 % of the area under the normal distribution curve. Hence, for this example, there is a 95.46 % probability of the gyroscope bias being within $\pm2°$/hour and a 99.73 % probability of the bias being within $\pm3°$/hour.

In an inertial navigation system, a number of sensors and components are required to operate together, each producing random errors. It is usual to assume that each error component is unrelated to each of the other error components, i.e. the error components are said to be independent. In combining the effect of many error sources, simply summing the individual contributions to the error budget arithmetically would give a very pessimistic prediction of system performance. Where a number of independent sources need to be combined, a more correct prediction of overall system performance is obtained by summing the individual 1σ errors quadratically, i.e. by taking the root sum square (RSS) error. Hence, if σ_1, σ_2, ... σ_n represent a number of independent errors which combine to produce an overall error, then the total or overall error is obtained using:

$$\sigma_{RSS} = \sqrt{\left(\sigma_1^2 + \sigma_2^2 + \ldots + \sigma_n^2\right)}$$

where σ_{RSS} is the 1σ root sum square error. In an inertial navigation system, an RSS position error can be calculated for both the along-track or down range position error (σ_x) and the cross-track position error (σ_y). Assuming all error contributions to both along- and cross-track errors to be Gaussian, then the behaviour of each is also described by a Gaussian distribution curve. When σ_x and σ_y are combined, a probability ellipse is used to described the behaviour of the total error. For the special case where $\sigma_x = \sigma_y$, the probability ellipse reduces to a probability circle. In mathematical terms, the radial probability distribution is given by:

$$p(r) = \frac{r}{\sigma^2} \exp\left(-r^2/2\sigma^2\right)$$

where σ is the standard deviation of the error in x and y. This is known as the Rayleigh probability density function. The probability of r lying between 0 and R, $P_r(<R)$, is given by:

$$P_r(<R) = \int_0^R p(r)\,dr$$

$$= 1 - \exp\left(-R^2/2\sigma^2\right)$$

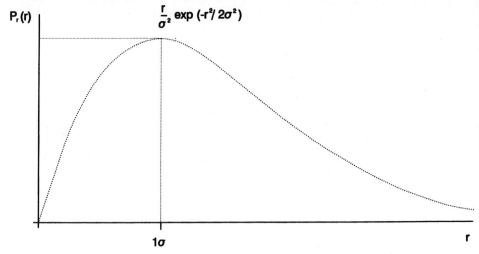

Figure B2   Rayleigh distribution

$P_r(<R)) = 0.5$ defines the 50% probability circle which occurs when $R/\sigma = 1.17710$. The radius of this circle is the 50% CEP referred to earlier.

Hence, 50% CEP = 1.1774s.

The reader is referred to any standard text on probability theory, Reference 1 for instance, for a more detailed discussion of the topics outlined in this Appendix.

Reference

1 LATHI, B. P.: 'An introduction to random signals and communication theory' (International Textbook Company, 1968)

List of symbols

Scalars

a_x	magnitude of acceleration acting along x axis of body
a_y	magnitude of acceleration acting along y axis of body
a_z	magnitude of acceleration acting along z axis of body
$B_{fx,y,z}$	g-independent (fixed) bias of inertial sensor
$B_{gx,y,z}$	gyroscope g-dependent bias coefficients
$B_{axz,yx}$	gyroscope anisoelastic bias coefficients
c	speed of light
c_{ij}	i, j element of direction cosine matrix
f	magnitude of specific force, the non-gravitational force per unit mass, acting on a body
f_N	magnitude of north component of specific force
f_E	magnitude of east component of specific force
f_D	magnitude of vertical (down) component of specific force
g	magnitude of acceleration due to Earth's gravity
h	height above ground
H	magnitude of angular momentum
I	moment of inertia
λ	longitude
L	latitude
M_x, M_y, M_z	inertial sensor cross-coupling coefficients
n_x, n_y	zero-mean random biases of inertial sensor
p	roll rate
q	pitch rate
r	yaw rate
R_0	mean radius of curvature of the Earth
S_x, S_y	scale factor error of inertial sensor

t	time
v_N	magnitude of north velocity
v_E	magnitude of east velocity
v_D	magnitude of vertical (down) velocity
δt	time increment
$\delta\phi$	rotation increment in roll
$\delta\theta$	rotation increment in pitch
$\delta\psi$	rotation increment in yaw
ξ	meridian deflection of the local gravity vector
η	deflection of the local gravity vector perpendicular to the meridian
ϕ	roll Euler angle
θ	pitch Euler angle
ψ	yaw Euler angle
π	pi
ω_x	roll rate of body with respect to navigation frame
ω_y	pitch rate of body with respect to navigation frame
ω_z	yaw rate of body with respect to navigation frame
ω_R	gyroscopic nutation frequency
ω_s	spin speed of gyroscope rotor
Ω	Earth's rate

Vectors

$\mathbf{a}_i$	acceleration with respect to inertial reference frame
$\mathbf{f}$	specific force vector
$\mathbf{g}$	mass attraction gravitation vector
$\mathbf{g}_l$	local gravity vector
$\mathbf{H}$	angular momentum vector
$\mathbf{I}$	unit vector in x direction of co-ordinate frame
$\mathbf{j}$	unit vector in y direction of co-ordinate frame
$\mathbf{k}$	unit vector in z direction of co-ordinate frame
$\mathbf{q}$	four element quaternion vector $[a\ b\ c\ d]$
$\mathbf{r}$	position vector
$\mathbf{T}$	torque acting on the body
$\mathbf{v}_i$	velocity with respect to inertial reference frame
$\mathbf{v}_e$	velocity with respect to the Earth, the ground speed
$\boldsymbol{\sigma}$	angle vector having components $\sigma_x, \sigma_y, \sigma_z$
$\boldsymbol{\omega}$	gyroscopic precession rate
$\boldsymbol{\omega}_{ie}$	turn rate of the Earth with respect to inertial reference frame
$\boldsymbol{\omega}_{ib}$	turn rate of body with respect to inertial reference frame

ω_{eb}	turn rate of body with respect to the Earth frame
ω_{in}	turn rate of navigation frame with respect to inertial reference frame
ω_{en}	turn rate of navigation frame with respect to Earth frame
ω_{ew}	turn rate of wander frame with respect to Earth frame

Matrices

$\mathbf{B}$	direction cosine matrix which defines transformation from true reference axes to estimated reference axes
$\mathbf{C}_b^i$	direction cosine matrix relating body frame to inertial reference frame
$\mathbf{C}_b^e$	direction cosine matrix relating body frame to Earth frame
$\mathbf{C}_b^n$	direction cosine matrix relating body frame to navigation frame
$\mathbf{C}_e^w$	direction cosine matrix relating earth frame to wander azimuth frame
$\mathbf{\Omega}_{ib}$	skew symmetric matrix defining turn rate of body with respect to inertial reference frame
$\mathbf{\Omega}_{eb}$	skew symmetric matrix defining turn rate of body with respect to Earth frame
$\mathbf{\Omega}_{nb}$	skew symmetric matrix defining turn rate of body with respect to navigation frame
$\mathbf{\Omega}_{ew}$	skew symmetric matrix defining turn rate of wander azimuth frame with respect to Earth frame
$\mathbf{\Psi}$	skew symmetric matrix defining incremental rotation

Other symbols

$\times$	denotes a vector cross product
$\sim$	denotes a measured quantity
$\wedge$	denotes an estimated quantity
$\cdot$	denotes a quaternion product
$*$	denotes the complex conjugate of a vector

Additional symbols are defined where they are used in the text.

Index